Introduction to
Microcontrollers and their Applications

Introduction to
Microcontrollers and their Applications

T.R. Padmanabhan

Alpha Science International Ltd.
Oxford, U.K.

T.R. Padmanabhan
Professor Emeritus, School of Engineering
Amrita University,
Ettimadai, Coimbatore
Tamilnadu, India

Copyright © 2007

Alpha Science International Ltd.
7200, The Quorum, Oxford Business Park North
Garsington Road, Oxford OX4 2JZ, U.K.

www.alphasci.com

Printed from the camera-ready copy provided by the Author.

ISBN 978-1-84265-399-9

Printed in India

Dedicated
to
Her Holiness Sri Mata Amritanandamayi Devi

PREFACE

"All the world is a stage, and all the men and women merely players. They have their exits and their entrances. And one man in his time plays many parts."

William Shakespeare

Then He spake through Shannon – an MS Student – that turning ON a motor is TRUE; turning it OFF is FALSE. The Mad Hatter declared "that means digital". Tweedledee said, "analog". Tweedledum said, "digital". The Rabbit gave Alice a stern look through its pince-nez and commanded "Count". Alice counted hesitantly, "4004, 4040, 8008, 8080, 8081, 8085, 8086, 8087, 8088, 8089, 8096, 80186, 80187, 80188, 80286, 80386, 80486, . .80586, . . 80686, . . 80786, . . . ". The Cheshire Cat intervened, "6502?, 6800?, 6801?, 6805?, 6811?, . . " and smiiiii . . . .led. The Dormouse pronounced "Yes, Z80, Z800, Bit Slice, Bitter Slice, . . it is all the same" and went back to sleep. The Queen gave the verdict, "Cut off their heads". The tea party ended – almost.

With the magic wand, He swept the stage for the next act.

A breed of digital designers descended with their 7400 series and CMOS 4000 series. They worked magic (at least that is what they said!) with these blocks and made their intricate digital system designs. The braggarts talked of Timing, Interface, Loading, Pick up, Noise headache, PCB layout, B(l)ack p(l)ane, Reflections, and Matching. They – armed with their Holy Book – warded off all these devil's temptations.

He said "enough" and dumped them. They murmured; but He pointed to the Swiss precision watch designers in the same bin – good company that was!

Then the 'Microprocessor Walla' descended – his basket full of Processors, Memories, Multifunction circuits, add-ons for Algebra, Communication, and DMA. He added to the cacophony with his Memory Access and Mapping, Linear and Decoded allocations, Internal and External Tasks, Local and Extended actions.

Yesterday once again He said "enough" and wrung the curtain. But the show continues – now in different stages concurrently.

Stage 1 has computer designers – a handful of them – talking of Multi-processing, Parallel processing, Multi-threading: They work at a higher layer (level) but do not soil their hands with circuits and the like.

Stage 2 has another handful; they claim inheritance from the SSI, MSI and LSI lineage; they carry the Crosses of Noise, Interference, Loading, Timing, Layout, and so on. They design VLSIs and claim to offer packaged solutions to all – called SOCs. They are hungry for problems, which they can chew, and offer zipped total solutions.

Stage 3 has the biggest crowd. The 'act of µC' is being played; embedded solutions are on offer. It is like asking "tell me what you want?" and almost doling out a µC and saying "this does the job"!

'Wake up! You are here!'

A substantial number of digital designs are done with off-the shelf items like µCs. Little of component level design is called for. But the need to form logic functions to minimize them, define flip-flop based sequences and design them, or build programmable circuits using SSIs and MSIs no longer exists. It calls for a radical change in the approach to teaching of digital design – already evident in the best of universities in the world. The need is to offer an intensive course in µCs and applications. At the end of it the student should be in a position to translate the needs of an application into functions that can be carried out by µC. He should be able to select a µC and directly relate application requirements with the features of peripherals and program needs. Doing program and going through the motions of debugging and downloading completes prototyping.

The approach implies three more things – Firstly there is the need for a course in 'Advanced Digital Design' – basics of digital design leading to VLSI through HDLs. A text book like [2] aims at this. Secondly there is no need to go through the motions of studying vintage 8-bit and 16-bit processors. µPs have only a niche market now – design of computer boards or similar items. Expertise with µCs and peripherals in them is more relevant and well sustained for the coming decades. Thirdly the approach here directly opens the window to embedded designs and applications. A variety of tools and approaches are on offer and under development; 16-bit and 32-bit µCs and RTOS are of this category. These can be taken up for study at senior levels. At Amrita University campuses, the µC course has been on offer for electronics and related streams at the second year level. The student community has enthusiastically received it.

The present book is an outgrowth of this effort. Following faculty members have contributed their mite for the program and the book:

From Coimbatore Campus:

Mr P V Sunil Nag of Electronics and Instrumentation Engineering Department; Mr S Krishnakumar, Ms S Malligadevi, Mr S Sooraj, MS K Divya, Ms V Vanita, and MS Vidya Lavanya of Electronics and Communication Engineering Department; MR G Jagadanand, Mr R Jayakrishnan, and Mr N Sivaramasubramanian of Electrical and Electronics Engineering Department

From Bangalore Campus:

Ms Sangita Khare of ISE Department; Ms Kusuma Keerthi of Computer Science and Engineering Department; Ms K Sudha of Electronics and Communication Engineering Department

From Amritapuri Campus

Ms K Miziya of Information Technology Department; MS V Nitha of Computer Science and Engineering Department; Ms Karthi Balasubramanian of Electronics and Communication Engineering Department.

I have been instrumental in 'linking and compiling' of their scripts. My role has essentially been in reediting the matter, ensuring meaningful continuity, and retaining focus.

The book is organized as follows:

The first chapter is essentially motivational in nature. Chapters 2 to 5 deal with the basics required to build a µC, understand its functions, and work with it. If the student has already gone through a course in digital design [9], these can be omitted or dealt with briefly to maintain continuity. The components are put together to form a basic µP, in Chapter 6. Concepts of instruction, instruction sequences forming programs, program memory and program counter, fetch and execute sequences are introduced. The following two chapters deal in detail with the features of the mid-range series of Microchip. Their worldwide and widespread use, market dominance, variety on offer, and the need for minimal additional hardware have been the primary reasons for the choice. From an instructional point of view a RISC processor like the PIC® mid range series has two advantages: Firstly one can easily understand the threads of programming and build up expertise in flowcharts, pseudo codes and programming *per se*. Secondly, enough attention can be given to peripherals and applications. The following chapters are devoted to the peripherals – that transform the µP to a µC. Their features, their programming, and application aspects are dealt with in detail. Typical (real life) applications are also dealt with. Chapter 13 is an introduction to the popular 8051 series. It also forms an example of a CISC processor.

A detailed discussion of possible projects – perhaps unconventional – forms the theme of chapter 14. Their variety is to bring out the potential and expanse of µC applications. Sufficient detail is given for each to enable the student to understand / appreciate the task on hand. Implementation specifics offered in some of the cases obviate the need for bridging the gap between application and µC features. Others introduced but confined to application details, are more demanding of the students. Many of the applications considered can be expanded substantially into innovative and fascinating products.

For teachers following the book, additional resources are available at the site **www.amrita.edu/downloads/micro.ppt**.

ACKNOWLEDGEMENTS

(Or ten tips to write a book)

Swami Abhayamrita Chaitanya runs
and with the help of the Ashramites RUNS the Institute.
Uma runs the home front well-oiled.
My colleagues prepare metal.
Ramachandran wades through my scrawls to elicit the manuscript.
The students edit and edit.
Prashant Nair pecks off the specks.
Harini paints (off) (non) English.
Easwaran gives his touch.
The Narosa team of Mehra, Anupama, and others do the rest.
I – Padmanabhan – shall take the 'Anantha' prefix;
and shall recline and relax all the while;
and allow all to do what they are to do for the book sans my interference.

'Mona Lisa, let me take a seat by your side –
rather a reclining pose – of course with mine own Smile'.

'Frown not dear; let the beholder say which is the SMILE!'

CONTENTS

CHAPTER 1

A CASE FOR MICROCONTROLLERS

1.1 INTRODUCTION

Microcontroller (µC) is in wide use today; the ubiquitous cell phone, calculator, sub-assemblies of automobiles, industrial instrumentation schemes, bio-medical instrumentation schemes, . . . thus goes the target list. What does the microcontroller do in these? How does it do the assigned functions? Given a potential application, how do you choose a microcontroller? What do you do with it to make it do the expected functions? We shall try to answer these questions in a graded manner in this book. Examining a few representative applications is a good starting point for us; through these let us get a feel of the microcontroller and its functions.

1.2 CURRENT MONITORING IN A HIGH TENSION LINE

Power transmission takes place in cross country terrain at high voltages – 110KV, 220KV and 400KV; higher voltages are not uncommon. Measurement of the current in the transmission line helps in knowing how much power is being transmitted; it is required for purposes of power flow control and co-ordination, for metering and allied commercial transactions, and for protection of the system. The conventional practice is to insert a current transformer in the power line and measure the truly reflected current in the secondary side of the transformer. Measurements at such high voltages are hazardous; the current transformer design and manufacture have to be such as to ensure sufficient, reliable and long lasting electrical insulation between its primary and secondary sides. The secondary is the low voltage side physically accessible to (human) operators. The activities mentioned above are carried out based on the secondary side current measured. The scene is likely to change radically for the better, thanks to the following recent developments:

- Low cost plastic fibre optic cables have become available. They provide enough electrical insulation and isolation between the terminations. They can carry data (information) in digital form at high speeds. The need for a costly current transformer with enough insulation between its primary and secondary sides is no longer there.
- Microcontrollers of today are quite powerful as measured by the activities they can carry out. They are of low cost and function with very low power to sustain then. A microcontroller at the HT end can carry out the current measurement and allied activities. It can transform this into meaningful digital data to be transmitted and couple it to the fibre optic cable. All these can be carried out with nominal insulation level.
- At the LT (ground) end, the fibre optic cable can couple the data into a local microcontroller. The microcontroller here can process it and extract from it the full information regarding the current. It can also carry out allied supervisory and monitoring functions.

- Rechargeable cells with capacity of the order of one ampere-hour have become available and affordable. They can be used to supply power to the μC at HT end by deriving the power (~ 10 MW) from a local CT through its secondary side. The rechargeable cell can be kept charged and used as a backup power source. It need supply power to the μC at HT end only in case of a power outage.

Let us consider the scheme in more detail; the aim is to understand the role of the μC which is of direct interest to us here. The scheme is shown in block diagram form in Fig. 1.1 It has three parts; the transmitter, the fibre optic cable and the receiver. The transmitter (by itself a low voltage unit) is kept insulated and close to the HT line. The receiver is kept at ground level. I is the current in the HT line; it is sensed by the current transformer CT. The circuit block B facilitates generation of power to the whole unit through the Power Supply Unit (PSU). It has a battery back-up. The circuit block B inputs the measured current to the μC. The μC converts the sensed current into a data bit stream; the same is coupled into the fibre optic cable through the transmitter Tx as a corresponding sequence of light pulses. The fibre optic cable carries these light pulses to the receiver. The cable also provides necessary electrical insulation and isolation. The receiver transforms the received light pulses into an electrical pulse sequence. The μC in the receiver accepts these pulses. It reconstructs the current signal and extracts necessary information from it. It is locally displayed and also made available to supervisory units. The μC in the transmitter does the following:

- Sample the current in the HT line at regular intervals and convert the same into an equivalent number
- Generate a 0-1 pulse sequence representing the above number conforming to a predefined protocol and transmit the pulse sequence through the fibre optic cable
- Keep the local battery charged; in case of power outage the battery takes over the power supply. When power is restored, battery is to revert to charge mode.

The μC in the receiver does the following:-

- From the received pulse sequence using the protocol used in the scheme, reconstruct the numbers representing the current samples
- Generate information from the sample sequence: this can include rms value of current, nature of variation of rms current with time, possibility of any overload in the terminal equipment, identification of hazards like short circuit etc.
- The information generated is displayed locally. It is also sent to a supervisory unit in the desired format.

Both the μCs carry out assigned functions continuously. They have to be pre-programmed for the same. They do not call for any intervention in their operations. Neither do they need to respond to manual inputs at any time in their regular operation. Both μCs function as 'blind Monitors'.

1.3 INFRA RED THERMOMETER

Conventionally temperature measurement is carried out by a sensor in contact with the source whose temperature is to be measured. There are situations where the source may not be accessible to have the sensor kept on it. There are situation where the sensor cannot put up with the source in contact with it; the sensor may be corroded or burned up. In all such cases the temperature measurement has to be based on non-contact type sensor – it is a case of 'judgment' based on observation. Infra Red (IR) sensors are available for the purpose. With the use of a suitable lens system, the radiation over a specified and narrow wavelength range is made to fall on the sensor; the sensor generates a corresponding voltage signal which has to be suitably interpreted and source temperature computed. Different factors can affect the accuracy and reliability of the scheme:

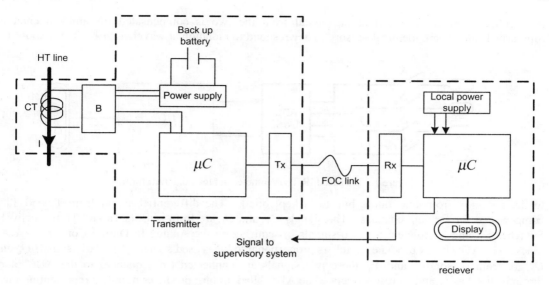

Figure 1.1 μC based HT current measurement

- Emissivity of the source
- An intervening medium like moisture has its own absorption characteristics
- Change in field of view of the sensor
- Unit to unit variation in the sensor characteristics

All these imponderables contributed to the relegation of IR thermometry to a novel idea without any practical use. But the μC has made a change to the scenario. One can account for all the above types of imponderables by proper selection of a measurement scheme and carrying out a few compensatory computations. Consider the two colour method of IR thermometry [15]. Radiations at two wavelengths – λ_1 and λ_2 – are related by the equation

$$\frac{W_1}{W_2} = \left(\frac{\lambda_1}{\lambda_2}\right)^{-5} \exp\left\{-\frac{C_2}{T}\left[\frac{1}{\lambda_1} - \frac{1}{\lambda_2}\right]\right\} \tag{1.1}$$

where
- T is the source temperature
- W_1 is the radiant intensity at wavelength λ_1
- W_2 is the radiant intensity at Wavelength λ_2
- C_2 is a known constant

One can select wavelengths λ_1 and λ_2 and also two sensors responding to these two wavelengths. The sensors can receive the radiations; their outputs will be functions of W_1 and W_2 respectively. Using Equation (1.1), one can do necessary computations and get the source temperature. The sensors used are sensitive to ambient temperature variations too. Compensation for this requires each sensor to be exposed to source temperature and ambient temperature alternatively. An alternative is to use two sensors – T_a and T_b – for wavelength λ_1 – and two sensors – T_c and T_d – for wavelength λ_2. The measurement scheme is shown in block diagram form in Figure 1.2. T_a is exposed to the radiation being received from the source. It responds to radiation at wavelength λ_1. T_b is an identical sensor but kept 'blind'. It receives no radiation. T_a and T_b are identical sensors. Ambient changes affect them in an identical manner. The differential signal from the two is compensated for ambient changes. Similarly T_c and T_d are identical sensors. They respond to radiation at wavelength λ_2. T_c is exposed

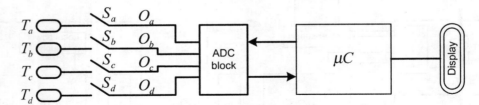

Figure 1.2 μC based IR Thermometer in block diagram form

to the radiation from the source but T_d is kept blind. The differential signal from T_c and T_d is compensated for ambient changes. The signals O_a and O_b are from sensors Ta and Tb respectively. Switches S_a and S_b connect the two signals in sequence to the Analog to Digital Converter (ADC) block. The ADC block produces numbers proportional to the sensed signals. Similarly signals O_c and O_d are from sensors T_c and T_d; these two signals are connected (in sequence) to the ADC block through switches S_c and S_d respectively. The ADC block in turn produces numbers representing these signals. The μC accepts all these number sequences, computes the temperature value of the source and displays it. Role of the μC in the scheme is as follows:

- Switch S_a and S_b in sequence at regular intervals and get sample values of signals O_a and $_{Ob}$.
- Switch S_c and S_d in sequence at regular intervals and get sample values of signals O_c and O_d.
- From sample values of O_a and O_b compute the compensated value of W_1 to be used in Equation (1.1).
- Similarly from sample values of O_c and O_d, compute compensated values of W_2 to be used in Equation (1.1).
- Using W_1 and W_2 values, calculate the value of temperature using Equation 1.1
- Display the temperature value at the display unit

The μC is assigned a specific set of tasks to be done in a sequence. The sequence includes intensive computational activities too. The μC is 'blind' in operation – in the sense it receives no further commands for its operation, from any source. The μC is a blind monitor cum computer.

1.4 CALCULATOR

The hand-held calculator has a numeric display and a set of 15 to 20 keys. It responds to each key stroke and carries out computations in a pre-arranged manner and displays the results. It functions in a structured manner on the following lines:

- Any number key pressed is understood by the calculator as a corresponding decimal digit
- A sequence of number keys pressed is interpreted as a composite decimal number.
- A function key like '+' pressed is interpreted as a corresponding operation to be carried out.
- A sequence of key strokes, '2', '4', '+', '9', '7' and '=' is interpreted as 'add number 24 with number 97 and display the result. Same holds good for other similar sequence too.

The calculator is built around a μC. It has a set of built-in programs each to identify the meaning of a key stroke. It has a program to display a desired number or character set. Another program – you may call it a supervisory program – calls a select set of these built-in programs and executes them in a sequence. The μC has to respond to the user commands. It has to do a display which the user sees and responds. The structure of the supervisory program is decided and set within the calculator; only the slots within are filled at the time of use of the calculator. Here the μC operation is interactive. The operation is in 'real time' in the sense that if the user wants an addition to be done, the μC does it 'immediately', at least as perceived by the user.

1.5 CASH REGISTER – POINT OF SALE TERMINAL

The Cash Register is in wide use in many commercial establishments. Let us examine its features in some detail. It has a keyboard with keys for the digits 0 to 9, keys for algebraic operations like addition and multiplication, and a few dominant functional keys for operation like – 'Enter', 'Total', 'Abort', and so on. It has two displays – one for the operator and the other for the user – both are identical. It has a dot matrix printer with the print out conforming to a specific format. Often the Cash Register has an electronic weighing machine attached to it. Once an object is kept on the weighting machine and weighed, through an operator command, the weight value can be transmitted to the Cash Register. In turn it computes the amount to be paid for the item weighed and adds it to the list being prepared for the customer.

The Cash Register can have two operational modes – one for the supervisor and the other for the operator. In the supervisory mode one can enter the names of different items in the shop in abbreviated form, the rates for each and freeze the information. The supervisor can access information regarding cumulative sales and collection. He can reset all such information at the end of the day. He can edit and change the rates for different items, add or delete items. The Cash Register can be programmed to show stock details and information regarding stock movement. The operator mode is designed for easiness of operations to prepare the bill of transaction. Item names can be entered through their numeric codes or abbreviated names; quantity sold to the customer has to be entered. On a command the Cash Register completes the line of entry in the bill. The line carries details like item name, rate, quantity and the total amount to be paid for the item. Similar entries are made for all the items. On a 'Total' command, the Cash Register totals the amount and prints out the bill in the specified format.

The Cash Register is built around a μC. Like the Calculator it has a set of built-in programs; it functions like a calculator. It has many other functions like multiplying the quantity entered by the rate for the item specified, doing totaling, updating stock information, providing distinct modes for operator and supervisor, and so on. The user interaction is more intensive and detailed than with a calculator. It has the provision to accept item weight information over the serial line. It has a separate program mode where entries regarding item name and rate can be updated. Again as in the calculator, the operation is in 'real time'; the operator gets instant response as perceived by him.

1.6 EMBEDDED APPLICATIONS

Today the μC is used in a variety of applications. We have examined a representative set here to understand the role of the μC. The μC and its role are distinctly different from that of a PC or other computer. In fact the μC does not do any visible computer work as we understand a computer of today. It receives no program to be executed. It does not offer any application platform like Visual Basic or Word Processor. The operator need not involve any such application program in his role as the user. Minimal skill or expertise suffices for the operation of the device; this is in contrast with any reasonable use of a PC. For example use of a commonly used platform like Word Processor requires the user to be quite skilled and adept at working with it.

In the applications of μC, the μC itself is only in the background. It may have a substantial amount of program built in. All that is well thought out in advance by the manufacturer, prepared as a program, loaded into the μC and kept frozen. It cannot be changed during subsequent use of the device. A program of this nature which is built into the μC at the time of manufacture of the device is called

'Firmware' (Software firmed up once for all). The μC itself is said to be 'embedded' into the device – embedded in the sense that the user does not perceive its presence directly [20].

The question naturally arises here; 'What should the μC do? What facilities or capabilities should it possess?' We attempt to answer these questions in the ensuing chapters. Nevertheless we are in a position to identify the skeletal features of a μC:

- It should have the capability of storing a program and running it.
- It should be able to carry out or execute elementary instructions; such instructions woven with dexterity form the program in the device.
- It should have a variety of peripheral circuits as part of itself; it should be able to support other peripherals easily.
- It should have a structured development support to facilitate development in terms of programming, testing and debugging the program, configuration of peripherals to suit the application and so on.

Through the following chapters, we shall see how to craft an elementary μC. We shall examine the features of representative and widely used μCs. We shall get acquainted with them to the extent necessary to embark upon development of comprehensive applications.

The flow of the rest of the book is broadly on the following lines:

Chapters 2 to 5 deal with the basic digital building blocks; their features, use and interconnection to build more complex blocks form the focus. Building up of a simple microprocessor from the basic blocks is illustrated in Chapter 6. In chapter 7, it is invoked to understand a representative and widely used processor – the PIC mid-range series. The chapters following are devoted to the hardware features, the instruction set and programming of the PIC® series. Program development support provided by the Integrated Development Environment (IDE) is discussed in brief – to the extent necessary for the text flow. The wide and varied set of peripherals offered by the manufacturer is discussed subsequently. In today's scenario, understanding the peripherals in all their dimensions is essential to the development work. Sometimes it turns out to be more important than understanding the processor itself! Use of the processor and the peripherals is illustrated through simple examples. The aim is to introduce the μC to the student and build up the motivation in him/her to use it (Taste blood?).

Amongst the wide spectrum of μCs, the PIC® series stands out with its simple and compact instruction set – being a RISC processor. In contrast a processor like 8051 has a comprehensive instruction set – being of the CISC category. 8051 and its variants offered by different manufacturers are in wide use due to their varied supplier base and variety of offerings. In the last chapter 8051 architecture and instruction set are discussed – as a representative CISC offering.

CHAPTER 2

NUMBER SYSTEMS AND ALGEBRA

2.1 INTRODUCTION

The decimal number system – where a set of ten digits are used to represent any number – is universal. There are other number systems with different sets of digits. Binary, Ternary, Octal, Hexadecimal are examples of such number systems. Digital systems and digital computers use binary numbers. All algebraic manipulations are carried out by them in the binary form. There is the need for us to interact with digital systems and computers. It requires working with the different number systems concurrently and calls for expertise in converting numbers from one number system to another. Here, we focus on various number systems and conversion of numbers from one system to another [9].

2.2 THE DECIMAL NUMBER SYSTEM

The decimal number system is built around ten unique symbols – 0, 1, 2, 3, 4, 5, 6, 7, 8, and 9 .The term 'decimal' signifies these ten. Any number is represented in terms of these same 10 symbols or digits. Consider the integer sequence

0123456789
1011 …192021…   293031…394041…99
100101…199200201…999
10001001…199920002001…   9999
10 000 10 001. . .

Any number is formed as a combination of the 10 basic symbols or digits. After all single digit representations are exhausted, 2 digit representations are used starting with 10 and going up to 99. The range is extended further by using 3 digit representations – from 100 to 999. Extending the logic as large a number as desired can be formed. Every number has a value implicitly attached to it. Thus the number represented as 2197 implies that

$$2197 = 2000 + 100 + 90 + 7$$
$$= 2 \times 10^3 + 1 \times 10^2 + 9 \times 10^1 + 7 \times 10^0.$$

In the number every digit position has a weight attached to it. The digit 2 – which is 4 positions to the left of the digit 7 – has a much larger weight attached to it than what 7 has. The position of a digit with respect to the left most digit in the integer representation, determines its weight.

With the accepted convention of integer representation, any digit combination has a unique position in the integer sequence; the weight system decides the position. The concept can be extended further with the use of decimal point to represent non-integer numbers. The number 2197.3208 can be seen to be equivalent to

$$2197.3208 = 2 \times 10^3 + 1 \times 10^2 + 9 \times 10^1 + 7 \times 10^0 + 3 \times 10^{-1} + 2 \times 10^{-2} + 0 \times 10^{-3} + 8 \times 10^{-4}.$$

With the convention adopted any number can be represented uniquely in terms of the 10 basic digits. In general a number Y can be represented as

$Y = d_n \times 10^n + d_{n-1} \times 10^{n-1} + ..d_1 \times 10^1 + d_0 \times 10^0 + d_{-1} \times 10^{-1} + d_{-2} \times 10^{-2} + .. + d_{-m} \times 10^{-m}$.

Here Y is the value of the entire number and d_i is the ith digit. The quantity 10 is called the base or 'radix' of the number system. It forms the pivot to determine the weight to be attached to a digit in the number representation.

2.2.1 Addition

It was mentioned earlier that any digit combination representing a number implies a unique value for the number in terms of the weights to the digits; implicitly it implies a unique magnitude for the number. Thus the number 32 is larger than the number 31. The number 32.14 is smaller than the number 32.15 and so on. Assigning magnitudes to numbers in this manner easily lends itself to the concept of addition. Let us consider a few examples.

Figure 2.1a shows the addition of two single digit numbers; the sum is also a single digit number – its magnitude is the sum of the magnitudes of the individual numbers. The numbers are represented as lines at the left of the figure with their lengths representing magnitudes. Addition is easy to visualize here as the addition of magnitudes. Figure 2.1b shows the addition of two single digit numbers – their sum being greater than 9 – the largest single digit number. The graphical magnitude representation and addition process are identical to the above case. Generation of the carry with addition of digitally represented values adds a new dimension to the process. The procedure can be generalized to multi-digit numbers as shown in Figure 2.1.c. It can be extended further to non-integers with decimal point representation as shown in Figure 2.1.d. Note that the decimal point is used essentially to align the numbers to be added; it has no further role in the addition process.

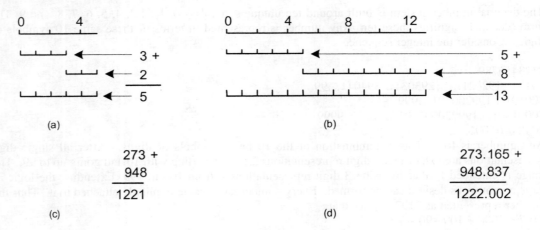

(a)

(b)

(c)

(d)

Figure 2.1 Examples of decimal addition

2.2.2 Negative Numbers and Subtraction

The numbers considered so far are implicitly assumed to be positive. Negative numbers are represented with a negative sign in front. Subtraction can be looked upon as the addition of a negative number to a positive number. A few illustrative examples are shown in Figure 2.2. Figure 2.2a shows a case where the subtrahend is less than the minuend and the result is positive. On the right the numbers are represented as lines with the magnitudes being shown as equivalent lengths. The result is a line of length unity in the positive direction. In Figure 2.2b the minuend is greater than the subtrahend and the result is a negative quantity. Figure 2.2c and 2.2d show the subtraction of non-integral numbers with positive and negative results. The graphical representation is similar to the respective integer examples in Figure 2.2a and Figure 2.2b respectively. Once again the decimal point can be seen to be used to align the numbers; it has no further significance in the subtraction process.

2.2.3 Complement Representation of Negative Numbers

Negative numbers can be represented in an alternative and useful manner – called 'complement' representation. A positive number uses zero as a reference and the magnitude represents its value above zero. In an analogous manner any negative number can use a positive value as reference and its value below the reference can be its magnitude. A single digit negative number can use 9 as a reference; -3 can be represented as 6; it is called the '9's complement' of 6. Two digit numbers can use 99 as reference; 23 can be represented as 76 which is the 9's complement of – 23. In all these cases, 9, 99, and so on, stand for the largest possible number with the same number of digits and digit positions as the negative number under discussion.

Other examples:
- 245   has its 9's complement representation as 754
- 24.5   has its 9's complement representation as 75.4
- 2.45   has its 9's complement representation as 7.54
- 0.245 has its 9's complement representation as 0.754
- 2450   has its 9's complement representation as 7549

Formation of 9's complement involves a digit to digit subtraction; no carry or borrow is involved; to this extent it involves only simple and direct algebra.

2.2.4 Subtraction Using Complements

Consider the subtraction of 245 from 452. We have

$$452 - 245 = 452 + (999 - 245) - 999$$
$$= 452 + 9\text{'s complement of } 245 - 999$$
$$= 452 + 9\text{'s complement of } 245 + 1 - 1000$$
$$= 452 + 754 + 1 - 1000$$
$$= 1207 - 1000$$
$$= 207$$

The step by step procedure of subtraction using 9's complement is as follows:
- Form the 9's complement of the subtrahend.
- Add it to the minuend
- Add 1 to the sum at the least digit position
- We get a carry bit of 1. Ignore it.

Often the last two steps are combined and put as 'take the carry around, and add it at lowest digit position'.

Figure 2.3 is a graphical representation of the subtraction process using complements. OA is the minuend. OB is the subtrahend. BR represents the 9's complement of the subtrahend. If we add OA with BR we get OR + BAa as result. We subtract the reference OR from it to get BAa which is the desired result. From the figure it can be seen to be the same as BbAb. For non-integral numbers, the procedure of subtraction using 9's complement can be generalized as follows:
- Decide the largest size of number to be handled – in terms of digits to the left as well as to the right of the decimal point. Let it have r digits to the right and l digits to the left of the decimal point.

- Form the 9's complement of the subtrahend using a number having l 9's to the left of decimal point and r 9's to its right.
- Add the 9's complement of the subtrahend to the minuend
- Take carry digit (usually 1) and do 'end –around-carry'.

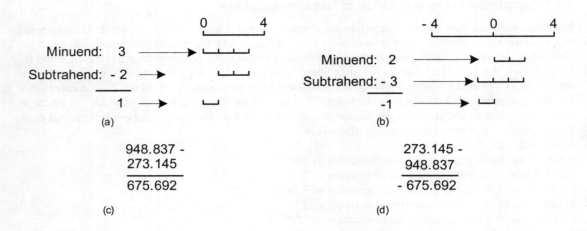

Figure 2.2 Examples of decimal subtraction

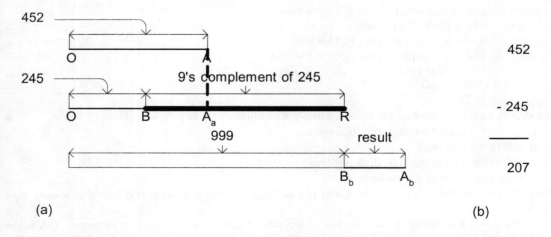

Figure 2.3 Illustration of decimal subtraction (a) Graphical representation (b) Conventional representation

If the subtrahend is larger than the minuend, the result is negative. Subtraction using complement representation does not yield a carry; it is an indication that the result is negative; further the result is directly available on the 9's complement form. The subtraction procedure outlined above can be generalized to include negative results; the last step in the procedure is to be modified as follows:

- If there is a carry, the result is positive; do-end-around carry to get the result. If there is no carry, the result is negative; it is available in 9's complement form.
- The 9's complement representation of a negative number uses the largest number within the specified range as reference. The algebra demands an end-around-carry operation. The 10's complement representation of a negative number uses the smallest number in the next range as reference.

Examples:

-245 has its 10's complement representation as 1000-245 = 755. We are dealing with 3-digit integers; the smallest 4 digit integer is 1000. It is taken as reference.

-24.5 has its 10's complement representation 100.0 – 24.5 = 75.5. The reference number is 100.

Continuing in the same fashion we get

7.55 as the 10's complement representation of – 2.45,

0.755 as the 10's complement representation of –0.245, and

7550 as the 10's complement representation of – 2450.

Identification of the reference for 10's complement formation is easier than that with 9's complement formation; one need not count the number of digits to the right of the decimal point. In digital systems, the number of digits to be used for algebra and position of decimal point (in fixed point algebra) are decided at the outset and do not change; the advantage claimed is trivial. Subtraction for formation of the complement can involve borrow bit at one or more digit positions (unlike the 9's complement formation where, there is no need for borrow); to this extent 10's complement is marginally different from that of 9's complement. Consider the earlier example.

$$452 – 245 = 452 + (1000 – 245) – 1000$$
$$= 452 + 10\text{'s complement of } 245 – 1000$$
$$= 452 + 755 – 1000$$
$$= 207$$

Note that subtraction of 1000 from the result is achieved by (consciously) ignoring the carry. The step by step procedure for subtraction using 10"s complements is as follows:

- Form the 10's complement of the subtrahend.
- Add it to the minuend.
- Ignore the carry bit generated.

The procedure is equally valid for non-integers. The graphical representation of formation of the complement and carrying out subtraction are identical for the 10's complement case as with the 9's complement case. If the subtrahend is larger than the minuend, the result of subtraction is negative; Absence of the carry bit in the result signifies this. Further the negative result is available in 10's complement form. Making allowance for the possibility of a negative result, the procedure for subtraction using 10's complements can be generalized as follows:

- Form the 10's complement of the subtrahend.
- Add it to the minuend.
- If there is a carry, the result is positive; it is obtained by ignoring the carry bit.
- If there is no carry, the result is negative; it is available in 10's complement form.

2.2.5 Floating Point Representation of Numbers

The discussion so far concerns decimal numbers that are integers or numbers represented in fixed point form. In the latter case the number is represented as a digit sequence with a fixed decimal point. As long as the numbers lie within a restricted range, the fixed point representation is acceptable and makes the algebra simple. For numbers over a wider range, the floating point representation is more common. The number representation is in three distinct fields.

- The first field stands for the sign of the numbers – positive or negative.

- The second field is for the mantissa – similar to that in a book of logarithmic tables. Depending on the precision desired, the mantissa may range from 4 to 10 digit
- The third field is for the exponent. It has a sign and (normally) two digits.

A few numbers represented in floating point mode are shown in Table 2.1. Respective number values are also given.

Table 2.1 Examples of number representations in floating point form

Representation	- 0.23456 + 08	+ 0.9876 - 12	- 0.53791 – 21
Number Value	- 0.23456 $\times 10^8$	0.9876×10^{-12}	-0.53791×10^{-21}

Table 2.2 Examples of number representations in floating point form with the convention of using only positive numbers for the exponent

Representation	- 0.23456 + 58	+ 0.9876 - 38	- 0.53791 – 29
Number Value	- 0.23456 $\times 10^8$	0.9876×10^{-12}	-0.53791×10^{-21}

In a slightly altered representation, the sign field for exponent is avoided. The exponent is given a range – –50 to + 49. +50 is added to it and the added value used in the representation. With this convention, the number representation in table 2.1 changes to those in table 2.2

Numbers in floating point mode can be multiplied and divided easily; the multiplication or division operation is carried out on the mantissa. The exponents are added or subtracted as the case may be. The sign of the result is decided following convention. Addition and subtraction operations are more cumbersome with floating point numbers. The exponents are to be made equal and then the addition or subtraction operation is carried out on the mantissa.

2.3 THE BINAR NUMBER SYSTEM

The binary number system uses only two digits – 0 and 1. It has a radix of 2. The binary digit is called a 'bit'; it can take the value '0' or the value '1'. The first few integers starting from 0 are given in decimal form as well as binary form below:

Decimal   01234 567891011

Binary 01 1011100101110111100010011010101011

The convention of forming numbers in the binary number system is similar to that followed with decimal numbers. A bit in a position has a weight attached to it – in terms of powers of 2. Non-integral binary numbers too can be formed in a similar manner. The integral part is to the left of the radix point (similar to decimal point). The fractional part is to its right. As an example consider the binary number 101. 0101

Considering the positional weights, the decimal value that the number signifies can be obtained as in Figure 2.4. The concepts of negative numbers, algebra and floating point representation used with decimal numbers, can be directly extended to the binary numbers.

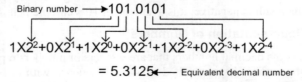

$$1 \times 2^2 + 0 \times 2^1 + 1 \times 2^0 + 0 \times 2^{-1} + 1 \times 2^{-2} + 0 \times 2^{-3} + 1 \times 2^{-4}$$

= 5.3125 ←—— Equivalent decimal number

Figure 2.4 obtaining the decimal value of a binary number

2.3.1 Addition

The numbers are aligned first. The binary bits at the left most position are added first to form the sum bit and carry bit. The bits at the immediate right position are added next; the carry bit from the previous bit addition (on the left side) is added along with these. The procedure is continued in a similar manner and completed.

Examples

$$
\begin{array}{r}
101\ + \\
10 \\
\hline
111 \\
\end{array}
$$

$$
\begin{array}{r}
1011\ + \\
1101 \\
\hline
11000 \\
\end{array}
$$

$$
\begin{array}{r}
1011.10\ + \\
1101.11 \\
\hline
11001.01 \\
\end{array}
$$

$$
\begin{array}{r}
1011.01\ + \\
1101.11 \\
\hline
11001.00 \\
\end{array}
$$

2.3.2 Subtraction

The numbers are aligned and subtraction carried out starting with the left most bits of minuend and subtrahend. Whenever the need arises a minuend bit is borrowed from the left. The procedure is continued until the right most bit position.

Examples

$$
\begin{array}{r}
111\ - \\
10 \\
101 \\
\end{array}
$$

$$
\begin{array}{r}
1101\ - \\
1011 \\
10 \\
\end{array}
$$

$$
\begin{array}{r}
1101.11\ - \\
1011.10 \\
10.01 \\
\end{array}
$$

If the minuend is smaller than the subtrahend, the result is negative. It is obtained by subtracting minuend from subtrahend and attaching a negative sign to the result.

Examples

$10 - 111$ 　　　　　　　　$=$ 　　　　　$- (111 - 10)$
　　　　　　　　　　　　　$=$ 　　　　　$- 101$

$1011 - 1101$ 　　　　　　$=$ 　　　　　$- (1101 - 1011)$
　　　　　　　　　　　　　$=$ 　　　　　$- 10$

1011.10 – 1101.11   =   - (1101.11 – 1011.10)
                                                    =   - 10.01

The subtraction procedure is more involved than the addition procedure. One has to compare the magnitudes of minuend and subtrahend and decide the sign of the result. If it is positive, subtrahend is subtracted from minuend. If the result is negative, the minuend is subtracted from subtrahend and a negative sign is attached to the result.

2.3.3 Complement Representation and Subtraction Using Complements

Negative numbers can be represented by their compliments in a manner analogous to the complements of decimal numbers. 1's complement and 2's complement are in wide use. 1's complement of binary numbers corresponds to the 9's complement of decimal numbers; method of forming complements and doing algebra is similar in both the cases. Since the binary bit can take only one of two values formation of 1's complement of a number is an easy and straightforward affair. Complementing every bit of the number, results in 1's complement of the number itself.

The 1's complement of 1101 is 0010
The 1's complement of 111 is 000
The 1's complement of 1101.11 is 00 10 . 00

Subtraction using 1's complement of the subtrahend follows the steps below:

- Form the 1's complement of the subtrahend.
- Add the 1's complement of the subtrahend to the minuend.
- If a carry bit is generated with the addition at the right most position, the result is positive; it is obtained by ignoring the carry bit and adding a 1-bit at the left most position. It is as though the carry bit at the right end is bodily taken out and added at the leftmost end. Hence the procedure is called 'end around-carry'.
- If a carry bit is not generated at the right-most position, the result is negative; it is already available in the complement form.

Examples

111 – 010 = 111 + 1's complement of 010 & end-around-carry if need be
   = 111 + 101 & end-around-carry if need be
   = 1100 & end-around-carry

The result has a carry bit; the result is positive. End-around-carry gives

$$111 – 010 \quad = \qquad 101$$
$$010 – 111 \quad \rightarrow \qquad 010 +$$
$$\underline{000}$$
$$010$$

The carry bit is absent; the result is negative and 010 is its 1's complement.

$$1011 – 1101 \quad \rightarrow \qquad 1011 +$$
$$\underline{0010}$$
$$1101$$

The result is negative; 1101 is the 1's complement of the result

$$1101.11 – 1011.10 \quad \rightarrow \qquad 1101.11 +$$
$$\underline{0100.01}$$
$$10010.00$$

The result is positive; end-around-carry gives the result as 0010.01

$$1011.10 – 1101.11 \quad \rightarrow \qquad 1011.10 +$$
$$\underline{0010.00}$$
$$1101.10$$

The result is negative; 1101.10 is its 1's complement.

Two observations are in order here.

- Subtraction using complements does not require prior comparison of numbers and changing the subtrahend and minuend positions before subtraction. The gain out of this observation appears trivial for decimal numbers, especially with manual subtraction. But it is a definite gain for (binary) numbers used in digital systems and computers.
- The visualization of complement representation and algebra in terms of corresponding length is equally valid here; it is not repeated.

The procedure for formation of 2's complement of binary number and subtraction using 2's complement are similar to 10's complement and its use for subtraction with binary numbers. Consider a binary number with r bits to the left of the radix point. The binary number of r + 1 bits, formed with r zeroes following a 1-bit, is the smallest number of r+1 bits size. Subtract the former from the latter to form its 2's complement. Alternatively the one's complement of the number can be formed and a 1-bit added at the right most position. The 1's complement and 2's complement of a few numbers are given below by way of examples:

Binary number	1's complement of the number	2's complement of the number
1 1 0 1	0 0 1 0	0 0 1 1
1 1 1	0 0 0	0 0 1
1 1 0 1.1 1	0 0 1 0.0 0	0 0 1 0.0 1
1 0 0.0 0 0	0 1 1.1 1 1	1 0 0.0 0 0

Subtraction through 2's complements involves the following steps:

- Form the 2's complement of subtrahend and add it to the minuend.
- The result is positive if the carry bit is present; the result is obtained by ignoring the carry bit
- The result is negative if the carry bit is absent. It is available in 2's complement form.

Examples

$$111 - \qquad\qquad \rightarrow \qquad\qquad 111 +$$
$$010 \qquad\qquad\qquad\qquad\qquad \underline{110} \leftarrow \text{2's complement of 010}$$
$$1101$$

→The result is positive and is equal to 101.

$$010 - \qquad\qquad \rightarrow \quad 010 +$$
$$111 \qquad\qquad\qquad \underline{001} \leftarrow \text{2's complement of 111}$$
$$011 \rightarrow \text{The result is negative; 011 is its 2's complement}$$

$$1101.11 - \qquad\qquad \rightarrow \quad 1101.11 +$$
$$1011.10 \qquad\qquad\qquad \underline{0100.10} \quad \leftarrow \text{2's complement of 1011.0}$$
$$10010.01$$

The result is positive; it is 10.01.

$$1011.10 - \quad \rightarrow \quad 1011.10 +$$
$$1101.11 \qquad\qquad \underline{0010.01} \leftarrow \text{2's complement of 1101.11}$$
$$1101.11 \rightarrow \text{The result is negative; its 2's complement is 101.11.}$$

2.3.4 Multiplication

For the multiplication of single bit numbers we have

$0 \times 0 = 0$

$0 \times 1 = 0$

$1 \times 0 = 0$

$1 \times 1 = 1$

Multiplication of multi-bit numbers is carried out on the same lines as with decimal numbers. In the decimal system at every stage a partial product is formed by multiplying the multiplicand by a digit in the multiplier. The partial product is shifted and added to the product term; the process is continued until completion. In the binary system, the value of any multiplier bit is either 0 or 1; in turn the partial product is zero or the multiplicand itself. Hence, the multiplication process simplifies to repeated shifts and additions of the multiplicand. The procedure comprises of the following steps:

- Start with a zero as the product.
- If bit b0 of the multiplier is 1, add multiplicand to the product;
- Shift multiplicand to left by one bit
- If bit b1 of multiplier is 1, add the shifted multiplicand value to the product.
- Repeat the shift and conditional add procedure until all the bit values of the multiplier are examined.
- The position of radix point for the product is decided separately.

Example

Multiplicand → 1 0 1 1   ×   1· 1 0 1   Multiplier   Multiplier bits are the indices
                  1 0 1 1   ——————— 1   to shifted position and
                                    ——————— 0   corresponding partial product.
              1 0 1 1   ———— 1
          1 0 1 1   — 1
        ———————————
        1 0 0 0 1 1 1 1   Final product

2.3.5 Division

Division is carried out by successive subtraction and shifting. The process is continued until the dividend becomes less than the divisor. At that stage, the value of the dividend left over, is the reminder. The step by step procedure is as follows:

- Assign zero value to the quotient.
- Align the LSB of division to the LSB of dividend.
- If the divisor is less than the MS part of dividend at the aligned position, subtract the division from the aligned dividend part. Add 1 to the quotient at the MSB position.
- If the divisor value is more, no such addition is carried out.
- Shift divisor to the right by one bit; again compare it with the 'aligned part' of the dividend.
- Continue the shifting and conditional addition to quotient until the divisor is aligned to the LSB end of the dividend. This completes division.
- The quantity remaining in the dividend is the reminder. The accumulated quantity obtained in the quotient position is the final quotient.

Example

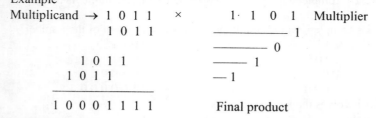

Repeatedly shifting the divisor left until it is aligned with the left most bit of the dividend is equivalent to multiplying it by power of 2: it implicitly adds an equal number of zero bits to the right of the divisor. Thus in the above case aligning 11 to 1011 at the right end and comparing the two numbers, is equivalent to comparing 1100 (divisor $\times 2^2$) with the dividend. Once thus aligned, division is carried out by a repeated sequence of compare –subtract-shift or compare-shift operations. The sequence is carried out until the left-over part of the dividend is less than the divisor; it is the reminder.

A slightly modified version of the sequence of operations to achieve division is available. It is more elegant. The improvement becomes perceptible for numbers involving a large number of bits. The procedure is as follows:

- Align the divisor with the dividend at the left
- Assume the dividend to have one more bit at left, its value being 1 to start with. Let it be designated P
- If P is 1, subtract the divisor from the dividend once, add 1 to the quotient
- Shift divisor to right by one bit position. Shift quotient left by one bit position
- If P = 1, repeat steps (3) and (4)
- If P = 0, add divisor to the dividend once, subtract 1 form the quotient
- Shift divisor to right by one bit position. Shift quotient left by one bit position
- Repeat the steps until the last position. Ascertain that at the end P = 1 and the reminder is less than divisor.

In the procedure here, the shift operation remains the same. Depending on whether P = 1 or P = 0 divisor is subtracted (and quotient incremented) or divisor is added (and quotient decremented). Consider the above example; let us carry out the division using the modified procedure:

- Align divisor to the dividend at the left and insert P = 1 to the left of dividend.

$$[P=1]1\ 1\ 0\ 1\ 1\ - \qquad\qquad\qquad 0\ 0\ 0$$
$$1\ 1 \qquad\qquad\qquad\qquad\qquad\qquad \uparrow$$

Since P = 1, subtract the divisor and increment   Quotient
$$[P=0]0\ 1\ 1\ 1\ 1 \qquad\qquad\qquad 1\ 0\ 0$$
$$1\ 1 \qquad\qquad\qquad\qquad\qquad\qquad \uparrow$$
$$\qquad\qquad\qquad\qquad\qquad\qquad\qquad\qquad \text{Quotient}$$

- Shift divisor to right by one bit position; shift quotient left by one bit position.

$$[P=0]0\ 1\ 1\ 1\ 1\ + \quad 1\ 0\ 0$$
$$1\ 1 \qquad\qquad \uparrow$$
$$\qquad\qquad\qquad \text{Quotient}$$

- P = 0; add divisor to dividend; decrement quotient at the bit position indicated.

$$[P=0]0\ 1\ 1\ 1\ 1\ + \quad 1\ 0\ 0$$
$$1\ 1 \qquad\qquad \uparrow$$
$$[P=1]1\ 0\ 1\ 0\ 1 \qquad 0\ 1\ 0$$
$$\qquad\qquad\qquad\quad \text{Quotient}$$

- Shift divisor by one bit to the right; shift quotient by one bit position to the left

$$[P=1]1\ 0\ 1\ 0\ 1\ + \quad 0\ 1\ 0$$
$$1\ 1 \qquad\qquad \uparrow$$
$$\qquad\qquad\qquad \text{Quotient}$$

- Since P=1, subtract divisor from the dividend; increment quotient at the bit position indicated

$$[P=1]1\ 0\ 1\ 0\ 1\ -\ \ 0\ 1\ 0$$
$$1\ 1\qquad\qquad \uparrow$$
$$[P=1]1\ 0\ 0\ 1\ 0\qquad 0\ 1\ 1$$
$$\text{Quotient}$$

- Now the divisor is at its final position. P=1; the dividend is less than the divisor. Hence we have
Quotient = 011
Remainder = 10

In the procedure detailed above, when we say dividend, it is implicitly assumed to be what is left of the dividend. The procedure can be applied with equal ease to numbers of many number of bits as well as non-integer numbers.

2.3.6 Algebra with Floating Point Numbers

The procedure for algebra with floating point numbers is evident from the representation itself. It is not elaborated further here.

2.4 OCTAL NUMBER SYSTEM

Octal Number System has a radix of 8. Since the radix value is a power of 2, octal numbers are closely related to binary numbers; this also explains their common use in digital systems and digital computers. The octal number system uses the eight digits – 0, 1, 2, 3, 4, 5, 6, and 7. All other numbers are represented in terms of these 8 digits. The sequence of integers from 0 is
0 1 2 3 ....7 10 1112... 1720 21. . . 27 30...77 100
The value of any integer can be obtained using the position based weights. When using numbers to different radices, the suffix to a number signifies its radix. With this convention
123_8 is the octal equivalent of $(1\times8^2+2\times8^1+3\times8^0)_{10}$ $= 83_{10}$
Similarly
1.23_8 is the octal equivalent of $(1\times8^0+2\times8^{-1}+3\times8^{-2})_{10} = (1+2\times0.125+3\times0.015625)_{10}$
$$= 1.296875_{10}$$

2.4.1 Addition

Addition of octal numbers is done on the same lines as with decimal numbers. Individual digits are added; if the sum exceeds 7, a carry is generated and taken to the digits to the left.
Examples –single digits

3+	3+	3+	7+
4	5	7	7
7	10	12	16

Examples – multiple digit numbers

1 2 3 +
6 5 4
7 7 7

1 2 3+
4 5 6
6 0 1

```
1 2 3 . 4 5 6+
6 5 4 . 3 2 1
7 7 7 . 7 7 7
```

```
1 2 3 . 4 5 6+
5 4 3 . 2 1 6
6 6 6 . 6 7 4
```

2.4.2 Subtraction

The radix point of the subtrahend is aligned to that of the minuend; subtraction is carried out starting with the digits at the right end.   Whenever necessary, a 'borrow' is done from the digit to the right.
Examples

```
6 5 4 –
1 2 3
5 3 1
```

```
  1 2 3 –
  6 5 4
- 5 3 1
```

```
6 5 4-
3 4 5
3 0 7
```

```
654.321-
123.456
530.643
```

```
 123.456-
 654.321
-530.643
```

2.4.3 Complement Representation of Negative Numbers

Negative octal numbers can be represented as 7's complements or 8's complements; the procedure to obtain them is similar to that with the complements in the binary and decimal systems. Subtraction can be carried out using either the 7's or the 8's complement.

2.4.3.1 Subtraction Using 7's Complements
Examples

654 –	$\rightarrow$	654 +
123		654← 7's complement of 123
		1530
Result is positive; do end around carry		531← Result
123 –	$\rightarrow$	123 +
654		123← 7's complement of 654
Result is negative and available in 7's complement form.		246← Result

654.321 −
123.456 → 654.321 +
654.321   7's complement of 123.456
1530.642
Result is positive; do end around carry 530.643   Result

123.456 −
654.321 → 123.456 +
123.456   7's complement of 654.321
Result is negative and available in 7's complement form. 246.134   Result

2.4.3.2 Subtraction Using 8's complements
Examples
654 −
123 → 654 +
655   8's complement of 123
1531
Result is positive; ignore carry 531   Result

123 −
654 → 123 +
124   8's complement of 654
Result is negative and available in 8's complement form. 247   Result

654.321 −
123.456 → 654.321 +
654.322   8's complement of 123.456
1530.643
Result is positive; ignore carry 530.643   Result

123.456 −
654.321 → 123.456 +
123.457   8's complement of 654.321
Result is negative and available in 8's complement form. 246.135   Result

2.4.4 Multiplication

Multiplication is carried out by forming the partial products of the multiplicand with individual digits of the multiplier and adding the partial products with appropriate shifts. The radix point (octal point) has to be inserted suitably.

654 × 123	654 × 3	2404	
	654 × 2	1530	
	654 × 1	654	
		105304	product

65.4×12.3	65.4×0.3	24.04	
	65.4 × 2	153.0	
	65.4 × 1	654	
		1053.04	product

2.4.5 Division

The long division procedure is similar to that with decimal numbers. It is detailed through an example.
*   Consider the division of 654_8 by 12_8.

- Align the division to the dividend at the right most octal digit position - 12_8 is aligned to '65'.

$$12 \overline{\smash{\big)}\ 654}$$
$$12$$

- Subtract from 65_8 the partial product of 12_8 with the octal digit that leaves a remainder less than 128: This forms the most significant digit of the (octal) quotient.

$$\begin{array}{r} 5 \\ 12 \overline{\smash{\big)}\ 654} \\ 62 \end{array}$$

- Carry out the subtraction and form new dividend.

$$\begin{array}{r} 5 \\ 12 \overline{\smash{\big)}\ 654} \\ \underline{62} \\ 34 \end{array}$$

- Shift divisor to next position and repeat procedure – subtract partial product from dividend; form the next significant octal digit of the quotient.

$$\begin{array}{r} 52 \\ 12 \overline{\smash{\big)}\ 654} \\ \underline{62} \\ 34 \\ \underline{24} \\ 10 \end{array}$$

- What is left of the dividend is less than the divisor; it forms the remainder.

The quotient is 52_8 and the remainder is 10_8.

Consider the example of non-integer type octal numbers. The steps involved are clear. The division is carried out here without elaborating the steps involved.

$$65.4_8 \div 12.3_8 \rightarrow \quad \begin{array}{r} 5.1202 \\ 12.3 \overline{\smash{\big)}\ 65.4} \\ \underline{63.7} \\ 1.50 \\ \underline{1.23} \\ 0.251 \\ \underline{.246} \\ .003 \\ \underline{.00246} \end{array}$$

Stopping at this stage and ignoring the remainder, we get

$65.4_8 \div 12.3_8 = 5.1202_8$

An algorithm involving sequential subtraction and addition of partial products was detailed with binary numbers earlier. It is applied here with octal numbers. The above two examples are considered for illustration; the steps involved are not detailed.

$$654_8 \div 12_8 \quad \begin{array}{r} 6 \\ \overline{\smash{\big)}\ [P=1]1654-} \\ \underline{74} \\ [P=0]0714 \end{array} \quad \leftarrow \quad \text{Subtract } 6 \times 12 \text{ from 65 and add 6 to quotient at the leftmost position.}$$

$$654_8 \div 12_8 \quad \begin{array}{r} 52 \\ \overline{\smash{\big)}\ [P=1]1654-} \\ \underline{74} \end{array}$$

$$\frac{[P=0]0714+}{74}$$
$$[P=1]1010$$

Shift divisor alignment to the right by one digit and quotient to the left by one digit. Add 6×12 to left-over part of dividend (to make P=1); subtract 6 from quotient at the aligned position.

What is left of the dividend is 10_8; it is less than the divisor and hence forms the remainder. The quotient is 52_8.

$$
\begin{array}{r}
6\\
\hline
\end{array}
$$
$$65.4_8 \div 12.3_8 \;\big|\; \frac{[P=1]165.4-}{76.2}$$

Subtract 6×12.3 from 65 and add 6 to quotient at the leftmost position

$$\frac{[P=0]067.2}{5.1}$$
$$65.4_8 \div 12.3_8 \;\big|\; \frac{[P=1]165.4-}{76.2}$$
$$\frac{[P=0]067.2+}{11.05}$$
$$[P=1]100.25$$

Shift divisor alignment to the right by one digit and quotient to the left by one digit. Add 7×12.3 to left-over part of dividend (to make P=1); subtract 7 from quotient at the aligned position.

$$
\begin{array}{r}
5.13\\
\end{array}
$$
$$65.4_8 \div 12.3_8 \qquad \big|\; \frac{[P=1]165.4-}{76.2}$$
$$\frac{[P=0]067.2+}{11.05}$$
$$\frac{[P=1]100.25-}{0.371}$$
$$[P=0]077.757$$

Shift divisor alignment to the right by one digit and quotient to the left by one digit. Add 3×12.3 to left-over part of dividend (to make P=1); subtract 3 from quotient at the aligned position.

Stopping at this stage, the procedure yields
$65.4_8 \div 12.3_8 = 5.13_8$
Proceeding two more steps further, we get 5.12638 as the quotient.

2.5 HEXADECIMAL NUMBER SYSTEM

The hexadecimal ('hex' in short form) number system has a radix of 16. Since $16 = 2^4$, the hex numbers are closely related to binary numbers as well as octal numbers. Hex numbers require 15 distinct digits apart from 0, for the number representation. It is customary to take these as the nine decimal digits (1, 2,.., 8, and 9) and the alphabetic characters a, b, c, d, e, and f ; the corresponding capital letters are also used interchangeably. These characters are used with equivalent decimal values of 10, 11, 12, 13, 14, and 15 respectively. Some hex numbers and their equivalent decimal counterparts are given below.

Hex number 123..89ab cde f1011....1920

Equivalent
Decimal Number 123..891011121314151617..31   32

Considering the positional weights, the decimal value of any hex number can be ascertained. Examples:

The hex number 123_{16} has the decimal equivalent value $1 \times 16^2 + 2 \times 16^1 + 3 \times 16^0 = 291_{10}$

The hex number fe_{16} has the decimal value $15 \times 16^1 + 14 \times 16^0 = 254_{10}$

The hex number 12.3h has the decimal value $\quad 1 \times 16^1 + 2 \times 16^0 + 3 \times 16^{-1} = 18.1875_{10}$

The hex number f.eh has the decimal value $15 \times 16^0 + 14 \times 16^{-1} = 15.875_{10}$

Formation of complements and algebra with hex numbers, are similar to those with octal umbers. These are not discussed further here.

2.6 NUMBER CONVERSIONS

Numbers are represented in digital systems in binary, octal or hex form. But we deal with numbers in decimal form. There is a need to convert a number in one system to one in another system. Some of the commonly used methods of such conversions are reviewed here. The methods are of use to realize corresponding hardware as well as to do conversions in software. All the procedures of conversion are based on the fact that every digit position in a number has a weight attached to it.

2.6.1 Binary to Decimal Conversion

The conversion is illustrated through examples:

Examples

To convert the binary number 1101_2 to decimal form, we write the value of the number considering the positional weights

$$1101_2 \quad = \quad 1 \times 2^3 + 1 \times 2^2 + 0 \times 2^1 + 1 \times 2^0$$
$$= \quad 8 + 4 + 1$$
$$= \quad 13$$

which is its decimal equivalent. As another example consider the non-integral number 1101.1011_2. Its decimal equivalent is obtained as

$1101.1101_2 \quad = \quad 1 \times 2^3 + 1 \times 2^2 + 0 \times 2^1 + 2^0 + 1 \times 2^{-1} + 0 \times 2^{-2} + 1 \times 2^{-3} + 1 \times 2^{-4}$
$\quad = \quad 8 + 4 + 1 + 0.5 + 0.125 + 0.0625$
$\quad = \quad 13.1875_{10}$

2.6.2 Decimal to Binary Conversion

Consider a binary integer N represented as the bit sequence $b_k\ b_{k-1}\ b_{k-2} .. b_1\ b_0$. Taking into account the positional weights, its decimal equivalent is obtained as

$$N \quad = \quad b_k\ 2^k + b_{k-1}\ 2^{k-1} + .. + b_2\ 2^2 + b_1\ 2^1 + b_0\ 2^0$$

Any decimal number can be represented in the above form. The conversion of a decimal number into binary form reduces to the evaluation of bits b_k, b_{k-1}, ..b_1, and b_0. The evaluation procedure for that is as follows:

- Divide N by 2; b_o is the remainder
- Divide the quotient (N/2) by 2; b_1 is the remainder
- Divide the quotient (N/4) by 2; b_2 is the remainder
- . . .
- . . .
- Proceed until b_k is obtained as the last remainder

Example:

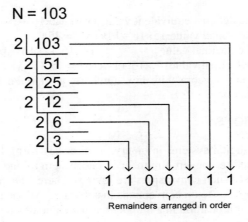

$$N = 103$$

Remainders arranged in order

Check:
$$2^6 + 2^5 + 2^2 + 2^1 + 2^0 \qquad = \qquad 64 + 32 + 4 + 2 + 1$$
$$= \qquad 103$$

Following a similar argument, one can convert the fractional part of a decimal number into binary form by repeated multiplication by 2.

Example

$$M = 0.372 \longleftarrow \text{Decimal value}$$

0.372 X 2
.744 X 2
.488 X 2
.976 X 2
.952 X 2
.904 X 2
.808 X 2
.616 X 2

0 1 0 1 1 1 1

Converted bits arranged in order

$$M = 0.0101111 \longleftarrow \text{Binary value}$$

The error in conversion is the order of 2^{-8}. It can be reduced as much as desired by continuing the repeated division by 2. The procedures to convert the integer and fraction can be combined to convert any non-integral number

Example

$$P = 103.372_{10}$$

From the two forgoing examples

$$P = 1100111.\,0101111_2$$

2.6.3 Octal Number Conversions

The conversion can be carried out in a manner similar to that with binary to decimal conversion
Example:

$N = 123.765_8$
$= 1 \times 8^2 + 2 \times 8^1 + 3 \times 8^0 + 7 \times 8^{-1} + 6 \times 8^{-2} + 5 \times 8^{-3}$
$= 83.97851_{10}$

Decimal numbers can be converted into octal form, as was done with binary numbers: Convert the integer and fractional parts separately and combine the results

Example
$P = 103.372_{10}$

$$
\begin{array}{ll}
8 \mid 103 \\
\quad\quad\quad 12 \quad\quad \rightarrow \quad 7 \\
\quad 8 \\
\quad\quad\quad 8 \mid 1 \quad\quad \rightarrow \quad 4 \\
\quad\quad\quad\quad\quad \rightarrow \quad 1 \\
\quad 0
\end{array}
$$

$\therefore 103_{10} = 147_8$

$$
\begin{array}{lll}
 & & 0.372 \times 8 \\
2 & + & 0.976 \times 8 \\
7 & + & 0.808 \times 8 \\
6 & + & 0.464 \times 8 \\
3 & + & 0.712 \times 8 \\
5 & + & 0.696
\end{array}
$$

$\therefore 0.372_{10} = 0.27635_8$

The error in conversion is of the order of 8^{-6}. Combining the two
$103.372_{10} = 147.27635_8$ (error $\sim 8^{-6}$)

2.6.4 Conversions with Hex Numbers

Hex numbers can be converted to decimal numbers using the positional weights
Examples:

$59\ C.F2_{16} = 5 \times 16^2 + 9 \times 16 + 12 \times 16^0 + 15 \times 16^{-1} + 2 \times 16^{-2}$
$= 1436.9453_{10}$ (accuracy $\sim 10^{-5}$)

Decimal number is converted into hex form by combining the converted values of integer and fractional parts –

Example
$N = 1436.9453_{10}$

$1436_{10} \rightarrow$

$$
\begin{array}{lll}
16 \mid 1436 \\
\quad\quad 16 \mid 89 \quad\quad \rightarrow \quad 12[C] \\
\quad\quad\quad\quad 16 \mid 5 \quad\quad \rightarrow \quad 9 \\
\quad\quad\quad\quad\quad\quad 0 \quad\quad \rightarrow \quad 5
\end{array}
$$

$\therefore 1436_{10} = 59C_{16}$

$$
\begin{array}{rl}
 & \underline{0.9453 \times 16} \\
15[F] \quad + & 0.1248 \times 16 \\
1 \quad + & 0.9968 \times 16 \\
15[F] \quad + & 0.9488 \times 16 \\
15[F] \quad + & 0.1808 \times 16 \\
2 \quad + & 0.8928
\end{array}
$$

$\therefore 0.9453_{10} = \text{F1FF2}_{16}$ (accuracy $\sim 16^{-6}$)

Combining the two parts

$1436.9453_{10} = 59\text{C. F1FF2}_{16}$

2.6.4.1 Conversions amongst Binary, Octal and Hex Numbers

The conversion procedures here are much simpler than those with decimal numbers – due to the fact that all the three radices are powers of two. The procedures are illustrated through examples. A hex number is converted into binary form as follows:

$59.\text{F2}_{16} = (1 \times 2^2 + 0 \times 2^1 + 1 \times 2^0) \times 16^1 + (1 \times 2^3 + 0 \times 2^2 + 0 \times 2^1 + 1 \times 2^0) \times 16^0 + (1 \times 2^3 + 1 \times 2^2$
$+ 1 \times 2^1 + 1 \times 2^0) \times 16^{-1} + (0 \times 2^3 + 0 \times 2^2 + 1 \times 2^1 + 0 \times 2^0) \times 16^{-2}$

$\qquad = 1011001.11110010_2 \quad$ binary equivalent

An octal number is converted into binary form as follows:

$73.64_8 = 7 \times 8^1 + 3 \times 8^0 + 6 \times 8^{-1} + 4 \times 8^{-2}$
$\qquad = (1 \times 2^2 + 1 \times 2^1 + 1 \times 2^0) \times 8 + (1 \times 2^1 + 1 \times 2^0) \times 8^0 + (1 \times 2^2 + 1 \times 2^1) \times 8^{-1} + 1 \times 2^2 \times 8^{-2}$
$\qquad = 111011.1101_2$

A binary number is converted into octal form by segregating the bits on either side of the radix point into groups of 3 bits each and assigning octal positions to the groups. The following example illustrates the procedure:

Example:

$1\,0\,1\,1\,0\,0\,1\,.\,1\,1\,1\,1\,0\,0\,1\,0_2 = 1\,(011)\,(001)\,.\,(111)\,(100)\,(10)$
$\qquad\qquad\qquad = 131.744_8$

A similar segregation into groups of 4 bits each is carried out for conversion into hex form

Example:

$1\,0\,1\,1\,0\,0\,1\,.\,1\,1\,1\,1\,0\,0\,1\,0_2 = 1\,0\,1\,(1001)\,.\,(1\,1\,1\,1)\,(0\,0\,1\,0)_2$
$\qquad\qquad\qquad = 59\,.\,\text{F2}_{16}$

2.6.5 Conversions of Numbers with Other Bases

A number to one base can be converted into a number to any other base using the positional weights. But our familiarity with decimal numbers makes the conversion process easier if we use the decimal number as an intermediary. Let us illustrate it through the conversion from a number with base 7 into one with base 5.

Example:

$123.45_7 = 1 \times 7^2 + 2 \times 7^1 + 3 \times 7^0 + 4 \times 7^{-1} + 5 \times 7^{-2}$
$\qquad = 66.6734693_{10}$

66_{10}

$$
\begin{array}{l}
5 \,\underline{|\,66} \\
\quad 5\,\underline{|\,13} \quad \rightarrow \quad 1 \\
\qquad 5\,\underline{|\,2} \quad \rightarrow \quad 3 \\
\qquad\quad 0 \quad \rightarrow \quad 2
\end{array}
$$

$\therefore\ 66_{10} = 231_5$

0.6734693_{10}

$$\begin{array}{r} 0.\ 6734693{\times}5 \\ \hline \end{array}$$

3	+	0.3673465 ×5
1	+	0.8367325 ×5
4	+	0.1836625 ×5
0	+	0.9183125 ×5
4	+	0.5915625

$\therefore \quad 0.6734693_{10} = 0.31404_5 \ (\sim 5\text{-}6)$

Combination of the two parts yields

$\therefore \quad 66.6734693_{10} = 231.31405_5$

and

$123.45_7 = 231.31405_5.$

2.7 EXERCISES

1) A number system to the base 3 is called a ternary system. It uses 3 digits – 0, 1, and 2. Convert the following decimal numbers into equivalent ternary numbers.

  11; 23; 91; 283; 0.12; 0.17; 0.333; 0.666; 0.363; 2.3; 2.6666

  Convert the following ternary numbers into decimal form.

  11; 21; 101; 222; 0.12; 2.12; 0.1

  Convert the following binary numbers into ternary form.

  11; 101; 1111; 1000; 0.11; 0.101; 0.1111

  Convert the following ternary numbers into binary form.

  11; 101; 121; 211; 0.12; 0.11; 0.1111; 0.021021021; 11.22; 22.11

2) Convert the following binary numbers into numbers to the base 4.

  1; 101; 101101; 10100111; 0.101; 0.1010101; 101.01011

3) Convert the following numbers to the base 4 to binary numbers.

  1023; 3201; 3; 0.303; 0.30213021; 12.21; 321.123

4) Convert the following octal numbers into hex numbers.

  723; 67; 126; 0.723; 0.67; 723.67; 126.723

5) Convert the following hex numbers into octal numbers.

  AB; A2; 2A4; 0.A2; 0.2AB; AB.BA; A2.A2

6) Convert the following octal numbers into decimal numbers.

  632; 236; 25; 77; 0.25; 0.71; 632.25; 25.77

7) Convert the following decimal numbers into octal numbers.

  632; 236; 25; 77; 0.25; 0.71; 632.25; 25.77

8) Convert the following decimal numbers into numbers to the base 5.

  632; 236; 25; 77; 0.25; 0.717171; 632.25; 25.77

9) Convert the following numbers to the base 5 into decimal numbers.

  234; 432; 42; 0.4444; 0.1111; 0.2222; 432.2222

10) Do the following binary algebra.

  1101+101; 101.1101+11.001; 101-10; 0.0111-0.0101; 101-111; 0.0101-0.1001; 101.01-1.11

11) Do the following subtractions using 1's complement. Express the subtrahend in each case in 1's complement form and do addition.

  101-10; 0.0111-0.0101; 101-111; 0.0101-0.0100; 101.01-1.11

12) Repeat the above using 2's complement form.

13) Do the following subtractions using 9's complements.

  1234-234; 4321-1234; 8888-1111; 1111-8888

14) Do the above subtractions using 10's complements.
15) Carry out the following binary multiplications.

    11x11; 111 x111; 111.011 x0.01101; 1.0011 x0.01101

16) Carry out the following multiplications of octal numbers.

    111.011 x110.101; 1.0011 x0.01101; 73.26 x124.652

17) Obtain the log to the base 2 of the decimal number 5.5. Convert it into a binary number
18) Obtain the log to the base 2 of the binary number 101.1. Carry out the full algebra in binary form.
19) Carry out the following binary divisions by repeated subtraction.

    11.10÷10.1; 101.1÷10.01; 11.01÷110.01

20) Carry out the above divisions by repeated subtraction; use 1's complement form of representations of the numbers
21) Carry out the divisions by repeated subtraction; use 2's complement form of representation of the divisor.
22) Carry out the division of the octal numbers as below:

    11.01÷10.1; 101.1÷10.01; 11.01÷110.01; 3.5÷2.5; 5.5÷2.25; 3.25÷6.25

23) Carry out the above octal divisions by repeated subtraction. Represent the divisors in 7's complement form.
24) Carry out the subtractions in above by repeated subtraction. Represent the divisors in 8's complement form.

CHAPTER 3

BOOLEAN ALGEBRA AND COMBINATORIAL CIRCUITS

3.1 INTRODUCTION

The quest for truth starts with simple statements of apparent facts. Each statement is examined in depth through 'whys' and 'what's. Through elementary arguments Aristotle and followers in succeeding centuries, established the basis for the quest; something was true and another was false. In the middle of 19th century, George Boole gave a new meaning to such logic through an algebraic abstraction. It evolved into the Boolean algebra – algebra of variables with two distinct states [21]. The variables are often called 'Boolean' variables, or 'logic' variables.

A Boolean variables has only two states – the states are called 'True' and 'False' ' High' and 'Low' or 'Yes' and 'No'. Boolean algebra is the algebra of such variables. Shannan – in the 1930s discovered that a circuit being 'ON' or 'OFF' is equivalent to a logic variable being true or false. With that all the results of Boolean algebra could be applied to study and analyze the behavior of circuits. This has led to a close parallel and identity between Boolean algebra and algebra with binary numbers. The philosopher may grumble and sneer at the fall of Boolean logic from the pristine heights of his quest for truth to the pedestrian level of 0's and 1's of 'the digital laity'. But the hijacking over, it is on our lap.

3.2 BOOLEAN VARIABLES

A variable which can take only two possible values is a Boolean variable. The values are variously called:

True (T)   and   False (F)
High (H)   and   Low (L)
Yes   and   No
'0'   and   '1'
ON   and   OFF

Boolean algebra is the algebra of such two valued variables. Every binary number is represented in terms of digits having two possible values – 0 and 1. Algebra involving such binary digits has a close parallel in Boolean algebra. The theorems and results of Boolean algebra are directly used in binary algebra.

3.3 BOOLEAN FUNCTIONS

Any Boolean function can be represented as an algebraic function. Alternately it can also be expressed in the form of a table called the 'Truth Table'. The truth table gives the values of the function for all possible combinations of values of the variables of the function. Hence the truth table is a complete representation of the function.   A function can be looked upon as a black box which generates the dependent variable from the given set of independent variables. With this perspective, the independent variables are also called input variables and the dependent variable is called the output variable. One can have various types of functions with the Boolean Variables [17, 18]. A few of them are basic ones and all others can be expressed as combinations of the basic functions.

3.3.1 NOT Function

Let A be the independent (input) variable and Y the dependent (output) variable of a Boolean function. A NOT function has the functional relation given in Truth Table 3.1. In the table, possible input values are shown as 0 and 1. They can equally effectively be shown as F and T, H and L. NO and YES or OFF and ON. Since our focus here is in digital systems, the states of Boolean Variables are taken as 0 and 1. The same convention is followed later too, unless specifically mentioned. The Not function is also called the invert function. The output is the universe of the input. The electronic circuit which realizes the NOT function is represented in any one of the forms shown in Figure 3.1. The circuit is also called the NOT gate. In all cases the 'bubble' signifies the inversion operation. The algebraic representation of the NOT function is

$Y = \overline{A}$

Table 3.1 Truth Table of NOT function

Input A	output Y
0	1
1	0

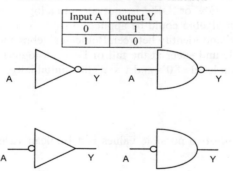

Figure 3.1 Electronic circuit representation of NOT function

3.3.2 AND Function

The AND function implies the statement 'Output is true only if all the inputs are true.' Compare it with the principle – 'The judgment is valid only if the decision of the jury is unanimous'. Truth table for a two-input AND function is given in Table 3.2.   An alternate and more compact form of the truth table is given as Table 3.3. Here X implies 'whatever be the value of the variable concerned'. Thus the first row of the table implies that 'Y = 0 as long as A = 0, whatever be the value of B'. A similar interpretation holds good for the second row. Use of X simplifies truth table representation in many cases. The electronic circuit symbol of the AND function is given in Figure 3.2. Often it is also called the AND gate.

Table 3.2 Truth table of AND function

Inputs		Output
A	B	Y
0	0	0
0	1	0
1	0	0
1	1	1

Table 3.3 Truth table of AND function – alternate representation

Inputs		Output
A	B	Y
0	x	0
x	0	0
1	1	1

Figure 3.2 Electronic circuit representation of AND function

The AND function and AND gate concept can be extended to AND function of more inputs. The algebraic representation of the two input AND function is

$$Y = A . B$$

The '.' symbol signifies AND operation. The 3 input AND function is expressed as

$$Y = A . B . C$$

where A, B, and C are the input logic variables. The concept can be directly extended for AND function of more inputs.

3.3.3 OR Function

The OR function represents the statement
'Output is true if at least any one of the inputs is true'. Compare it with 'The judgment is valid if at least one of the jury favours it'.
Table 3.4 is the truth table for OR function of two logic variables. The electronic circuit symbol of the two input OR function is shown in Figure 3.3. The OR function can be represented as an algebraic relation

$$Y = A + B$$

The '+' symbol signifies OR operation. The OR function concept can be directly extended to more than 2 inputs. With A, B, and C as 3 inputs,

$$Y = A + B + C$$

where Y represents the OR function of A, B, and C.

Table 3.4 Truth table of OR function

Inputs		Output
A	B	Y
0	0	0
x	1	1
1	x	1

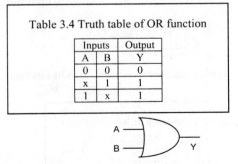

Figure 3.3 Electronic circuit representation of OR function

3.3.4 NAND and NOR Functions

NOT, AND and OR functions are the basic logic functions. All other logic functions can be realized in terms of these, However NAND and NOR functions are more commonly used for reasons to be explained shortly. NAND function is a combination of AND and NOT functions. The two input NAND function can be expressed as

$Y = \overline{A.B}$

It implies the statement 'The judgment is invalid if the jury unanimously says yes'. The two input electronic NAND gate has the symbol in Figure 3.4; it   can be seen to be a combination of the AND and NOT gates. The truth table is given in Table 3.5. The idea of a NAND gate can be extended directly to 3 or more input cases.

The NOR function of two logic variables is expressed as

$Y = \overline{A + B}$

It signifies the statement 'If at least one of the inputs is true, the output is false'. Table3.6 gives the truth table of the NOR function with two inputs. The symbol for the 2 input NOR gate circuit is shown in Figure 3.5. The concept of NOR gate can be directly extended to the case of 3 or more inputs.

Figure 3.4 Electronic circuit representation of NAND function

Table 3.5 Truth table of NAND function

Inputs		Output
A	B	Y
0	X	1
X	0	1
1	1	0

Table 3.6 Truth table of NOR function

Inputs		Output
A	B	Y
0	0	1
X	1	0
1	X	0

Figure 3.5 Electronic circuit representation of NAND function

Figure 3.6 Electronic circuit representation of XOR function

Table 3.7 Truth table of XOR function

Inputs		Output
A	B	Y
0	0	0
0	1	1
1	0	1
1	1	0

3.3.5 Exclusive OR Function

The exclusive OR function of two logic variables is represented by the symbolic equation
$Y = A \oplus B$

The symbol $\oplus$ signifies the 'exclusive OR' – often shortened as XOR – Operation. The function signifies the statement 'Output is true only if the two inputs are unlike each other' or the output is false only if the two inputs are identical.' The electronic circuit symbol of XOR gate is shown in fig. 3.6. The truth Table is given in Table 3.7. XOR function is basically a function of two logic variables.

3.4 LAWS and THEOREMS

There are some commonly used laws and theorems of Boolean algebra. They are useful in the simplification of Boolean expressions, establishing the identity between apparently different expressions, as well as identifying alternate representations of a given expression. The statements are given here with minimal explanations accompanying wherever necessary. All such statements involve only one, two, or three Boolean Variables. The veracity of each such statement can be established by forming the related truth tables [10].

Commutative laws state that

$A+B \qquad\qquad = \qquad B+A$

and

$A.B \qquad\qquad = \qquad B.A$

The law implies that in an AND or OR operation, the order in which the Boolean Variables are considered in an operation, is immaterial.

Associate laws for OR and AND operations are

$A+ (B+C) \qquad\qquad = \qquad (A+B) + C$

and

$A.(B.C) \qquad\qquad = \qquad (A.B).C$

The law implies that the order in which an operation is carried out is immaterial. Distributive law states that

$A.(B+C) \qquad\qquad = \qquad A.B + A.C$

The distributive law means that the order in which the two types of operations are carried out is immaterial. Invoking the commutative law one can state that

$A.(B+C) \qquad\qquad = \qquad (B+C).A$
$\qquad\qquad\qquad\qquad = \qquad A.B + A.C$

from the distributive law. Again applying the commutative law we get

$(B+C)A \qquad\qquad = \qquad BA + CA$

This is an alternate statement of the distributive law

The commutative, associate, and distributive laws are illustrated through corresponding digital circuit blocks in Figure 3.7

3.4.1 Laws of OR and AND Operations

There are four laws related to the basic operations involving logic states. For any logic variable A we have

$A . 0 \qquad\qquad = \qquad 0$
$A . 1 \qquad\qquad = \qquad A$
$A + 0 \qquad\qquad = \qquad A$
$A + 1 \qquad\qquad = \qquad 1$

Assigning possible values to A, the laws also imply the following:-

$$0 . 0 \qquad\qquad = \qquad 0$$
$$1 . 1 \qquad\qquad = \qquad 1$$
$$0 + 0 \qquad\qquad = \qquad 0$$
$$0 + 1 \qquad\qquad = \qquad 1$$
$$1 + 1 \qquad\qquad = \qquad 1$$

We can also confirm the following through truth table verification.

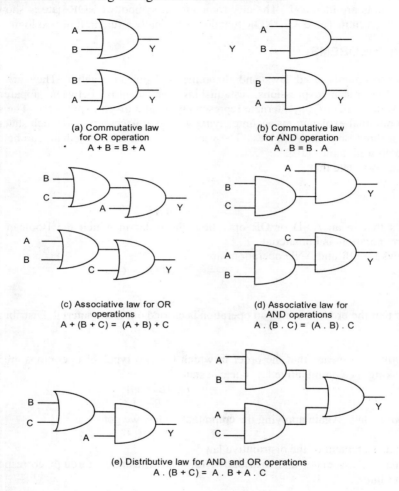

(a) Commutative law
for OR operation
A + B = B + A

(b) Commutative law
for AND operation
A . B = B . A

(c) Associative law for OR
operations
A + (B + C) = (A + B) + C

(d) Associative law for
AND operations
A . (B . C) = (A . B) . C

(e) Distributive law for AND and OR operations
A . (B + C) = A . B + A . C

Figure 3.7 Electronic circuit representations of laws of Boolean algebra

$$A . \overline{A} \ = 0$$
$$A + \overline{A} = 1$$
$$\overline{\overline{A}} = A$$

3.4.2 De Morgan Theorem

De Morgan theorem can be stated as

$$\overline{A + B} = \overline{A} \cdot \overline{B}$$

The theorem can be proved by forming the truth tables for the expressions on either side and confirming that they are identical. An alternate and equally valid statement of De Morgan theorem is the dual of the above statement.

$$\overline{A \cdot B} = \overline{A} + \overline{B}$$

The equivalence statements of De Morgan theorem are illustrated for the corresponding electronic circuit blocks in Figure 3.8

(a)   (b)

Figure 3.8 Electronic circuit representations of De Morgan theorems: (a) $\overline{A \cdot B} = \overline{A} + \overline{B}$ and (b) $\overline{A + B} = \overline{A} \cdot \overline{B}$

De Morgan theorem implies that the bubbles at the input side of an operation are equivalent to a corresponding bubble at the output side of the dual operation. Here the AND and OR operations are the implied duals.

3.4.3 Duality Theorem

The duality theorem is the generalization of a result we have used above in specific cases. The theorem states that any Boolean relation has a corresponding dual relation; the dual relation is obtained by carrying out the following changes to the relation.
- Change each AND operation to an OR operation
- Change each OR operation to an AND operation and
- Complement any 0 or 1 appearing in the expressions of the relation

As an example consider the relation (De Morgan theorem)

$$\overline{A + B} = \overline{A} \cdot \overline{B}$$

By duality theorem we get another relation

$$\overline{A \cdot B} = \overline{A} + \overline{B}$$

which is the alternate statement of De Morgan theorem.

3.4.4 Two Tier Representations of Logic Functions

A logic function may involve AND, OR, and Invert operations in combinations and sequences. But by a systematic application of the above laws and theorems all such functions can be simplified to a 2-tier format. It may take one of the following two forms.
- A set of OR functions, with their outputs ANDed to form the final output. It is often referred as the 'Product of-Sums' form.
- A set of AND functions, with their outputs ORed to form the final output. It is often referred as the 'Sum of Products' form.

As an example consider the function Y of Boolean variables A, B, and C

$$
\begin{aligned}
Y &= ((A\overline{B} + AB)C + A\overline{C})(\overline{B} + C) \\
&= (A\overline{B}C + ABC + A\overline{C})(\overline{B} + C) \\
&= A\overline{B}\,\overline{B}\,C + AB\overline{B}\,C + A\overline{B}\,\overline{C} + A\overline{B}\,CC + ABCC + A\overline{C}\,C \\
&\quad A\overline{B}\,C + A\overline{B}\,\overline{C} + ABC
\end{aligned}
$$

This is the sum of products representation of Y.

Taking the complement of the right side twice,

$$Y = \overline{\overline{\overline{A}BC + A\overline{B}\overline{C}\ ABC}}$$

$$= \overline{\overline{\overline{A}BC}.\overline{A\overline{B}\overline{C}}.\overline{ABC}}$$

$$= (\overline{A}+B+\overline{C}).(\overline{A}+B+C).(\overline{A}+\overline{B}+\overline{C})$$

This is the product of sums representation of Y. Either form can be used for circuit realization. Logic functions can be simplified and represented with minimum number of AND and OR operations. Such minimal functions can be used for circuit realizations. Minimization problems are dealt with in text books on digital design [4, 5].

3.5 UNIVERSAL GATES

IC realization of any logic function requires all the basic functions to be realized using respective logic gate circuits. They can be interconnected to realize the overall function. But use of one standard circuit element to realize all the function types is preferred. NAND and NOR gates are examples of such standard elements. Any logic function can be realized by repeated and judicious use of NAND gates. A 2-input NAND gate is the simplest such element – for this reason, it is often called a 'universal' gate. For the same reason a 2-input NOR gate is also called a 'universal' gate. Use of 2-input NAND gates to realize *different* functions is illustrated in figure 3.9. All such functions can be realized using 2-input NOR gates also.

3.6 APPLICATIONS OF LOGIC CIRCUITS

Logic circuits are used for a variety of applications. In each case one has to examine the application requirements, formulate the logic expressions required and realize them in terms of the gates discussed so far [7]. In many cases three categories of circuits are in frequent use; they are discussed in more detail here. The categories are
• Multiplexers and demultiplexers – more often referred as 'mux' and 'demux' circuits
• Data encoders and decoders
• Algebraic circuits

3.6.1 Multiplexers and Demultiplexers

Consider the circuit of Figure 3.10. S is a logic input called a 'selector bit'. I_1 and I_2 are two data input lines; each carries data represented by values – 0 and 1. When $S = 1$, output of AND gate G_1 is I_1 itself; that is as long as $S = 1$, output of G_1 is 1 if $I_1 = 1$ and 0 if $I_1 = 0$. It is as though S has opened a gate for the data on I_1. Since $S = 1$, $\overline{S} = 0$. Output of AND gate G_2 is zero. It does not change even if the data on input line I_2 changes. It is as though gate G_2 is closed and does not allow the data on line I_2 to go through. The OR gate R combines the outputs of G_1 and G_2 and makes it available on the output line O. Since output of G_2 is zero, O reflects the data on I1. When $S = 1$, data on input line I1 is selected and made available on the output line 0. By a similar argument one can see that if $S = 0$, data on input line I_2 is selected and made available on the output line. Combining both cases one can see that the circuit as a whole multiplexes the data on the two input lines I_1 and I_2 and makes it available on output line O. Selection is decided by the logic input on selector line S. Figure 3.11 shows typical input waveforms for I_1 and I_2 and the change in output as the selector line value changes. During time interval A to B, $S = 1$ signal I_1 is selected and connected to O through gate G_1. From instant B to C, $S=0$ and gate G_1 is closed; gate G_2 is open and signal I_2 is connected to output O. The concept can be extended to multiplex 4, 8, or 16 inputs to one output. A 4-to-1 mux circuit is shown in Figure 3.12. It has two select lines and four input lines. If $S_1 = 0$ and $S_0 = 0$, data on line I_0 is made available on output line O. Similarly, for values of 01, 10 and 11 on $S_1 S_0$ pair, data on input lines I_1, I_2, and I_3 respectively are made available on output line O. If only 3 inputs are to be multiplexed, line I_3 may be omitted. $S_1 S_0$ need not be assigned the value I_1. If the number of channels is more than 4 but less than 8, a 7-to-1 mux can be used. The unconnected channels need not be activated.

Consider the circuit of Figure 3.13 (a). It has one input select line S. If $S=0$, gate G_0 is enabled, data on input line I is made available on output line O_0. Output of gate G_1 is zero – gate G_1 is disabled. Similarly when $S=1$, Gate G_0 is disabled. Gate G_1 is enabled and data on input line I made available on output line O_1. The circuit is a '1-to-2 demultiplexer' – demux. The demux operation with a square wave input is shown in Figure 3.13 (b).

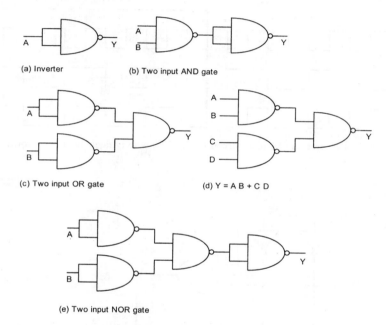

(a) Inverter

(b) Two input AND gate

(c) Two input OR gate

(d) Y = A B + C D

(e) Two input NOR gate

Figure 3.9 Realization of a few basic Boolean functions using the 2-input (universal) NAND gate

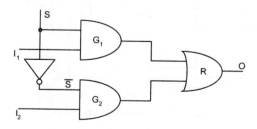

Figure 3.10 Circuit of a 2-to-1 mux

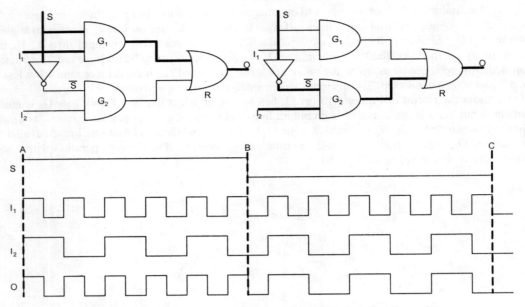

Figure 3.11 Operation of 2 to 1 mux of Figure 3.10 with typical signal waveforms

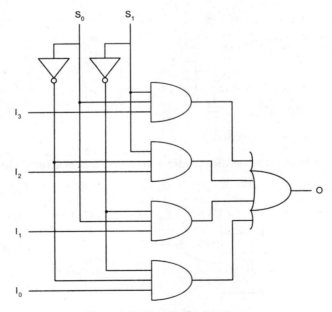

Figure 3.12 Circuit of a 4-to-1 mux

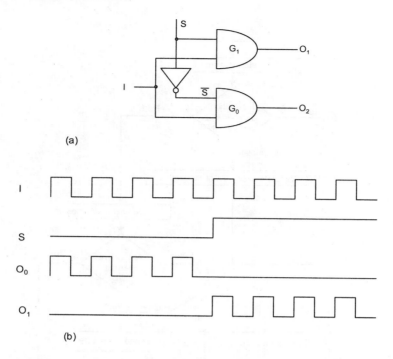

(a)

(b)

Figure 3.13 2-to-1 demux (a) Circuit (b) Operation with typical signal waveforms

The concept of demux can be directly extended to demux data on one data line to multiple output lines. Figure 3.14 shows a 1-to-4 demux. Path selection and control is through a set of two select lines S_1 and S_0. When $S_1 S_0 = 00$, gate G_0 is enabled. Data on line I is steered to output line O_0. All the other three gates are disabled. Similarly for $S_1 S_0$ values of 01, 10 and 11 data on input line I is steered to outputs O_1, O_2 and O_3 respectively. A demux of 3 output lines can be obtained by deleting gate G_3 and output line O_3. Select lines S_1 and S_0 are not given the input value 11. The concept can be directly extended to demux data on one data line into 8 or 16 output lines; 3 or 4 path select lines are needed for this. By suitably deleting gates and data combinations on the select lines demux of other sizes can be accommodated.

3.6.2 Circuits for Algebra

Consider the addition of two binary bits – a and b. In general the sum is 2 bits wide. If comprises of a sum bit S and a carry bit C. All combinations of values of a and b are considered and corresponding values of S and C bits are given in Table 3.8.
A cursory examination of the table shows that S and C can be generated as the logic functions of a and b:

$$S = a \oplus b$$
$$C = a . b$$

The adder circuit in terms of corresponding gates is shown in Figure 3.15 (a). The adder circuit to add two bits is called a 'Half Adder'. The half adder as a block is shown in Figure 3.15 (b). The half adder accepts two input bits and generates a sum bit and a carry bit. In a multi-bit adder, the adder unit used at any intermediate bit position is called a 'Full Adder'. Consider a full adder at the ith bit position. It has to accept a carry bit input – C_{i-1} generated from the adder in the previous stage. C_{i-1} is

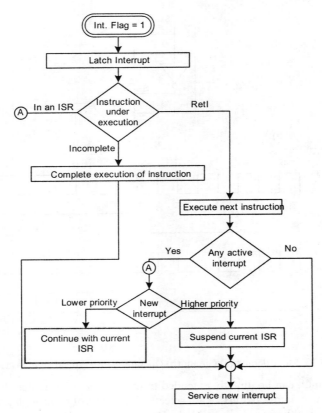

Figure 13.7 Flowchart for interrupt service

Table 13.3 Arithmetic instructions of 8051 Series ('*' signifies availability of the instruction in the mode. The execution time is given in terms of the number of machine cycles)

Mnemonic	Operation	Addressing modes				Execution time
		Dir.	Ind.	Reg	Imm	
Add A, <byte>	(A)+<byte>→(A)	*	*	*	*	1
ADDC A, <byte>	(A)+<byte>+C→(A)	*	*	*	*	1
SUBB A, <byte>	(A)–<byte>–C→(A)	*	*	*	*	1
INC A	(A) + 1 → (A)	Accumulator only				1
INC <byte>	<byte> + 1 →<byte>	*	*	*		1
INC DPTR	(DPTR) + 1 → (DPTR)	Data pointer only				1
DEC A	(A) – 1 → (A)	Accumulator only				1
DEC <byte>	<byte> – 1 →<byte>	*	*	*		1
MUL AB	(B) × (A) → (B):(A)	ACC & B only				4
DIV AB	Int[(A)/(B)] → (A) Mod[(A)/(B)] → (B)	ACC & B only				4
DA A	Decimal adjust	Accumulator only				1

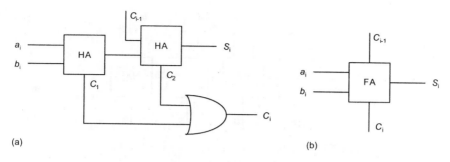

Figure 3.16 Full Adder (a) Circuit (b) Block diagram representation

Formation of a full adder in terms of 2 half adders requires 6 levels of gating. The scheme is useful from the point of view of easy understanding. A preferred approach is to form the truth tables for S and C in terms of a, b, and c and realize their functions in sum-of-products or product-of-sums form. With either form, the functions can be realized in an optimized manner. Minimization procedures are discussed in books on digital design.

Figure 3.17 shows a nibble adder formed with one half adder and a set of three full adders. The adder chain can be expanded as much as desired. In all such multi-bit adders, the carry bit is generated after propagation of the addition process through all stages; it causes maximum delay. To circumvent the delay problem, quality adders generate the carry through a circuit dedicated for it.

Figure 3.18 shows the circuit of a half subtractor. It generates a difference bit and borrow bit as outputs (Verify circuit function using truth table). Two half subtractors can be combined to form a full subtractor. Full subtractor chains can be used to subtract one multi-bit number from another. These are left as exercises.

Subtraction can be done by adding 2's complement of the subtrahend to the minuend. Figure 3.19 shows such a subtractor for two nibbles. Nibble b_3 b_2 b_1 b_0 is complemented (1's complement). It is added to nibble a_3 a_2 a_1 a_0. Addition of 1 at right end is equivalent to forming the 2's complement of b_3 b_2 b_1 b_0 and adding it to the minuend a_3 a_2 a_1 a_0. If the carry bit $C=1$, result represented by d_3 d_2 d_1 d_0 is positive. If $C=0$, the result is negative. d_3 d_2 d_1 d_0 represents its 2's complement value. The subtractor size can be increased by adding inverters and full adders to the nibble subtractor. In all such multi-bit subtractors the carry bit generation occurs with maximum delay and puts an upper limit on the speed of operation of the subtraction.

3.6.3 Multiplier

Binary numbers are multiplied by repeating the multiply, shift and add sequence. The circuit to multiply two binary nibbles is shown in Figure 3.20. The steps involved in the multiplication process are as follows:

- From the partial product of multiplicand $a_3a_2a_1a_0$ with $b0$. This is the first partial sum.
- Form the partial product of a_3 a_2 a_1 a_0 with b_1. Shift it left by one bit position. Add it with the above partial sum to form the new partial sum.
- From the partial product of a_3 a_2 a_1 a_0 with b_2. Shift it left by one bit position. Add it with the partial sum to get the new partial sum.
- Form the partial product of a_3 a_2 a_1 a_0 with b_3. Shift it left by one bit position. Add it with the partial sum. The partial sum at this stage is the final product.

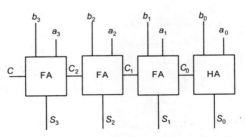

Figure 3.17 A nibble adder in terms of Half and Full adders

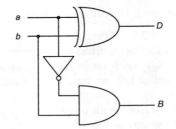

Figure 3.18 Half subtractor circuit

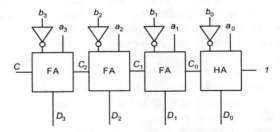

Figure 3.19 A nibble subtractor in terms of Half and Full adders

The sequence of forming partial products, shifting and forming partial sums is shown in Figure 3.21. Its one to one correspondence with the circuit of Figure 3.20 eases understanding of the multiplication operation. Multiplication is an important and widely occurring activity in any digital system. As such the following observations are in order here:

- The nibble multiplier can be looked upon as a multiple-input (8 bit) multiple-output (8 bit) Boolean function set
- The function set realization shown in Figure 3.20 is perhaps the most elaborate and least optimized? One can realize it in terms of basic logic functions – after optimizing it. The types, number, and size of the gates can be used as variables for optimization. The optimization aims at compactness of the circuit (minimization of Si area in an IC) and speed of operation.
- Byte multipliers can be realized by a direct extension of the concept. The same can be done for multiplication of 16 bit and 32 bit words. However the process becomes cumbersome and unrealistic, beyond 16 or 32 bits. Nevertheless the whole multiplication is carried out at 'one go'.
- Multipliers of larger sizes are realized in 2 ways in ICs.
    * Repeat the nibble or byte multiplier block in the IC.

* Use one nibble, byte or word multiplier unit repeatedly. After every addition of shifted partial product, store the partial sum in a storage location (to be discussed in the following chapter). Repeated use of the multiplier unit requires the activity to be carried out in a sequential manner – in consecutive time slots of operation of the digital system

- The choice between the two approaches to realization of the multiplier discussed above is essentially a hardware–software trade-off problem. It shows up at different stages and levels in any digital system design.

3.6.4 Encoders and Decoders

A variety of encoders and decoders are in use. They transform a set of bits into another set on a one to one basis. An encoder does the transformation to make the set more compact by reducing the number of bits, the reduction being carried out conforming to a defined rule. Alternatively, the encoder may transform a bit combination into another with a larger number of bits; the aim may be to mask it behind a veil or to build redundancy and hence reliability. A decoder receives an encoded bit set and retrieves the original bit set. The general procedure of encoding (or decoding) is to use the rule of conversion and express the output bits as Boolean functions of the input bits [5]. The functions are realized through the standard gates we are familiar with. A representative set of encoders and decoders is studied here.

3.6.5 2-to-4 Decoder

Consider the circuit of Figure 3.22 – a_1 and a_0 are two input data lines. Using these, 4 output lines O0, O_1, O_2, and O_3 are generated. The truth table for O_0, O_1, O_2, and O_3 in terms of a_1 and a_0 is given in Table 3.9. We can see that if a_1 a_0 is treated as a 2-bit binary number, O_0 = 1 when the binary number is zero,

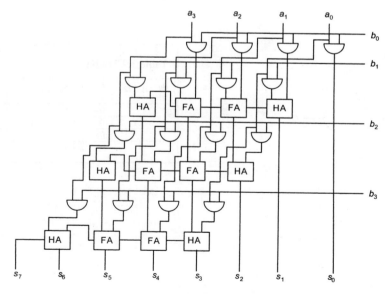

Figure 3.20 A nibble multiplier in terms of Half and Full adders

		a_3b_0	$a_2 b_0$	$a_1 b_0$	$a_0 b_0$
	a_3b_1	$a2 b_1$	$a_1 b_1$	$a_0 b_1$	
S_5	S_4	S_3	S_2	S_1	S_0
a_3b_2	$a_2 b_2$	$a_1 b_2$	$a_0 b_2$		

S_6	S_5	S_4	S_3	S_2	S_1	S_0
a_3b_3	$a_2 b_3$	$a_1 b_3$	$a_0 b_3$			
S_7 S_6	S_5	S_4	S_3	S_2	S_1	S_0

Figure 3.21 Multiplication of two binary nibbles by forming partial products and partial sums

$O_1 = 1$ when it is 1,
$O_2 = 1$ when it is 2, and
$O_3 = 1$ when it is equal to 3.
The 2-to-4 decoder can be directly expanded to form a 3-to-8 decoder and a 4-to-16 decoder. All such decoders can be realized using AND gates and inverters. The 7-segment decoder is commonly used for numerical displays. It is discussed in Section 10.5.2.

3.6.6 BCD Encoder

In some applications decimal numbers are retained and processed without changing their decimal identity. It requires each decimal digit to be represented separately using binary bits. One common approach is to use the Binary Coded Decimal (BCD) representation. Here each decimal digit is represented by 4 binary bits; the binary values 0000, 0001, 1000, and 1001 are used to represent digits 0, 1, 2, ... 8, and 9 respectively – The values 1010, 1011, 1100, 1101, 1110 and 1111 are not used. A few decimal numbers and their BCD equivalents are shown in Table 3.10 as examples. One can see that each set of 4 bits has a separate identity. The bit combinations cannot be directly related to their equivalent binary values.

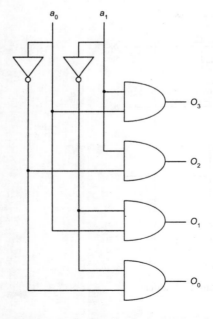

Figure 3.22 The circuit of a 2-to-4 decoder

Table 3.9 Truth table of 2-to-4 decoder					
a_0	a_1	O_0	O_1	O_2	O_3
0	0	1	0	0	0
0	1	0	1	0	0
1	0	0	0	1	0
1	1	0	0	0	1

Table 3.10 Examples of decimal numbers and their equivalent BCD representations

Decimal number	Equivalent BCD representation
15	0001 0101
151	0001 0101 0001
191	0001 1001 0001
909	1001 0000 1001

Figure 3.23 shows a decimal to BCD encoder. It has 10 input lines – each represents a decimal digit value. If the digit value is 0, the zero line is at 1 state. All the other 9 lines are at 0 state. If the digit value is 1, the 1-line is at high state and all others are at 0 state and so on.

b_0 is high if the digit value is 1,3,5,7 or 9.
b_1 is high if the digit value is 2,3,6 or 7.
b_2 is high if the digit value is 4,5,6 or 7.
b_3 is high if the digit value is 8 or 9.

In the encoder the values of these 4 bits are realized through a set of 4 OR gates.

3.6.7 Parity Bit Encoder

Digital systems store, transmit and retrieve data as part of their regular operations. Any one of these activities can corrupt the data. To make data secure against such corruption, redundancy is built into the data before storage or transmission. During retrieval the redundancy property can be revoked to check the authenticity of the data and retrieve the original data. Parity check is the most elementary form of such redundancy operations.

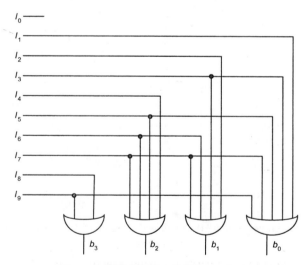

Figure 3.23 Decimal to BCD encoder

To implement parity checking data is segmented into groups, each group being of a definite and predetermined size. Such grouping into bytes is common. Each byte is examined for the number of 1-bits and an additional bit – called the 'parity bit' – is added to it. The parity bit is assigned the value 1, if the number of 1's in the byte is odd; it is assigned the value 0, if the number of 1 bits in the byte is even. With the addition of the 9th bit – bit b8 – the group of 9 bits always has even number of 1s. We are said to have implemented even parity. If the data gets corrupted, one bit may change state – from 0 to 1 or 1 to 0 (We implicitly assume here that such data corruption possibility is low for a size of 8 bits and at the worst only one bit can get corrupted). The parity condition is no longer satisfied. Through an 'even parity check' we can detect the error – called 'parity error'. If such a parity error is detected, the data byte can be declared as erroneous and discarded.

The parity checking and parity bit insertion process conform to the 'even parity' scheme, where we go by the dictum 'even number of bits is right'. Alternatively we can make each group of the 9 bits representing a byte of data conform to 'odd parity'. Parity checking and parity bit insertion can conform to the dictum 'odd number of bits is right'. Such a scheme is called an 'odd parity' scheme. For a given situation, one goes by even parity or odd parity throughout. Parity implementation is the basic form of error checking and building reliability. The principle can be used repeatedly and more redundancy built into the scheme. It improves reliability and allows correction of corrupted data. Adding a parity bit to a group (like a byte) is called 'parity encoding' [checking the encoded group (like a group of 9 bits) can be called 'parity decoding']. The simplest scheme to implement parity encoding is to use XOR operation. If we have two bits – a and b – generate bit c as

$$c \quad = \quad a \oplus b$$

The set of 3 bits a, b, and c together conform to even parity. The idea can be extended to a byte represented by – b_7, b_6, $\ldots$ b_1, and b_0 – to generate a parity bit b_8 as

$$b_8 \quad = \quad b_7 \oplus b_6 \oplus \ldots \oplus b1 \oplus b_0$$

The 9 bits together satisfy even parity. The scheme to generate b_8 from the 8 bits – b_7, b_6, $\ldots$ b_1, and b_0 – using 2-input XOR gates is shown in Figure 3.24(a). It is represented in a simplified form in Figure 3.24(b) as an 8 input XOR gate.

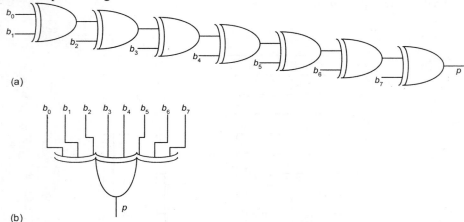

(a)

(b)

Figure 3.24 Circuit for parity bit generation (a) Circuit with 2-input XOR gates (b) Equivalent 8-input XOR gate representation

A similar scheme can be used to do parity checking with the 9 bits – b_7, b_6, $\ldots$ b_1, and b_0 – and a parity bit p be generated. One can see that

$$p \quad = \quad 0$$

if the even parity check is satisfied.

$$p \quad = \quad 1$$

if the even parity check is not satisfied. The parity bit – p – being equal to 1, is an indication that an error has occurred. It can be used as a flag to reject the erroneous byte of data or to seek a retransmission of the byte itself.

3.6.8 Gray Code

Representation of natural numbers as binary numbers in the form 0000 for 0001 for 1, 0010 for 2, 0011 for 3, and so on is one scheme. The approach here is similar to decimal system with weights for positions. Other representations are also possible; the Gray code is one such. Table 3.11 gives the first few integers and their binary and Gray code representations. Gray code has the characteristic that the number of bit transitions between adjacent numbers is always one; this is in contrast with the binary numbers where such transitions can take place in one or more bits. For example as we advance from 0111 to1000 all the bits change. When we advance from 0011 to 0100, three bits change. Displacement encoding is one of the common applications of Gray code. Whenever angular or linear displacement is to be converted into equivalent digital form, Gray code is the preferred choice. The fact that successive bit combinations differ only by one bit, minimizes the conversion error. However once a displacement value is available in such a digital form, it has to be converted into equivalent binary form; the binary form of data is easily amenable for further processing. Gray code is converted into equivalent binary form with the following procedure:

- For any number representation, Gray code and binary have the same number of bits. Let these be g_{n-1}, g_{n-2} ... g_1, and g_0 for the Gray code and b_{n-1}, b_{n-2}, .. b_1; and b_0 for the binary
- Make $b_n = g_n$
- For a bit b_i, we have the relation $b_i = b_{i+1} \oplus g_i$
- Repeat the procedure for all b_is down to b_0

Conversion of an *n* bit Gray Code into binary form requires n-1 XOR gates each of the 2-input type. Figure 3.25 shows the conversion scheme.

A number in binary form can be converted into Gray code through the following procedure:

- Make $g_n = b_n$.
- For a bit g_i, we have the relation $g_i = b_{i+1} \oplus b_i$.
- Repeat the procedure for all g_i down to g_0.

Figure 3.26 shows the conversion scheme.

The procedure requires n-1 number of XOR gates each of the two input type.

Table 3.11 Integers and their binary and Gray code representations

Decimal Number	Binary Equivalent	Gray code Equivalent	Decimal Number	Binary Equivalent	Gray code Equivalent
0	00000	00000	16	10000	10000
1	00001	00001	17	10001	10001
2	00010	00011	18	10010	10011
3	00011	00010	19	10011	10010
4	00100	00110	20	10100	10110
5	00101	00111	21	10101	10111
6	00110	00101	22	10110	10101
7	00111	00100	23	10111	10100
8	01000	01100	24	11000	11100
9	01001	01101	25	11001	11101
10	01010	01111	26	11010	11111
11	01011	01110	27	11011	11110
12	01100	01010	28	11100	11010
13	01101	01011	29	11101	11011
14	01110	01001	30	11110	11001
15	01111	01000	31	11111	11000

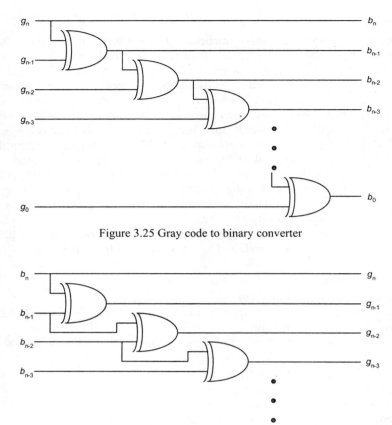

Figure 3.25 Gray code to binary converter

Figure 3.26 Binary to Gray code converter

3.7 ARITHMETIC LOGIC UNIT (ALU)

Relative to the size of circuits that we are discussing here, the size of an IC is practically infinite. One can have a more comprehensive approach to realization of the various functions discussed earlier. For example consider two bits a and b. One can have an adder, subtractor, AND, OR, XOR functions realized using the two bits. Any one of these individual functional outputs can be selected and made available; it requires a function selector to be provided. A circuit with a variety of functions built-in and the facility to select any one of them at a given time for implementation is called an ALU.

A single bit ALU is shown in Figure 3.27. The ALU has a and b as inputs and 7 outputs. The three function select bits S2S1S0 together select one function and steer its output to the common output bit through the 7 input OR gate. The selection is done through a 3-to-8 decoder whose outputs gate the desired functional output to the OR gate. A practical ALU is far less complex. The functions will not be realized through such separate functional units. All will be combined and an optimal functional realization will be done. Further practical ALUs will be 8 or 16 bit wide depending on the system. The function select inputs will be common for all of them.

An ALU is a key functional block in all μCs and many digital systems. It allows a sequence of different functions to be selected and executed. Programmability is built into it.

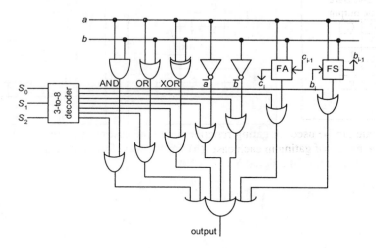

Figure 3.27 A single bit ALU

3.8 EXERCISES

1) 3.13.1 Express the following functions in sum of products form

$$D = (A + B + \overline{C})(A + C)(\overline{A} + \overline{B} + C)$$

$$C = (A + B)(\overline{A} + B)$$

$$D = (\overline{A} + \overline{B} + \overline{C})(A + B + B)(A + B + \overline{C})$$

2) Express the following functions in product of sums form

$$C = AB + \overline{A}\overline{B}$$

$$D = ABC + \overline{A}BC + \overline{A}\overline{B}\overline{C}$$

$$D = \overline{A}B + AB\overline{C} + \overline{A}B\overline{C}$$

3) Give the truth tables for each of the functions in Exercises 3.1 and 3.2 above

4) Truth tables of a few functions are given in Tables 3.12, 3.13, and 3.14. Express each function in sum of products form. In each case the last column represents the output.

5) Express each of the functions in Exercise 3.4 in product of sum form

6) Give the truth tables for the functions below

(a) $C = \overline{A}B + A\overline{B} + \overline{A}\overline{B}$ (b) $D = ABC + \overline{A}\overline{B}C + A\overline{B}C$ (c) $D = (A + B + \overline{C})(\overline{A} + \overline{B} + C)$

(d) $E = ABCD + \overline{A}BCD + \overline{A}B\overline{C}D$

7) Use the 2-input NAND gate as a basic building block. Realize each of the functions in Exercise 3.6 using the NAND gate

8) Use the 2-input NOR gate as the basic building block. Realize each of the functions in Exercise 3.6 using the NOR gate

Table 3.13A, B, and C are inputs; D is the output			
A	B	C	D
0	0	0	0
0	0	1	1
0	1	0	1
0	1	1	0
1	0	0	1
1	0	1	0
1	1	0	0
1	1	1	1

Table 3.14A, B, and C are inputs; D is the output			
A	B	C	D
0	0	0	1
0	0	1	0
0	1	0	0
0	1	1	1
1	0	0	0
1	0	1	1
1	1	0	1
1	1	1	0

Table 3.12A and B are inputs; C is the output		
A	B	C
0	0	0
0	1	1
1	0	1
1	1	0

9) A two input gate can be used for gating function. In all the cases in Figure 3.28 A is a gating input. (a) Explain the nature of gating in each case. (b) For the waveforms of A and B signals sketched in (j), sketch the waveform of the signal Y.

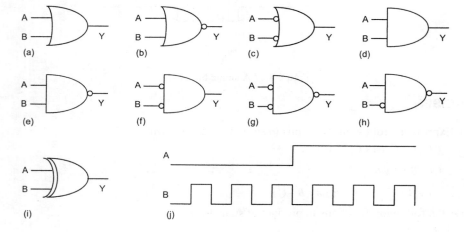

Figure 3.28 Use of 2-input gates for gating a signal

10) Two signals – A and B are given as inputs to an NAND gate. The output is C. A and B are square waves. Different cases of phase shift between A and B are considered as in Figure 3.29. In each case, sketch the waveform of C. The voltage levels for all signals are 0V (0 state) and 5V (1 state). Plot the average value of C against the phase shift between A and B

11) The signals A and B in Figure 3.29 are given as inputs to a 2-input XOR gate. Repeat the steps in Exercise 3.10 for this case.

12) Give the logic expressions for the following. Based on them, give the circuit to realize each. The circuits should be in terms of 2 input NAND gates

    4-bit binary to BCD code converter
    8-bit binary to 3 BCDs converter
    Three BCDs to 8-bit binary converter [The 3 BCDs together represent a number less than 256 in magnitude]
    4-bit binary adder
    4-bit binary subtractor

BCD adder

BCD subtractor

In the 4-bit adder, give the logic function for the carry bit. Generate the carry bit with a separate circuit

Binary to Grey Code converter – both of 8 bits

Grey code to binary converter – both of 8 bits

Decimal to 7 segment decoder

8-bit even parity generator

8-bit odd parity generator

8-bit even parity checker

8-bit odd parity checker

Half adder

Full adder

2-bit multiplier

4-bit multiplier in terms of 2-bit multiplier blocks, 4-bit adder blocks and other minimal necessary gates

8-bit multiplier in terms of 4-bit multiplier blocks, 4 bit adder blocks and other minimal additional gates necessary

13) For each case in Exercise 3.12, identify the signal output which requires the longest path to be traversed. Here longest is implied in terms of the number of basic gates – AND, OR or NOT type. Each such gate causes a delay of 2 ns for propagation. Obtain the worst case delay for the signal. Note that the worst case delay decides the maximum speed of operation of the circuit.

14) Give the circuit of a 3-to-8 decoder. Build a 4-to-16 decoder in terms of two 3-to-8 decoders.

15) Build a versatile 4-bit ALU. It is to have two sets of 4-bit inputs, a carry input bit and a set of function select input lines. All function select input lines are to be in encoded form (3 function select inputs can select one of 8 possible functions). The ALU can have 4 output bits, a carry output bit and if necessary an additional output bit with suitable functional assignment.

It may be instructive to prepare tables for representing output bits in terms of all input bits and then realize the ALU using them.

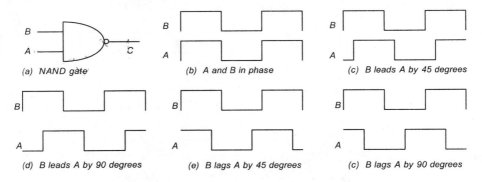

(a) NAND gate   (b) A and B in phase   (c) B leads A by 45 degrees

(d) B leads A by 90 degrees   (e) B lags A by 45 degrees   (c) B lags A by 90 degrees

Figure 3.29 NAND gate with 2 periodic signals at different phases

CHAPTER 4

FLIP FLOPS, COUNTERS, AND REGISTERS

4.1 INTRODUCTION

With the developments in the last two chapters, an identity between numbers, number systems and algebra of numbers on the one hand and logic variables, Boolean algebra and functions of Boolean variables on the other hand, has been established. The identity and correspondence between the two helps in realizing algebraic operations in terms of logic circuits. All such quantities – numbers or logic variables – and the operations – logic or arithmetic have one thing in common – an existence at the present. Values of variables and their relations are specified and studied on an 'as is where is' basis. None of them relate to the past or future. There is no room for questions like the following:

- 'A Boolean variable has a value now or a given set of values at specified past instants of time. What will be its value at a future instant?'
- 'A Boolean variable has a set of given values in the past. Another Boolean variable is dependent on this. Can the latter take on values now, depending on the past values of the former?'

The questions beg time to be discretized and brought into the picture. Further, values of Boolean variables are to be memorized or stored and retrieved later. Such storage and retrieval helps in relating the states or values of one Boolean variable at one instant of time to those of the same and/or other Boolean variables at other instants. Provision of storage and the possibility to relate Boolean variables at different times lead to the abstraction of a digital system as shown in Figure 4.1. A set of Boolean inputs to the model are represented by $I_1, I_2 \ldots$. Variable values in the past instants of time are stored in the memory in layers M0 (present), and past represented by $M_{-1}, M_{-2}, \ldots$ Their values are retrieved and used along with the inputs $I_1, I_2, I_3, \ldots$ in combinations. The combinatorial functions relate these. Such relations are specified in terms of Boolean functions discussed so far. The Boolean functions generate outputs. Some of these are stored in memory. Others are system outputs represented by $O_1, O_2, O_3, \ldots$ Schemes of the type in Figure 4.1 are modeled as sequential circuits and studied [21]. The aim here has been merely to establish the need for a memory and it's potential.

4.2 FLIP FLOPS

A flip flop can be looked upon as a binary storage element. It has two possible states which can be designated as 0 and 1. When it is in state 0, it is said to store (or memorize) the binary value 0. Similarly when it is in state 1, it is said to store (or memorize) the binary value 1. Further the flip flop will continue to be in its state and retain the value in it, until the stored value is changed externally at some instant of time.

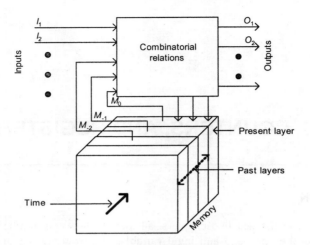

Figure 4.1 Abstract model of a digital system

4.2.1 Flip Flops with NOR Gate

A simple flip flop can be formed by connecting two NOR gates as shown in Figure 4.2. The truth table of the flip flop is given in Table 4.1. We can observe the following:

- If $R = 1$ and $S = 0$, the flip flop takes the definite state with $Q = 0$ and $\overline{Q} = 1$. The flip flop is said to be in 'reset' state.
- If $R = 0$ and $S = 1$, the flip flop takes the definite state with Q =1 and $\overline{Q} = 0$. The flip flop is said to be in 'set' state.
- If $R = 0$ and $S=0$, two possibilities exist for the outputs Q and $\overline{Q}$.
- If the input values were $R = 0$ and $S = 1$ prior to this state, the output continues to remain in the state Q = 1 and $\overline{Q} = 0$ which it had before.
- If the input values were R = 1 and S = 0 prior this state, the output continues to remain in the state $\overline{Q} = 1$ and $Q = 0$ which it had before.
- If $R = 1$ and $S = 1$, both outputs – Q and $\overline{Q}$ – can be at 0. The very purpose of the flip flop is to provide two outputs Q and $\overline{Q}$ complementary to each other. This state is prohibited.

Connection of the two NOR gates in a back to back fashion as shown in the figure has led to an output pattern not encountered so far: For $S = 0$ and $R = 0$, the flip flop retains its old output set. We have been able to achieve the property of memory using basic gates. All types of flip flops, registers, counters etc., build up on the memory property and exploit it in different ways. A basic RS flip flop of the type discussed here can be formed using any of the universal gates. Figure 4.3 shows such alternate realizations (See Exercise 4.1). The input signals can be R and S or $\overline{R}$ and $\overline{S}$.

The RS flip flop of Figure 4.2 has been modified and shown in Figure 4.4 (a). The input positions have been interchanged. With that the input combination S = 1 and R = 0 directly relates to the set state. Similarly $S = 0$ and R = 1 directly relates to the reset state of the flip flop. When $EN = 1$, the transitions in S and R are directly available at the input leads X and Y of the flip flop. The flip flop behaves in a manner identical to that of the flip flop in Figure 4.2. When $EN = 0$ the input leads to the flip flop X and Y are at zero state; they are cut off from R and S signals. Transitions in R and S signals do not affect the flip flop behavior. Since $X = 0$ and $Y = 0$, the flip flop continues to remain in the last state. The truth table for the flip is shown in Table 4.2. It is implicitly assumed here that the two outputs of the flip flop will be complements. Only one of them is included in the table. The same procedure will be followed hereafter too. The nomenclature used for the output states Q_n and Q_{n+1} warrant an explanation. The transitions in EN decide the flip flop behavior. Any transition in the flip flop state – from set to reset state or vice versa – can occur only after a positive transition in EN from 0 to 1. On a negative transition on EN- from 1 to 0 – the flip flop retains its previous state. It continues retention of the stored value until the following positive transition on EN. The time interval between two successive negative transitions on EN is referred as nth, n+1th, etc. The storage in the flip flop refers to these time intervals. The transition instants in the enable input decide the flip flop behavior; the flip flop is also known as a 'clocked flip flop'. The commonly used symbol of the clocked flip flop is shown in Figure 4.4 (b). It is customary to keep the duration of the state – $Q = 1$ (referred as the 'clock pulse') – short; this duration will be normally enough to sense the value of R and S inputs. As such the values of R and S are assumed constant during the short clock pulse. With this assumption a set of waveforms of input signals EN, R, and S are shown in Figure 4.5. Corresponding transitions in the output Q are also shown. The time instants at which the transitions in input signals occur are marked as $t_1, t_2, t_3 \ldots$ etc. Time instants $t_1, t_3, t_5, t_7, t_9, t_{11}$, and t_{13} represent the starting points of the clock pulses; the flip flop takes the new status at these instants. Time instants $t_2, t_4, t_6, t_8, t_{10}, t_{12}$, and t_{14} represent the end points of the clock pulses; the flip flop freeze the new status at these instants. In the intervals $t_2 - t_3, t_4 - t_5, t_6 - t_7, t_8 - t_9, t_{10} - t_{11}$, and $t_{12} - t_{13}$ the value of EN is zero and the output remains at the frozen value even though the R and S values may change.

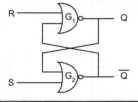

Figure 4.2 A flip flop formed using 2 NOR gates

Table 4.1 Truth table of the flip flop in Figure 4.2				
R	S	Q	$\overline{Q}$	Remarks
0	1	1	0	Set state
1	0	0	1	Reset state
0	0	Last state		Memory property
1	1	Not allowed		

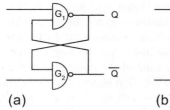

(a)

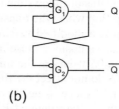

(b)

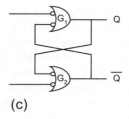

(c)

Figure 4.3 Flip flops formed with different universal gates

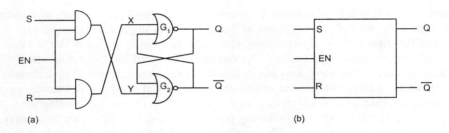

(a)   (b)

Figure 4.4 A clocked RS flip flop (a) circuit in terms of basic gates (b) Commonly used symbol

Table 4.2 Truth table of the clocked RS flip flop in Figure 4.4

EN	R	S	Qn+1
0	X	X	Qn
1	0	1	1
1	1	0	0
1	0	0	Last state
1	1	1	Not allowed

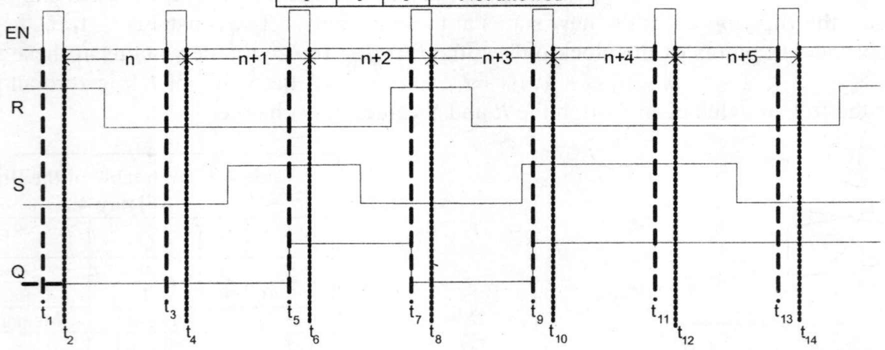

Figure 4.5 Waveforms of signals illustrating the operation of the clocked RS flip flop of Figure 4.4

4.2.2 Edge Triggered Flip Flops

The clocked flip flop either stores the past status or follows the inputs – all decided by the enable input. It is an improvement over the plain RS flip flop where the transitions in output always depend on those at the input. The edge triggered flip flop is a modified version of the clocked flip flop. Consider a positive edge-triggered flip flop. It generates a positive pulse inside the flip flop whenever the enable signal changes from 0 to 1 state. The R and S values at the time of the pulse decide the next state of the flip flop. Subsequent changes in R or S do not affect the flip flop status. The pulse should be long enough to allow the transitions inside to take place; it need not last any longer; typically the trigger edge is a few nanoseconds wide. A positive edge-triggered RS flip flop with its normal symbol, is shown in Figure 4.6 (a). All transitions in the output take place at the positive going edge of the clock. Operation of the edge-triggered flip flop can be understood from the waveforms shown in Figure 4.6(b). It is customary to use a periodic clock signal so that all possible transition instants are known and fixed beforehand. $t_1, t_3, t_5, \ldots$ are the possible transition instants – being the instants of positive going edges. Values of R and S are sampled at these instants and the flip flop output changed accordingly. The waveforms of EN, R, and S signals in Figure 4.6(b) are identical to those of Figure

4.5. At instant t_3, R and S are at 0; the flip flop output retains its last value. At all other positive clock edges the flip flop status is decided by whether $R = 1$ or $S = 1$.

Figure 4.7(a) shows the symbol used for a negative edge-triggered RS flip flop; the flip flop behavior is illustrated with the waveforms for EN, R, and S as in the last two cases [Figure 4.7(b)]. The transition instants for the output are marked as t_2, t_4, t_6, ... etc. At instant t_2, R and S are at 0; the flip flop output retains its last value. At all other negative clock edges the flip flop status is decided by whether $R = 1$ or $S = 1$.

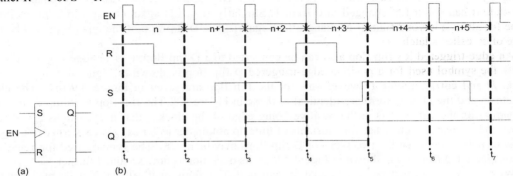

Figure 4.6 (a) Positive edge-triggered flip flop and (b) its operational waveforms

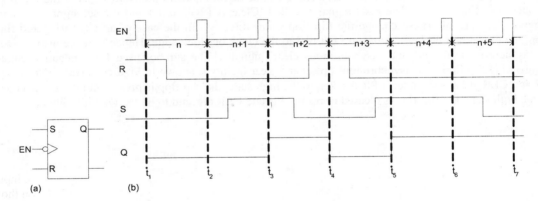

Figure 4.7 (a) Negative edge-triggered flip flop and (b) its operational waveforms

Digital systems have a number of flip flops connected in different configurations and functioning simultaneously. If all of them are selected as edge-triggered flip flops, transitions will take place simultaneously and synchronously. Such systems are referred as 'synchronous systems'. The predictability and orderly behavior of synchronous systems have made them more popular and more widely used than the 'asynchronous systems' (those not using edge-triggered flip flops as the basic element).

4.2.3 D Flip Flop

The RS flip flop can be modified in different ways to prevent it being in the forbidden state. The D flip flop is one such modification. The clocked RS flip flop of Figure 4.4 (b) has been modified to form a D flip flop and shown in Figure 4.8 (a). The input lead D is directly connected to the S input form a D flip flop and shown in Figure 4.8 (a). The input lead D is directly connected to the S input and $\overline{D}$ is connected to R; R and S inputs are always complements ($R = S$ condition is avoided). The symbol for D flip flop is shown in Figure 4.8 (b). As long as $EN = 1$, the flip flop is transparent. Any change in D is instantaneously reflected as a change in Q also. When EN goes to zero, the flip flop status is frozen at the value prior to the transition of EN to the 0 value. Typical waveforms of operation are shown in Figure 4.8 (c). In the interval $t_1 - t_2$, $EN = 1$; changes in D appear as corresponding changes in Q; same is the case with the interval $t_3 - t_4$. During interval t_2-t_3 $EN = 0$. Q remains frozen at 1 state it had when EN changed to 0 from 1. Similarly at t_3, EN again changes to 0, Q retains its value for all time beyond (until EN changes to 1 again). The D flip flop of the type discussed here, is more often called a latch.

An edge-triggered RS flip flop also can be converted to a D flip flop; it is an edge-triggered D flip flop – the symbol used for a positive edge-triggered D flip flop is shown in Figure 4.9 (a). A set of clock, D and corresponding Q waveforms for the flip flop are given in Figure 4.9 (c). The clock waveform and the D waveform are identical to those in Figure 4.8. The flip flop takes up a new state depending on the value of D at the positive going edge of the clock – that is at instants t_1 and t_3. D values and changes in it at other time instants of time do not matter as far as Q is concerned. Change in behavior pattern between the two types of D flip flops may be noted. The symbol used for a negative edge-triggered D flip flop is shown in Figure 4.9(c); its operation is not explained further.

When power is turned on to a digital system with flip flops in it, all the signals may not take predictable values; it may result in an erratic behavior of the system. It can be avoided by providing separate inputs to set or reset the flip flop. These can be direct inputs to the NOR gates of the flip flop as shown in Figure 4.10. The reset signal is called 'Clear (CLR)' signal and the set signal is called 'Preset (PR)' signal. These change the flip flop status directly. In the case of an edge triggered flip flop, the preset and clear functions are not tied to the clock or clock edge. Hence they are also known as 'asynchronous preset' and 'asynchronous clear' signals. The symbol used for a negative edge-triggered D flip flop with asynchronous preset and clear facility is shown in Figure 4.11(a). Normally PR and CLR are at low state; if PR is taken to the high state, the flip flop is preset; and if CLR is taken to the high state, the flip flop is cleared or reset. PR and CLR are said to be 'active high' inputs.

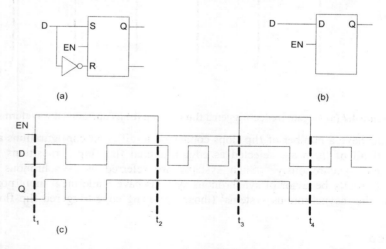

Figure 4.8 D Flip flop and its operation (a) Modification of clocked RS flip flop to form the D flip flop (b) Commonly used symbol of the D flip flop and (c) Operational waveforms

Another version of the negative edge-triggered D flip flop with asynchronous PR and CLR inputs is shown in Figure 4.11(b). Here normally PR and CLR are at the high state; if PR is taken to the low state, the flip flop is preset; instead if CLR is taken to the low state, the flip flop is cleared or reset. PR and CLR are 'active low' inputs.

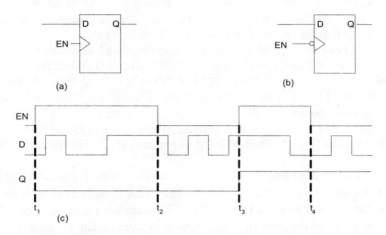

Figure 4.9 Edge-triggered D Flip flop and its operation (a) Symbol of positive edge-triggered D flip flop (b) Symbol of negative edge-triggered D flip flop and (c) Operational waveforms of the positive edge-triggered D flip flop

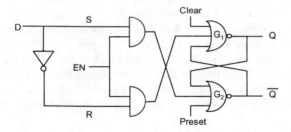

Figure 4.10 D flip flop with PR and CLR inputs

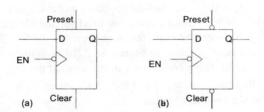

Figure 4.11 Symbols of negative edge-triggered D flip flops with PR and CLR inputs (a) PR and CLR are active high (b) PR and CLR are active low

4.2.4 JK Flip Flop

The JK flip flop is another modification of the RS flip flop to avoid the forbidden stake. It is formed by the connections shown in Figure 4.12 (a). The S input to the RS flip flop is formed by ANDing the $\overline{Q}$ state, clock pulse input, and an external logic input termed J. Similarly the R input to the RS flip flop is formed by ANDing the Q state, clock pulse input and an external logic input termed K. Operation of the JK flip flop for various input conditions is as follows:

- The flip flop has been configured as a positive edge-triggered one. The CLK input is used to generate a positive pulse P given to the two AND gates G_1 and G_2. When the CLK input is at state 1 or 0, P = 0. Outputs of gates G_1 and G_2 are 0; R = S = 0. The flip flop continues in its previous state. Any transition Q or $\overline{Q}$ can occur only when CLK changes from 0 to 1 state. Only these transition instants are considered below.

- Consider the case when $J = 0$ and $K = 0$. Outputs of gates G_1 and G_2 are 0; $R = S = 0$; when the pulse P occurs, the flip flop continues to be in its previous state. This corresponds to the storage mode.

- Consider the case when $J = 1$ and $K = 0$. If the flip flop is in reset state prior to the clock pulse, $\overline{Q}$ = 1 and $Q = 0$. Output of G_2=0. When the clock pulse occurs, output of G_1 goes to 1 state. Since S = 1and $R = 0$, the flip flop gets set. If the flip flop were already in the set state $\overline{Q}$ = 0 and Q = 1 prior to the occurrence of the positive pulse, Outputs of gates G_1 and G_2 are at 0, $R = S = 0$, when the clock pulse occurs; the flip-flop output continues in its old state; that is, it is in set state.

- Combining both the conditions, the flip flop can be seen to get set to $Q = 1$ state if $J = 1$ and $K = 0$ at the positive edge of CLK.

- Operation for the case $J = 0$ and $K = 1$ can be analyzed on the same lines as above. When the clock pulse occurs, the flip flop gets reset and Q becomes 0.

- Consider the fourth condition when $J = 1$ and $K = 1$. Two possibilities arise. Prior to the clock pulse, if $Q = 1$, at the positive edge of CLK, $S = 0$ and $R = 1$. The flip flop gets reset and Q becomes 0. In the alternate case Q = 0 prior to the clock pulse; when the positive edge of clock pulse occurs, $S = 1$ and $R = 0$ and the flip flop gets set. Combining the two cases, the flip flop can be seen to change state. Thus if $J = 1$ and $K = 1$, with every clock pulse, the flip flop changes state. It toggles; often the flip flop with these conditions is said to be a 'Toggle Flip Flop' – or a 'T flip flop'.

Truth table of the JK flip flop is shown in Table 4.3. The case when CLK is steady at 0 or 1 causes no change in the flip flop status. It is not included in the truth table. All the cases shown are implicitly taken as the instants of transition when the positive edge of the clock occurs. Q_{n+1} stands for the value of Q after the nth clock pulse; Q_n is the value before the clock pulse occurs. Preset and clear facilities can be added to the JK flip flop. The symbol commonly used for the clocked JK flip flop with Preset and clear facilities is shown in Figure 4.12 (b); the way they are shown, PR and CL are active low signals. The JK flip flop with the configuration in Figure 4.12 (b) is characterized by the following:

- It is the most versatile of the flip flops considered so far with all the facilities and modes.
- All its states are definite; there are no illegal and forbidden states.
- All the transitions occur at the (positive) clock edge.

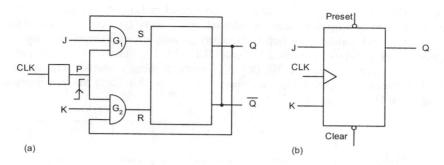

Figure 4.12 (a) Positive edge-triggered JK flip flop formed with RS flip flop (b) Symbol of Positive edge-triggered JK flip flop with active low PR and CLR inputs

Table 4.3 Truth table of JK flip flop: The last column gives the Q value after the active clock edge

J	K	Qn+1
0	0	Qn
1	0	1
0	1	0
1	1	$\overline{Q}_n$

The versatility and definiteness of operation of the JK flip flop makes it the automatic choice for building different digital circuits and systems. Hence it is popular and widely used. (Use of the negative clock edge for operation of the flip flop is equally acceptable. Similarly active high type of PR and CL can be used as inputs to preset and clear the flip flop). All subsequent discussions in the book use the JK flip flop in Figure 4.12 (b) as the basic one (exceptions are specifically indicated).

The Master-Slave flip flop (MS flip flop) is an alternative, versatile, and equally popular flip flop (See Exercise 4.21). As far as the discussions using flip flops wherever we use JK flip flop are concerned, the MS flip flop can be used equally effectively. MS flip flop is not discussed further here [10].

4.3 COUNTERS

Counters form one of the most common applications of flip flops. They help count and keep track of number of events of a repetitive nature. Numbers of people passing through a turn style, number of vehicles crossing a point on a road, number of packets being stocked in a carton, number of voltage pulses on a signal line are examples of such count requirements. Formation of counters using flip flops and counters of different types are discussed here. Due to its versatility the JK flip flop is taken as the basic unit to build all the counters here. Further all the JK flip flops considered are assumed to respond to the negative edge of the clock.

4.3.1 Ripple Counter

Consider a negative edge-triggered JK flip flop with $J = 1$ and $K = 1$ as shown in Figure 4.13 (a). A clock input of square wave type is considered as input clock to the flip flop. J and K being at 1 state, the output Q toggles at every negative edge of CLK. The CLK and Q waveforms are shown in the Figure 4.13 (b). It is evident from the figure that for every two pulses of CLK, Q gives out one pulse: This is independent of whether CLK is periodic or not; even if the pulses on CLK line occur at irregular intervals the transitions on Q take place faithfully at the negative edges of CLK signal.

Three JK flip flops – designated a, b and c – are connected in Tandem in Figure 4.14 (a). All the J and K inputs are at state 1. Flip flop a has an external square wave type clock signal as input to it. Its output Qa forms the clock input to flip flop b. Similarly Q_b output forms the clock input to flip flop c. The output waveforms are shown sketched in Figure 4.14 (b). The negative transitions on CLK are at instants $t_0, t_1, t_2, \ldots$ etc. Output of flip flop a changes at these instants. Transitions on Q_b are coincident with the negative edges of Q_a – at instants $t_1, t_3, t_5 \ldots$ etc. Similarly the transitions on Q_c are at the negative edges of Q_b – at instants $t_3, t_7, t_{11} \ldots$

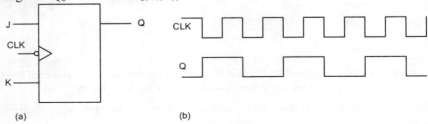

(a)   (b)

Figure 4.13 (a) Symbol of the negative edge-triggered JK flip flop and (b) CLK and Q waveforms with $J = 1$ and $Q = 1$

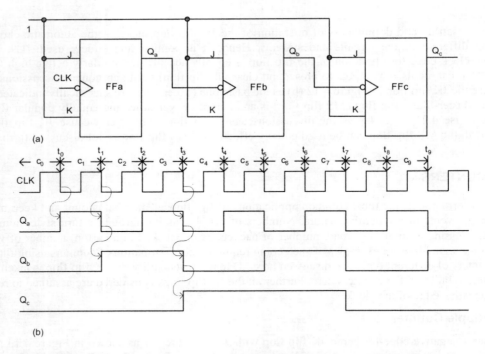

(a)

(b)

Figure 4.14 (a) A ripple counter with 3 flip flops and (b) The counting waveforms

Successive periods of CLK signal are designated $C_0, C_1, C_2, \ldots$ States of the three flip flops in successive intervals are tabulated in Table 4.4. If we take Q_a, Q_b, and Q_c together as a binary number, it can be seen to advance from 000 to 111 repeatedly. The Increment takes place at the negative edges of

CLK. The flip flop combination can be looked upon as a binary counter – counting the number of negative pulses on CLK. The following observations are in order here:

- The count value advances at the negative edges on CLK – whether the edges are spaced equally or not.
- An additional flip flop – d – can be added to the chain with Q_c as its clock input. With that the binary count will advance from 0000 to 1111 in 16 successive pulses. Generalizing with K flip flops connected in tandem, the counter chain counts up to 2K pulses.
- The transitions on Q_a occur at the negative edges of CLK; the transitions on Q_b occur at the negative edges of Q_a; those on Q_c occur at the negative edges of Q_b and so on. The transitions occur with one triggering that in the following flip flop – in a serial fashion. It imparts the name 'serial binary counter' to the counter chain. It is analogous to the ripples in a wave; the counter is also called a 'ripple counter'.
- At instants $t_0, t_1, t_2, \ldots$ the count value in the counter increments consistently. The counter is said to count up; it is called an 'Up Counter'.
- The transition in a flip flop occurs only after the transition in the preceding flip flop. It makes the counter operate in an 'asynchronous' manner. Hence the serial binary counter is also called the 'asynchronous binary counter'.

Table 4.4 States of flip flops in successive periods of the ripple counter in Figure 4.14 (a)

periods	Qc	Qb	Qa	Decimal equivalent of Qc Qb Qa	periods	Qc	Qb	Qa	Decimal equivalent of Qc Qb Qa
C0	0	0	0	0	C8	0	0	0	0
C1	0	0	1	1	C9	0	0	1	1
C2	0	1	0	2	C10	0	1	0	2
C3	0	1	1	3	C11	0	1	1	3
C4	1	0	0	4	C12	1	0	0	4
C5	1	0	1	5	C13	1	0	1	5
C6	1	1	0	6	C14	1	1	0	6
C7	1	1	1	7	C15	1	1	1	7

4.3.2 Synchronous Binary Counter

Consider the set of flip flops in Figure 4.15 with all the clock inputs being from a single source CLK. With any flip flop, the Q values of all the preceding flip flops are ANDed together and used as the J and K inputs. Hence a flip flop in the chain changes state (toggles) only if all the preceding flip flops are at one state and a negative transition occurs at its CLK input. With a periodic clock at CLK, outputs of the flip flops have waveforms similar to those in Figure 4.14 (b). The flip flop combination once again functions as a binary counter. The counting mode and count sequence are identical to those of the counter in Figure 4.14. The chain operates as a binary up-counter. Further all the transition instants in the counter are coincident with the negative edge of the clock (It will be violated with a flip flop in the chain if the cumulative delay of transitions in all the preceding flip flops exceeds the CLK clock period). Since the transitions in all the flip flops are in synchronism with those in CLK, the counter chain is called a 'synchronous counter'. It is also called a 'parallel counter' since all the transitions occur in parallel – no waiting for the transition in a previous flip flop.

The synchronous counter has a predictability and definiteness associated with its operation – unlike the asynchronous counter. Hence a synchronous counter is the preferred choice for applications. An asynchronous counter is used only if the application demands it on extraneous considerations. For example a synchronous counter needs more hardware compared to an asynchronous counter; where hardware is at a premium, the asynchronous counter may be the preferred choice – especially for counters of large number of flip flops.

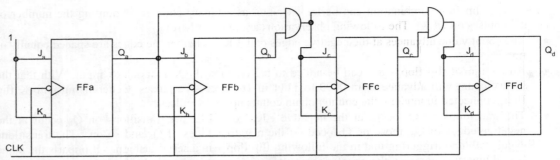

Figure 4.15 A synchronous binary up-counter of four flip flops

4.3.3 A Versatile Counter

The counter in Figure 4.16 is a modified version of that in Figure 4.15. J_b and K_b are connected to $\overline{Q}_a$. Similarly J_c and K_c are connected to the ANDed outputs – $\overline{Q}_a$ and $\overline{Q}_b$. Waveforms of Qa, Qb, and Qc for a square wave of periodic pulses on CLK, are shown in Figure 4.17. Values of the flip flop outputs after successive clock pulses are given in Table 4.5. With every negative transition on CLK, the count represented by $Q_c\,Q_b\,Q_a$ can be seen to decrement by 1. The counter counts down repeatedly from 1111 to 0001. It is a 'down counter'. Specifically it is a 'synchronous down counter' or 'parallel down counter'.

Observations

- Additional flip flops can be cascaded to the counter chain and a down counter of any desired size formed.

- The serial counter Figure 4.14 (a) can be changed to a serial down counter in an analogous manner. It requires the clock input to every flip flop to be connected to the $\overline{Q}$ of the preceding flip flop.

- In the up counter of Figure 4.15, the combination outputs $\overline{Q}_c\,\overline{Q}_b\,\overline{Q}_a$ always represent the 1's complement of the combination of outputs $Q_c\,Q_b\,Q_a$. It represents the down counter. In general in a counter if the Q outputs together do up-counting, the $\overline{Q}$ outputs together do the down-counting simultaneously.

- If necessary the counter output can be decoded and used for specific applications as explained in Section 3.6.4. Since every binary variable has its complement already available (Q and $\overline{Q}$), the decoder hardware is simpler.

The up-counter of Figure 4.15 and the Down-counter of Figure 4.16 have been combined and shown in Figure 4.18. It has a mode select input designed UP/DN. As long as UP/DN is high, the counter functions as an up-counter; if UP/DN is low the counter functions as a down-counter. The same counter can be made to count up for some time and then count down. The mode change is brought about by the control signal UP/DN. It is implicitly assumed so far that all the PR and CL signals are at 1 (inactive). With necessary additional hardware PR and CL inputs may be selectively used to set and reset the flip flops in the counter chain. Such setting and clearing have to be carried out and completed before the negative edge of the clock. With this addition the counter becomes more flexible; one can count up to a number and reset the counter; the counter can be made to count up from a preset number or count down from it. The count up can be up to one number and a count down to another number may follow.

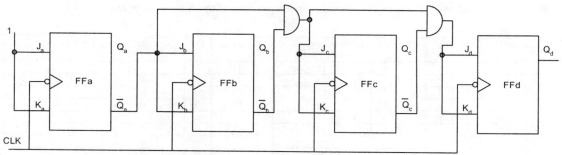

Figure 4.16 A synchronous binary down-counter of four flip flops

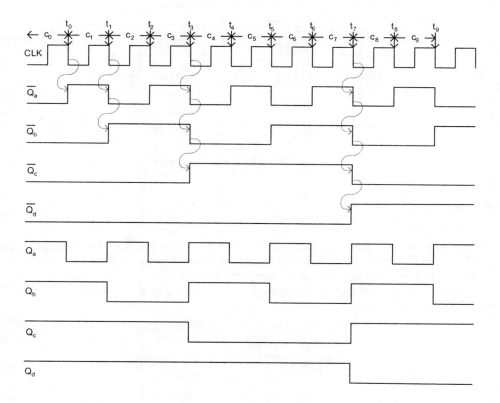

Figure 4.17 Waveforms of operation of the synchronous down-counter in Figure 4.16

4.3.4 Decade Counter

A decade counter counts from 0000 to 1001(in binary form) at successive clock pulses and then resets to zero; the corresponding count sequence is shown in Table 4.6. When the count value is 1001, the counter has to advance to 0000 at the following active clock edge. It requires the J and K values to be redefined as follows:

Table 4.5 States of flip flops in successive periods of the down-counter in Figure 4.17

periods	Qd	Qc	Qb	Qa	Decimal equivalent of Qc Qb Qa	periods	Qd	Qc	Qb	Qa	Decimal equivalent of Qc Qb Qa
C0	1	1	1	1	15	C8	0	1	1	1	7
C1	1	1	1	0	14	C9	0	1	1	0	6
C2	1	1	0	1	13	C10	0	1	0	1	5
C3	1	1	0	0	12	C11	0	1	0	0	4
C4	1	0	1	1	11	C12	0	0	1	1	3
C5	1	0	1	0	10	C13	0	0	1	0	2
C6	1	0	0	1	9	C14	0	0	0	1	1
C7	1	0	0	0	8	C15	0	0	0	0	0

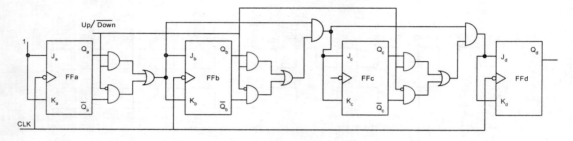

Figure 4.18 A 4-bit Up/Down counter

$$J_B = K_B = A. \overline{AD}$$

$$J_C = K_C = A.B.\overline{AD}$$

$$J_D = K_D = A.BC + \overline{AD}$$

The JA and KA values remain unchanged. Figure 4.19 shows the modified version of the 4-bit counter. The modification assures that the counter advances directly to 0000 state from 1001 state (Actual circuit implementation can be more compact). Decade counter modules can be cascaded and counting done with the combination in Figure 4.19. It eliminates the need to decode large binary numbers to get equivalent decimal values.

Table 4.6 Count Sequence of a decade counter

Qd	Qc	Qb	Qa	Decimal equivalent of Qd Qc Qb Qa
0	0	0	0	0
0	0	0	1	1
0	0	1	0	2
0	0	1	1	3
0	1	0	0	4
0	1	0	1	5
0	1	1	0	6
0	1	1	1	7
1	0	0	0	8
1	0	0	1	9
0	0	0	0	0

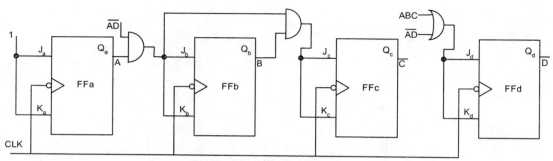

Figure 4.19 A decade counter

4.3.5 Sequence Generator

In a binary counter with every clock pulse the counter advances to the next number in the natural number sequence; the counter sequences through the natural number set repeatedly. The restriction of natural number sequence can be removed and any arbitrary number sequence specified in its place. The flip flops can be configured to advance through such a sequence cyclically. An example is considered here by way of illustration.

Consider a sequence shown in Table 4.7. The sequence is represented by 5 binary numbers. Each of the numbers is assigned an alphabetic character to represent it. With a set of three flip flops the sequence can be generated cyclically. The abstraction of cyclically sequencing through the numbers is represented in Figure 4.20. The need for 5 distinct states is met through three flip flops; these are designated as a, b, and c in Table 4.7, representing the bits of the sequence states in the same order. Consider flip flop a: When in state L (010), at the clock edge, flip flop a has to get set. It can take place if $J_a = 1$ when abc = 010 – irrespective of the value of K_a. Continuing the argument in the same manner, we get all the J_a values to ensure that the required transitions take place. Similarly all the J and K values can be specified. The truth tables for J and K values have been obtained in this manner and given in Table 4.8 with A, B, and C representing respective flip flop outputs. The corresponding Boolean functions for J and K values are as follows:

$$J_A = \overline{A}BC$$
$$K_A = 0$$
$$J_B = A\overline{B}C$$
$$K_B = \overline{A}BC$$
$$J_C = \overline{A}B\overline{C}$$
$$K_C = ABC + AB\overline{C}$$
$$= AB$$

Table 4.7 An arbitrary number sequence of 5 distinct states

Bit sequence			State designation
C	B	A	
0	1	0	L
1	1	0	M
1	0	1	N
1	1	1	P
0	1	1	R

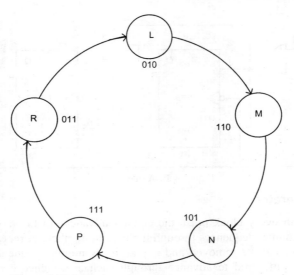

Figure 4.20 Representation of a 5-state cyclic sequence

Table 4.8 Truth tables for J and K values for the sequence in Table 4.7

State	C	B	A	JC	KC	JB	KB	JA	KA
L	0	1	0						
				1	X	X	0	0	x
M	1	1	0						
				X	0	X	1	1	x
N	1	0	1						
				X	0	1	X	X	0
P	1	1	1						
				X	1	X	0	X	0
R	0	1	1						
				X	1	X	0	X	1

The functions can be minimized and J and K values expressed as optimized functions. In fact with the judicious use of the don't-care cases in Table 4.8 the functions can often be simplified further [4]. By realizing the functions in hardware, the three flip flops – A, B, and C – can be made to function as the desired sequence generator.

Observations:

- A sequence generator of K states requires at least $\log_2 K$ flip flops for its realization. The closest larger integer is to be selected.

- A sequence generator can be more complex than what is shown in Figure 4.20 and involve external inputs [18]. By a proper analysis of the sequence the minimal number of states required can be identified and number of flip flops required decided.

- The states 000, 001, and 100 do not figure in the specified sequence; they are prohibited states. But at power on or due to some extraneous disturbance, the sequence generator may take one of these states; it can be forced to one of the states of the specified sequence by redefining the J and K values [See Exercise 4.9].

- The sequence generator can be also realized through a different but elegant route [See Section5.5].

4.4 REGISTERS

A group of flip flops functioning together with some common features can be termed a register. A register stores information or data in the form of a binary number. A register is characterized mainly by the facility offered to load data into it and the facility to shift data out of it. Either can be done in serial or parallel form. The facilities are discussed here with a set of three flip flops forming a register; however the discussion is equally valid for registers with larger number of flip flops.

Consider the connection in Figure 4.21. At the instance of the clock pulse, the data in data input line will get loaded into flip flop a, since the flip flop will get set or reset depending on whether this data is 1 or 0 respectively. Similarly, the state of Q_a before the clock pulse will decide the state of Q_b after the clock pulse; in other words, data in flip flop a will get shifted to flip flop b. Data in flip flop b will get shifted to flip flop c. The data which was originally in flip flop c will be lost. Continuing the argument, with very clock pulse, the data in the serial input line will get serially shifted into the register at the input side of the register (flip flop a). Similarly it will get shifted out at the output side of the register (flip flop c). Operation of the register is illustrated through the waveforms in Figure 4.22. Successive transition instants – negative edges of the clock – are designated t_0, t_1, t_2, t_3 . . etc. Successive clock intervals are designated C_0, C_1, C_2, C_3, . . etc. A typical data input waveform is shown. All transitions in the input are assumed to have taken place before the transition instants following. Before the first transition instant t0, all flip flops are assumed to be reset – $Q_a = Q_b = Q_c = 0$. With every clock pulse the data can be seen to be serially shifted into the register.

The serial data input to the register has to be synchronized to the clock pulses to ensure satisfactory operation of the shift operation. Synchronism implies two things; successive bits of data in the serial input line should occur in successive intervals of the clock and transitions in the data should occur well before the negative edge of the clock input. The register in Figure 4.21 is called a 'Serial –in–Serial–Out' type register. The Serial –in–Serial–Out register has been modified and shown in Figure 4.23. Normally, the Read Line is low and the outputs are at 0 state. When it is taken high all the AND gates are enabled. The data in the flip flops of the register are made available on the output lines O_a, O_b, and O_c simultaneously. The data in the register is read in parallel form. Loading of data into the register is still in serial form. The register is said to be of the 'Serial-in-Parallel-Out' type.

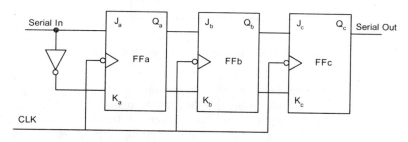

Figure 4.21 Serial –in–Serial–Out type shift register

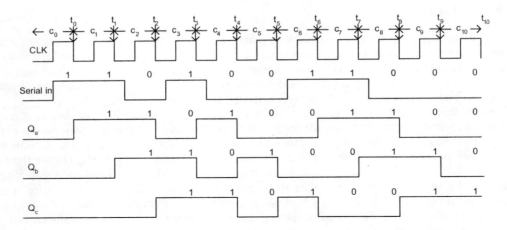

Figure 4.22 Waveforms of operation of the serial shift register in Figure 4.21

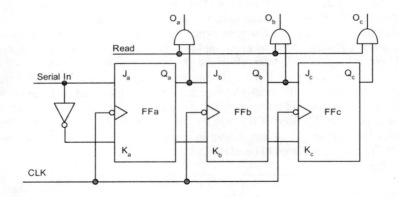

Figure 4.23 Serial –in–parallel–Out type shift register

The flip flops of the register have been modified on the input side and shown in Figure 4.24. The inverters – I_a, I_b, and I_c – at the input side ensure that the J and K inputs are complements of each other. S (for shift) and L (for load) are two input control lines. When $S = 1$ (and $L = 0$) output of the flip flop on the left is steered into the J and K leads of the flip flop. The flip flops of the register are connected as in Figure 4.21 and data is serially shifted in. When $L = 1$ (and $S = 0$), the flip flops of the register are isolated from one another; data on P_a, P_b, and P_c lines are connected to the respective J and K inputs. At the following negative edge of the clock the data is loaded in parallel into the flip flops and hence the register. With the modification here the register has the 'parallel-in-and serial-shift' facility. Serial-out facility is already built in.

The versatility of registers makes them useful for a variety of data transfer operations. As an illustration 4 registers – R_1, R_2, R_3, and R_4 – are shown in Figure 4.25. Each register is 8 bit wide and can store a byte. Consider a data byte stored in R_1. A typical set of operations that can be carried out with the register set is as follows:

- Load the data from R_1 into R_2. For this make Read1 high. Make S_2 input of R_2 low and L_2 input high (ready for parallel loading); Give a clock pulse to R2. After this Read1 control signal may be made 0.
- Make $L_2 = 0$ and $S_2 = 1$ for R_2; $L_3 = 0$ and $S_3 = 0$ for R_3.   Serial data movement is enabled from R_2 to R_3. Give 8 clock pulses to R_2 and R_3 simultaneously. The data in R_2 gets serially shifted into R_3.
- Transfer the data byte in R_3 in parallel into R_4. For this make Read3 high to read data in R_3 in parallel. Make $S_4 = 0$ and $L_4 = 1$ for R_4 and make R_4 ready for parallel loading. Give a clock pulse to R_4. It completes parallel data transfer into R_4. At the end of the transfer Read3 may be made low.

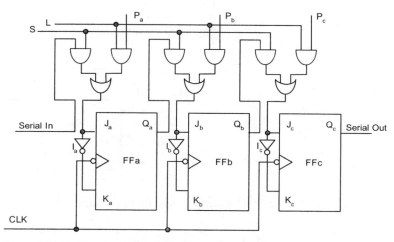

Figure 4.24 Shift register with the facility for serial shifting and parallel loading

The facilities of loading and shifting of data make registers very versatile. In general a register which has facilities to load data into it and to shift the data is called a 'shift register'. In this sense all the registers discussed so far are shift registers. With all the shift registers considered so far data is transferred from left to right. The register can be made more flexible by adding the facility to transfer data from right to left. Consider the modification to flip flop b as shown in Figure 4.26. A similar arrangement can be provided for flip flops a and c too. If R / $\overline{L}$ = 1, Q_a is connected to J_b and $\overline{Q}_a$ to K_b. Similar connection is made to flip flops a and c too. With this the three flip flops together form a shift register with data being serially shifted right. If R / $\overline{L}$ = 0, Q_c is steered to J_b and $\overline{Q}_c$ to K_b. Similar connection is effected to flip flops a and c as well. The three flip flops together function as a shift register with left shift facility.

4.4.1 Circular Counters

The shift register can be converted into an elegant counter by a simple additional connection as shown in Figure 4.27 (a). The connection resembles a ring (See Figure 4.27 (b)); hence it is called a 'ring counter'. Initially a number 001 is loaded into the ring counter (Figure 4.27 (c)) and the counter is continuously clocked. With every clock pulse the number shifts right by one bit; the bit in the rightmost flip flop gets shifted into the leftmost flip flop. If the flip flops are arranged in the form of a ring, the number can be seen to circulate in the ring.

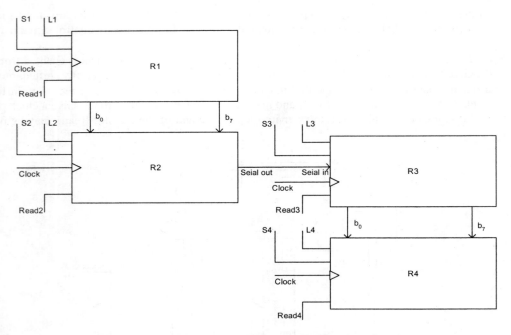

Figure 4.25 A group of 4 registers arranged for flexible data transfer

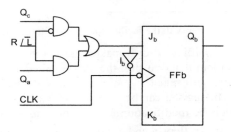

Figure 4.26 A flip flop in a shift register with facilities for right and left shifts

Observations:

- The ring counter has as many states as the number of flip flops used in the ring (contrast with the binary counter which has 2n states with n flip flops). But the outputs are already available in decoded form.
- Any binary number can be initially loaded into the ring counter; the same will continue circulating within the ring.
- Many applications require a well arranged sequence of operations to be repeatedly carried out (Flying lights, pick and place mechanisms, batch operations are examples). The ring counter is ideally suited to act as a master controller in such applications.
- Loading a 000 or a 111 into a ring counter does not serve any purpose. Hence the same is not done.

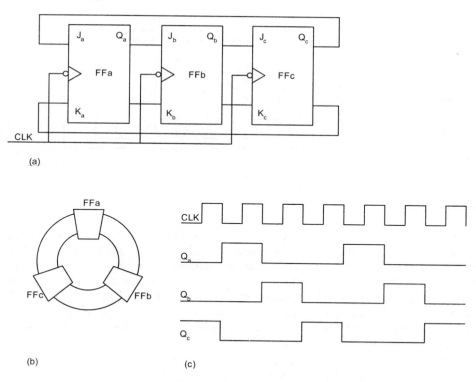

(a)

(b)   (c)

Figure 4.27 A 3-bit ring counter and its operation (a) Shift register connected as a ring counter (b)The unit arranged in the form of a ring and (c) Operational waveforms

4.4.2 Johnson Counter

The circuit of Figure 4.28 is a slightly modified version of the ring counter; it is called the 'Johnson Counter'. Connections to the J and K inputs of one of the flip flops of the ring counter are interchanged to form the Johnson counter. All the flip flops of the Johnson counter are assumed to be reset before embarking on the shift operations. The count sequence of the Johnson Counter is given in Table 4.9. At every clock pulse the bit in flip flop a shifts to flip flop b directly; similarly that in flip flop b shifts to flip flop c directly. But the bit in flip flop c is complemented and shifted to flip flop a. With a properly selected initial condition, the Johnson Counter sequences through twice as many states as a ring counter does with the same number of flip flops. The binary counter, binary counter modified to count to specific modulus, the ring counter and the Johnson Counter can be combined in different ways to get a variety of counters. These are more of academic interest.

Table 4.9 Count sequence of the 3-bit Johnson counter in Figure 4.28

Qa	Qb	Qc
0	0	0
1	0	0
1	1	0
1	1	1
0	1	1
0	0	1
0	0	0

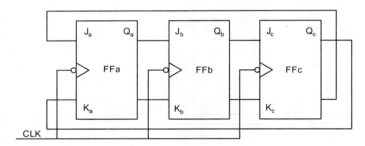

Figure 4.28 A 3-bit Johnson counter

4.5 ACCUMULATOR

An accumulator is a versatile register with a variety of facilities. The facilities – discussed above as count up, count down, shift register etc., – are selectable and changeable as required in the application. An accumulator can be built using JK flip flops by combining the required facilities. Figure 4.29 shows the flip flop b with a variety of such facilities combined. The J_b and K_b inputs are selected by assigning suitable values to the mode select inputs – R, L, P, and C / $\overline{S}$. Table 4.10 lists the assignments possible and the corresponding functions. The other flip flops of the register too can be modified accordingly and the whole set connected together to form the accumulator. However, the four mode-select inputs are common for all of them.

Observations :

- The logic input circuits used in Figure 4.29 have at least four tiers. By forming respective Boolean functions and carrying out optimization, they can be expressed in more compact and two tier forms. Going a step further, Q_b and $\overline{Q_b}$ can be expressed as Boolean functions of the present state and the inputs and realized optimally.

- The accumulator assembly here is suited for parallel or synchronous operation. CLK input is common for all the flip flops.

- The mode select inputs – in Table 4.10 – are not independent. Only with C/$\overline{S}$ = 1, P can take the value 1. At a time only R or L is high. When P is high R and L are low.

- Accumulators are integral parts of μCs. Each μC shall have at least one versatile accumulator. Some – like the 68XX series or Z80 – have two versatile accumulators. Many other μCs have one versatile accumulator and a number of additional ones that are not so versatile.

Table 4.10 Assignments and corresponding functions of the accumulator formed with flip flops of the type in Figure 4.20

C/$\overline{S}$	P	R	L	Function carried out
0	1	0	0	Parallel load of data on Db line
	0	1	0	Right shift type shift register
	0	0	1	Left shift type shift register
1	0	1	0	Up counter
	0	0	1	Down counter

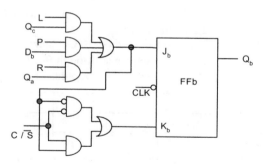

Figure 4.29 The circuit of one bit of an accumulator

4.5.1 Accumulator and ALU

Accumulator and ALU are integral to every µC. They operate in close co-ordination. Whenever an ALU does an operation involving two operands – like AND, OR … Add, Subtract, etc., – one of the operands is taken as the accumulator content. Result of the operation is often loaded into the accumulator by default. The control signals to the accumulator and the ALU may be organized to function in co-ordination.

4.6 EXERCISES

1) Analyze each of the flip flops in Figure 4.3 and prepare the truth table in each case.
2) Realize a mod-5 counter and a mod-2 counter (single flip flop). Cascade the two to form a mod-10 Counter.
3) The cascading in the above case can be done in 2 ways – the flip flop representing mod-2 counter following the mod-5 counter or preceding it. Give the scheme for both. In both cases provide decimal decoder and compare the hardware requirements.
4) Realize mod-7 and mod-11 counters.
5) The propagation delay of a single gate – AND, OR, NAND or NOR is 3 ns. The propagation time associated with a change in J or K value of a JK flip flop is 8 ns.

     What is the maximum acceptable clock frequency of a 4 bit ripple counter assembled using JK flip flops?

     What is the maximum acceptable clock frequency of an 8 bit ripple counter assembled using JK flip flops?

     What is the maximum acceptable clock frequency of a synchronous counter of 4 bits assembled using JK flip flops?

     What is the maximum acceptable clock frequency of the decade counter of Figure 4.19?

6) Consider the sequence generator of Figure 4.20. If it gets into any of the prohibited states, at the following negative edge of the clock, it has to be forced to the 111 state. Modify the truth table in Table 4.7 to accommodate this additional requirement. Form the corresponding modified functions for all the J and K values.
7) A ring counter of 4 flip flops is loaded with the binary number 0101 initially. How many states does it sequence through? What are the states?
8) A Johnson Counter of 4 flip flops is loaded with the binary number of 0101 initially. What are the states it sequences through? Repeat if the initial number loaded is 0000.

9) A set of 5 flip flops is connected in a ring form as shown in Figure 4.30. The set is initially loaded with 00000. What are the states it sequences through?

10) Have a 4 bit ripple counter. Modify it to form a mod-10 Counter.

11) A sequence generator of 3 flip flops is to sequence through $0 \rightarrow 3 \rightarrow 6 \rightarrow 1 \rightarrow 4 \rightarrow 0 \rightarrow 3 . . .$ Prepare the truth table for the Js and Ks. Obtain the respective Boolean functions.

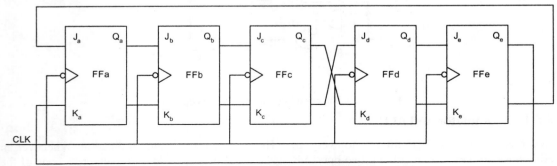

Figure 4.30 The ring counter considered in Ex. 4.12

12) Figure 4.31 shows a set of 4 JK flip flops connected as a shift register. Initially all the flip flops are reset. A data stream of bits 100111 is clocked into the shift register, the data being synchronous with the clock – the left most bit being the first. Sketch the output waveforms of each flip-flop for 10 successive clock periods

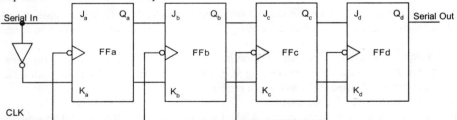

Figure 4.31 A shift register of four flip flops

After 10 clock pulses external input to Ja and Ka are removed. Qd and $\overline{Q}_d$ are connected to Ja and J_b respectively; the unit functions as a ring counter. Sketch the waveforms of all flip flop outputs for the next 10 clock pulses – clock pulse 11th to 20th.

After the 20th pulse, Q_a is connected to K_a and $\overline{Q}_d$ to J_a (Johnson Counter). Sketch the flip flop output waveforms for clock pulses 21st to 30th.

After the 30th pulse, the flip-flops are reconnected; Q_d and $\overline{Q}_d$ are connected to J_c and K_c. Q_c and $\overline{Q}_c$ are connected to J_b and K_b. Q_b and $\overline{Q}_b$ are connected to J_a and K_a. Q_a and $\overline{Q}_a$ are connected to J_d and K_d. Sketch the flip flop output waveforms for clock pulses 31st to 40th

- After the 40th pulse, Q_a is reconnected to K_d and $\overline{Q}_a$ to J_d. Sketch the flip flop output waveforms for the next 10 pulses 41st to 50th.

13) 4.16 Two flip flops are arranged as a ripple counter. Both are reset initially and clock pulses are given to the first flip flop A. Provide a decoder using Q_a, $\overline{Q}_a$, Q_b and $\overline{Q}_b$ with 4 outputs. Each

decoder output should be high during the corresponding pulse period and low during the other three pulse periods. Give the circuit of the decoder and sketch decoder output waveforms.

14) 4.17 Give the truth table of a clocked JK flip-flop. Consider a set of 3 JK flip-flops, their output being A, B and C respectively.

15) A sequence generator (SG) has a set of defined states, each state being represented by a binary number. When a clock input is given to it, it shifts to the next state in the sequence. Any arbitrary sequence can be defined and the sequence generator made to advance through it. Consider the sequence below:

> The SG sequences through 5 distinct states cyclically. Realize the SG with the 3 flip flops above. If ABC represents the bit sequence of the SG, to shift from state 011 to state 101, do the following and give a clock input.
>
> Similar functional assignments can be made in a general manner. Complete the definition of function J_a, J_b and J_c. Realize the sequence generator.

16) Show that a sequence generator of k states requires log2 k flip-flops to be used.

17) Give the outline of a 'flying light' type display scheme using a sequence generator. Note that the scheme should have the facility to define the sequence and dwell time in each state.

18) A set of 4 flip-flops are connected as a ripple counter. Modify it to count (a) from 0 to 9 (b) 0 to 10 (c) 0-11 (d) 0 – 12 (e) 0-13 (f) 0-14 (g) 0-15

19) Two flip-flops are connected to form a ripple counter. Prepare the truth table.

20) A Sequence generator has 5 states designated A, B, C, D, and E. The sequence to be generated is
A – B – C – D – E A – B – C – D – E – A – C – D – A – . . .
Realize the sequence generator (Hint: Realize an 8-state SG as A – B – C – D – E –. .).

21) Figure 4.32 shows the scheme of a common and widely used flip flop – called the 'Master-Slave' flip-flop [MS flip flop]. It is formed using two flip flops a master flip flop and a slave flip flop. The clock input is a square wave. Assume the M and S flip flops to be in reset state to start with. The clock is in zero state. With $J_M = 1$ and $K_M = 0$, made CLK=1. Then made CLK=Q. Obtain the values of Q_M and Q_M after each transition in CLK.

22) Repeat for various combinations of values J_M and K_M. Repeat for different sequences of J_M and K_M.

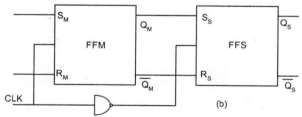

Figure 4.32 An MS flip flop formed using two RS flip flops

23) Using an MS flip flop, make a T flip flop.
24) Using an MS flip flop, make a D flip flop.
25) Using MS flip flops, make a 4-bit synchronous Up-counter.
26) Using MS flip flops, make a 3-bit synchronous Down-counter.
27) Using MS flip flops, make a 3-bit synchronous Up/Down-Counter.
28) Using MS flip flops, make a 3-bit synchronous Programmable Counter.
29) Using MS flip flops, make a 3-bit synchronous Ring Counter.

CHAPTER 5

MEMORY

5.1 INTRODUCTION

Digital systems require data to be stored and retrieved. Storage may be merely to maintain a record or for archival purposes. More often it is for use later. Extraction of information from a group of data or taking decisions using the data, are typical uses. A variety of storage facilities are available [3]. The discussion here is limited to the memory elements used with μC-based systems. Concepts of tri-state buffers and buses are introduced as a prelude to the study of memory.

5.2 TRI-STATE BUFFER

A buffer is essentially an electronic amplifier. It will reproduce the input signal at its output terminal. In a digital system scenario, if the input is at state 0, the output too will be at state 0; if the input is at state 1, the output too is at state 1. The output will have much higher signal drive capacity compared to the input. A buffer is inserted in a digital line whenever such enhanced drive capacity is called for. Often the buffer is provided with a control input. The control input transforms the buffer into a series switch. The symbol for the buffer is shown in Figure 5.1(a); it has x as its input, y as the output and C as the control input. When $C = 1$, $y = x$; when $C = 0$, y is isolated from the input and left floating. Its state is undefined – such a state of the signal is called its 'tri-state' and the output is said to be at tri-state. The buffer with the added facility is called a 'tri-state buffer'. An alternate form of the tri-state buffer is shown in Figure 5.1 (b). When $C = 0$ the buffer is on, and it is in the off state when $C = 1$. The bubble at the control input signifies inversion in the control signal.

The tri-state buffer makes connections and data transfer quite flexible. As an example, consider a 2-to-1 mux shown in Figure 5.2 (a). When $C = 1$, input L is steered to the output line N. When $C = 0$, input M is steered to the output line N. An identical mux is realized using the tri-state buffers in Figure 5.2 (b). The scheme in Figure 5.2 (b) has two added benefits.

- When $C = 1$, signal line M is disconnected from the output line. M can take on values of its own without affecting the status of line N. Similarly, with $C = 0$, L line is isolated and can take on values without affecting the output.
- We can keep both the tri-state buffers TS1 and TS2 in off state (by providing separate control inputs). It isolates the output from both the inputs. Line N can be connected anywhere else in the circuit and made to carry the corresponding signal.

The flexibility that tri-state buffers impart to a digital circuit can be realized from the hypothetical circuit (similar connections are common in digital circuits) in Figure 5.3. It has four tri-state buffers C_1, C_2, C_3 and C_4. Some possibilities are indicated below:

- With C_2 and C_3 in off state the two segments of the circuit – segment 1 and segment 2 are isolated. They function independently.
- With C_2 on (and C_3 off) a device in segment 1 can transmit a signal to another device in segment 2. With C_3 on (and C_2 Off) a device in segment 2 can transmit a signal to another in segment 1.
- Signal on line B can be made a function of the signal on line A, by turning on C_1 and keeping C_2 and C_3 in off state. Simultaneously C_4 can be turned on to make the signal on line D dependent on that on line E. These are illustrations of the two segments being isolated and functioning independently.
- With C_2 and C_1 on, signal on line A can be an input for the gate deciding the signal on line D. It is a case of a signal from one location in segment 1 being transmitted to a specific location in segment 2. The reverse transmission and functional dependence can be brought about by turning on C_3 and C_4.

A VLSI IC has about 100 million transistors in it; a million circuit blocks can reside in it. They may have to be connected and reconnected in different ways to do desired functions. It is achieved by providing long signal lines and tri-state buffers. Thus the tri-state buffer is a key player in the functioning of all VLSIs.

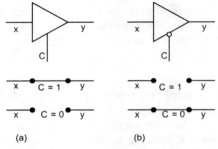

Figure 5.1 (a) Tri-state buffer with active control high (b) Tri-state buffer with active control low

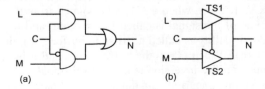

Figure 5.2 2-to-1 mux (a) with logic gates and (b) with tri-state buffers

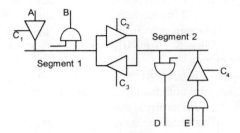

Figure 5.3 Illustration of digital circuit reconfiguration using tri-state buffers

5.3 BUS

In a digital system a collection of bits can have a common identity – like a byte or a 16-bit word. A byte may be stored in a register as a single entity. It may be added to another byte. It may be transferred as a byte to another register in a different location. All such operations – byte centered – require data to be transferred on 8 signal lines. The signal lines may be run as one entity, they may be connected, or disconnected *etc.*, all together. A set of 8 such signal lines is called a 'bus' – a bus of 8 lines. In a system one can have buses of different widths. Physically all lines belonging to one type of such an entity will run parallel and close to each other. To simplify representation in circuits a bus is represented by a double line as in Figure 5.4. Buses can 'turn'; they can have branches. In line with the double line representation, direction of data transfer is indicated by a double arrow; the arrow in front of register R2 signifies that the data is input to R2 from the bus. The number 8 shown by the side of the bus along with the 'cross line' signifies the bus width as 8 lines. As with tri-state buffers, the bus too is a key player in the functioning of all VLSIs. They facilitate transfer of a data amongst different functional units in different ways.

5.4 REGISTER FILES

A set of registers arranged in an orderly manner is called a register file. Register files can be of the serial or parallel types. In a serial register file data is loaded one behind the other and read in a similar manner. The register access is in serial form – one after the other. In a parallel register file, a register location is accessed to write data or to read data, in parallel from. Both types have their own applications.

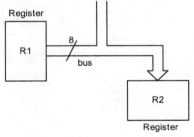

Figure 5.4 Illustration of typical bus connections and layout

5.4.1 Serial Register File

A serial register file scheme is shown in Figure 5.5. The register file is assembled using a set of 8 shift registers – each of 4 stages. The shift registers are arranged to operate in an 'up-down' fashion here rather than the 'right-left' type discussed in Section 4.4. The positions of these are designated as b_0, b_1, .. b_6, and b_7 respectively. Eight bits of data – one byte wide – can be handled by the register at a time. The top layer of 8-bits in the file is assigned the 'address' A_0; the one below has address A_1 and so on. The bottommost layer has address A_3. The register file is said to be one byte wide and 4 locations deep.

Consider the operation of the shift register set with shifting done in down direction. With every clock pulse, byte-wide data that enters at the top is loaded in parallel at location A0. Data at location A_0 moves to location A_1 and so on as shown in Figure 5.6. Byte-wide data is serially input at the top. After the fourth clock pulse, with every additional clock pulse, the data is output at the bottom – each byte in synchronism with a clock. The first byte of data input at the top, comes out first at the bottom (after 4 clock periods). The register file is said to be of the 'First Input First Output (FIFO)' type.

A FIFO file is possibly the simplest of register files. Bytes can be stacked in a FIFO one behind the other to be processed in the same order. Such a queue is a typical application of a FIFO. Storage and later retrieval allows a FIFO to function as a delay line for data. It is another common application of a FIFO. 'Move-up-move-down' facility can be added to the register file (akin to the 'move-right-move-left' facility discussed earlier). Whenever a byte is loaded the register file is in 'move-in' mode. Whenever a byte is read, the register file is in move out mode. With these provisions, the register file is referred as a 'Last Input First Output (LIFO)' type. A LIFO is also called a stack. A stack is an important component of every µC based system. The input and output are on the two sides for a FIFO. They are on the same side for a LIFO (See Figure 5.7); the function is decided by whether the LIFO is set to 'Move-in' or 'Move-out' mode.

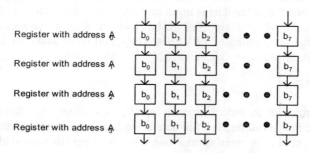

Figure 5.5 A serial register file of 4 registers

Register address	Status before 1st clock pulse	Status after 1st clock pulse	Status after 2nd clock pulse	Status after 3rd clock pulse	Status after 4th clock pulse	Status after 5th clock pulse
A_0		Byte 1	Byte2	Byte 3	Byte 4	Byte 5
A_1			Byte 1	Byte 2	Byte 3	Byte 4
A_2				Byte 1	Byte 2	Byte 3
A_3					Byte 1	Byte 2
					Output →	Byte 1

Figure 5.6 Serial Shifting of data through the FIFO file of Figure 5.5

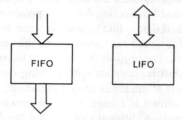

Figure 5.7 Serial register files (a) FIFO (b) LIFO

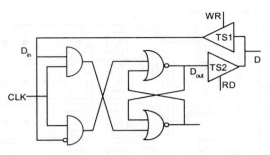

Figure 5.8 The basic memory cell built around flip flop

5.5 MEMORY

A parallel register file is more widely known as a memory. Memory can be of various types depending on the hardware, size, and function. Those relevant to µCs and their applications are discussed here in detail.

5.5.1 Scratch Pad Register

The circuit in Figure 5.8 is built around an RS flip-flop. It forms the basic memory all. It is characterized by two operations:

- Write: When a data bit is made available on the data line, Tri-state buffer TS1 can be turned on by exercising WR line – that is making it high. On a positive pulse on the clock line CLK, the data is transferred to the RS flip flop: If D is high, Q is set to 1 and if $D = 0$, Q is reset to 0. This constitutes the 'Write' operation. Subsequently until the next pulse on CLK, the data remains frozen in the RS flip flop, even after WR is taken low. The data is said to be stored in the flip flop.

- Read: When RD line is taken high, Tri-state buffer TS2 is turned on. The stored bit value in the flip flop is connected to the data line. This is the formal 'Read' operation. In short the circuit has the facility to write a data bit into it when desired, to store the bit as long as desired, and to read the bit as and when desired. It forms a basic memory cell. The memory cell requires only an RS flip flop as its core; JK or more involved flip flops are not necessary (more often, the cell core is even simpler than the RS flip flop).

Figure 5.9 shows a register file built around the memory cell. It has four rows of registers; each register is one byte wide. Thus the register file as a whole has (4×8=) 32 memory cells of the type in Figure 5.8. The b0 data lines are connected together and brought out as a d0 data line. Similarly every one of the other data bit lines is also brought out. Each of the cells has its clock input. All these 32 clock inputs are connected together to form a common clock input to the unit. Being common to all the cells, it has been left out in the figure.

The file has a set of two 'address lines' – A_0 and A_1; they are decoded using a 2-to-4 decoder to form output lines L_0, L_1, L_2, and L_3. If $A_0 A_1 = 00$, $L0$ is high. The WR and RD inputs to the register in the top row are enabled; in other words the byte location $L0$ is 'selected'. Data in the selected location can be 'read' by taking RD to high state; it makes all the eight bits of data available on the data lines. Similarly data can be written into the selected location. For this the byte-wide data is made available on the eight data lines, WR taken high and a clock pulse given to the unit. In an analogous manner – with the address lines A_0 and A_1 – any one of the 4 locations can be selected for reading and writing.

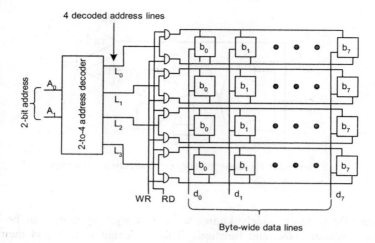

Figure 5.9 A register file of 4-byte locations shown with all the control and data input and output lines: The CLK signal is input to all the bit locations and hence not shown

The following are noteworthy:

- The WR, RD and CLK inputs are common to all the memory cells
- The decoded address line is common to all the bits of the selected location
- Data bit line d_0 is common to the 0th bits of all the registers (or locations). Similar is the case with all the other bit-lines.

With n address lines one can address a total of 2^n distinct locations. The unit is said to have a 'storage capacity' of 2^n bytes. Every one of the 2^n locations is equally accessible; in other words any address can be selected at random and accessed with equal ease. The memory unit is referred as 'Random Access Memory' (RAM). In μC systems and computers, a RAM of the type discussed here is used to store data and retrieve it 'at short notice'. The unit is called a 'Scratch Pad Register'. The scratch pad capacity in a system can vary from 32 bytes to 65 k bytes or more, depending upon the size and scope of the system. Normally (as mentioned earlier) the capacity is a power of 2.

A scratch pad type RAM is shown in Figure 5.10. The n address lines form an address bus. The Memory Address Register (MAR) stores the memory address location to be accessed next. With the provision of MAR, the location concerned is accessed first and subsequently the Read or Write operation executed. Further the address in MAR can be updated in 2 or more stages – by loading the address in segments of one byte at a time. This is in line with the digital system organization where all activities are carried out in bytes.

Chip Select (CS) is an additional address line common to the whole unit of Figure 5.10. Any address selection for memory read or write operation will be done only if CS is high. Such common CS type lines help in increasing memory capacity and in interfacing different devices to the system.

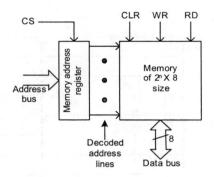

Figure 5.10 A scratch pad type RAM

5.5.2 RAM operation – Timing

Addressing, control and data line signals are to be properly sequenced for satisfactory operation of the RAM. The sequencing has to be taken care of in the digital system design. Manufacturers' data sheets give timing details elaborately. The aim here is to discuss the same in enough detail to bring out the constraints. Typical read timing waveforms are shown in Figure 5.11. The address lines have to take on the new values corresponding to the location accessed. CS and RD lines have to go high. After all these have changed, enough time has to be allowed for data to become available on the data lines in a reliable manner. The time delay (normally in ns) is marked as tp1 in figure. Any device using the Read data, has to wait for at least tp1 ns before the data can be taken as reliable.

Once the Read operation is over, CS and RD will go to 0 state. The address lines may take on a new address. With a delay of t_{p2} ns, data lines are turned off. Only after that they are available for the next operation. Manufacturers normally specify a range for t_{p1} and t_{p2} and guarantee that actual values lie within the specified range. Timing signal waveforms for the Write operation are shown in Figure 5.12. The address lines take on the values corresponding to the location to be addressed. The CS and WR lines go high. After this a minimum time of t_1 ns has to be provided after changes in all the lines mentioned above. Enough time has to be allowed for the data to settle to the new values, before the clock signal is activated. The time period is indicated at t_2 ns in Figure 5.12. Once the data is written, the system has to wait long enough before embarking on next memory access. The minimum wait time for this is t_3 ns after the clock transition to the low state. Manufacturers give ranges for all such delays and guarantee performance within the range.

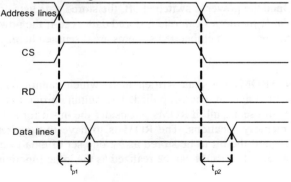

Figure 5.11 Read timing waveforms for RAM

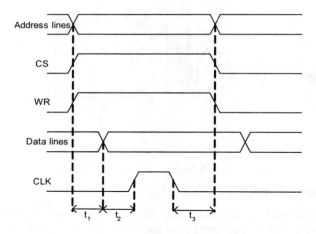

Figure 5.12 Write timing waveforms for RAM

5.5.3 Dynamic RAM

The RAM discussed above is built around the RS flip flop as the basic storage cell. Once a bit is stored in a cell, it stays put there until it is changed during a subsequent write operation at the same location. The static nature of the cell and storage has earned it the name – 'Static RAM' or SRAM

A 'Dynamic RAM (DRAM)' works on a different principle. The gate of a MOS transistor accumulates a charge and decides the status of the transistor as on or off; this constitutes the storage. The charge leaks through and the stored information is lost in tens of milliseconds. To sustain the storage, the charge has to be replenished and restored at regular intervals. Normally it is done by shifting the charge (or storage) by one location at regular intervals. Such shifting and eventual rotation of the storage is built into every DRAM. The replenishment and rotation will not hinder memory access for reading or writing. DRAM occupies much less space compared to the SRAM; or for the same IC size and price the dynamic RAM capacity is one order higher than the SRAM capacity.

5.5.4 Non-Volatile Memory

SRAM and DRAM function only when energized. They require the specified power supply to be available to store data. Once the power is switched off, the stored data gets erased. The memory is said to be 'Volatile'. Non-volatile memories can retain stored data even if the power supply is switched off. The non-volatile memories used with embedded systems are discussed here.

5.5.4.1 ROM

A 'Read Only Memory' (ROM) has data written in it, which cannot be erased. The memory is manufactured with the written data in it and supplied. The running programs in calculators, cell phones and the variety of devices in use are all in ROMs. Consider the outfit for a 4×1 ROM; it stores one bit of data with 4 distinct memory locations. The ROM is to have two address lines. Designating the address lines as A_0 and A_1 and the bit to be stored as b_0, one set of data to be stored in a ROM as an example, is given in Table 5.1. The ROM can be realized as the logic function.

$b_0 = \overline{A_1}\,\overline{A_0} + \overline{A_1}\,A_0$

Table 5.1 Data for a typical 4 × 1 ROM

A1	A0	b0
0	0	1
0	1	1
1	0	0
1	1	0

Multi-bit ROMs can be realized in a similar manner by expressing every output bit as a function of the address bits. An n×8 ROM has n address lines and 8 output bit lines. The eight bits $b0$, $b1$, ... $b6$, and $b7$ can be expressed as Boolean functions of the address line A_0, A_1,. . A_{n-2}, and A_{n-1}. ROM is a realization of this multiple output logic function set.

ROM manufacturers require ROM data to be supplied by the user in a specific format. User can make it available to the manufacturer and get the ROM supplied. With the fixed cost overheads of semiconductor manufacture, ROMs are economic only if the volume requirement is in tens of thousands. Examples are programs in Calculators and Cell Phones.

If the volume demand for a ROM is only the in thousands range, memory manufacturers provide the option of using a Programmable ROM (PROM). A PROM can be programmed only once and the program cannot be changed thereafter. Programming is done by executing a sequence of specified operations. The layout of an 8×3 PROM is shown in Figure 5.13. It has 3 address lines – designated A0, A1 and A3 – and 3 data output lines – designated b_0, b_1 and b_2. The address lines are decoded through a 3-to-8 decoder; the decoded outputs are available as L_0, L_1, . . L_6, and L_7. The factory-supplied version of the PROM has all the L-lines as inputs to the OR Gates forming the three bit outputs. User programming is essentially the process of disconnecting the unwanted L-inputs from each of the three bit lines. The junctions forming the connections are all fuse wires. Any one of these fuses can be selected and by passing a large enough pulse of current, it can be blown. With the PROM energized under normal operating conditions the fuse functions as an ordinary conductor. Programming is done, by accessing each memory location sequentially and blowing off the unwanted fuses using the specified current pulses.

Table 5.2 gives the data for Gray to binary encoder. It has been realized as an 8×3 PROM and shown in Figure 5.14. The dots in the cross-over matrix of lines represent connections which are left 'unprogrammed'. They can be seen to correspond to the data in the table. Commercial versions of PROMs are more often one byte wide. PC-based standard programmers are available. They provide PC interface. The program to be written is to be prepared as tabular data on the PC in a specified format and down-loaded into the programmer module. On a specified command the programmer programs the PROM IC.

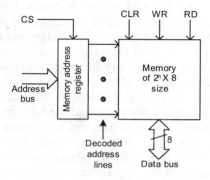

Figure 5.13 PROM based 3-bit Gray to Binary encoder

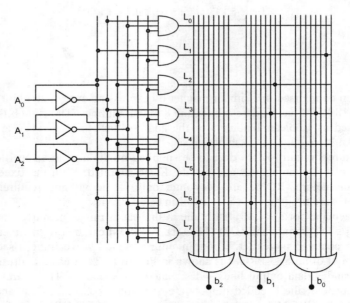

Figure 5.14 PROM based realization of 3-bit Gray to Binary encoder

Table 5.2 bit values of 3-bit Gray to binary encoder

A2	A1	A0	L	b2	b1	b0
0	0	0	L0	0	0	0
0	0	1	L1	0	0	1
0	1	1	L2	0	1	0
0	1	0	L3	0	1	1
1	1	0	L4	1	0	0
1	1	1	L5	1	0	1
1	0	1	L6	1	1	0
1	0	0	L7	1	1	1

5.5.4.2 EPROM

Erasable Programmable ROM (EPROM) is functionally an improvement over the PROM. It has the added facility of 'bulk erasure'. An EPROM is a MOS device. Data bit is stored in a location by storing a definite charge in the gate region of a MOS transistor. The programming involves accessing the location and applying an electrical pulse of a specified height and duration. Data to be written has to be prepared as sequence of bytes. Each of the memory locations is accessed and data written byte by byte. Once programmed, data remains frozen in the respective locations. The IC chip can be put in the circuit of interest and used as long as desired.

Once a data bit is written in a location, it is stored as 'trapped charge' in the concerned MOS device. The whole IC can be exposed to UV radiation of specified intensity and duration to erase the data; all trapped charges are released and the IC restored to the un-programmed state. The IC can be repeatedly programmed and erased in this manner a few thousand times (guaranteed life cycles of the device). Programming is done using standard programmer kits available commercially. Erasure is done with the help of UV light provided for the purpose.

Observations:

- A PROM chip is not reusable. If wrongly programmed, it has to be discarded. If minor alterations to the programmed data – often required in applications especially in development stage – are required, the programmed chip has to be completely discarded and a new un-programmed device used. An EPROM chip with its reprogramming facility scores over this.
- ROM, PROM, and EPROM – all are available with pin-to-pin replacement facility. During development a circuit can use EPROM. After completion of development, a PROM can replace the EPROM; it is used for test marketing and in small production volumes. A ROM need be used only if the volume justifies its use.
- The programming waveforms to be used as well as the UV dose for erasure of an EPROM, varies with manufacturers. Commercially available programmers have the facility to accommodate such variations.
- EPROMS are erased using UV lamps; they are sometimes referred as UV-erasable PROMs.
- The memory area of an EPROM is not encapsulated. It is left exposed with a transparent glass cover for protection. Erasure is done by exposing the device to UV radiation through this window. However, once the EPROM is programmed, a tape opaque to UV radiation is stuck over the window. It prevents UV exposure of the device and any inadvertent erasure.

5.5.4.3 EEPROM

Electrically Erasable PROM (EEPROM or E2PROM) is an improvisation over the EPROM. Erasure is carried out by applying prescribed electrical pulses. Erasure can be carried out on a byte-by-byte basis. However, erasure of a byte takes much longer time (~1ms) than reading (~ 20 ns). As such it cannot replace a RAM. But the selective programmability allows the data stored and to be altered when desired. It can be done in situ. The added facility of in situ programming, makes E2PROM more versatile than an EPROM.

Flash E2PROM is an improvisation of an E2PROM. It has improved memory density but requires bulk erasure; erasure cannot be done on a selective basis.

5.5.4.4 GENERAL

Running programs and permanent data used (universal constants like the value of π) remain normally invariant during the use of a digital system. ROM, PROM and EPROM are used to store them. E2PROM is used to store data changed infrequently (like calibration constants, measured values of a variable to be stored and retrieved for later use). RAM is used for scratchpad type storage. Thus the volatile RAM and the different types of non-volatile memory can co-exist in a digital system.

5.6 BULK STORAGE DEVICES

Bulk storage devices can store data of 1 MB size or more. All of them are of the non-volatile type. They store data as a serial bit stream. Access to data in a specified location is also in serial fashion (in contrast to the random access possible with devices so far). Access to a specific location can take one millisecond or a few seconds depending on the status of the storage device and the location specified. The large access time makes them unsuitable for direct use with μC-based systems. They are used only as back up storage devices in PCs, Servers or their applications. Often storage or retrieval is done one full file of data at a time or at least a bulk portion of it. 100 k bytes is a typical size of data handled at a time. Non-volatile nature of their operation and the relatively low cost of storage are the reasons for their continued popularity.

Perhaps the only situation where such a storage device can come into direct contact with a μC based system or application is for data logging. The μC may collect specific data and accumulate the same as part of its regular operation. Such data may be 'downloaded' into a bulk storage device at regular intervals. Subsequently a PC may reprocess such data off-line. Fluid flow rate in a pipe is a typical example. A µC-based flow meter may measure the flow rate at regular intervals. The accumulated flow rate data can be downloaded into a storage device at regular intervals. A PC can access it off-line separately for further processing. Such processing can be to calculate volume or weight of quantity delivered, decide commercial transactions, do model based analysis, decide control strategies and so on.

Bulk storage devices – not being of direct relevance to µC-based systems – are not dealt with in detail here [14]. However, they are briefly touched upon for the sake of completeness and establishing familiarity. Bulk storage devices in use are of three types; magnetic storage, optical storage and semiconductor storage.

5.6.1 Magnetic Storage Devices

All the devices have tiny fine-grained magnetic particles uniformly coated on a support base. When data is written, the particles get magnetized. The magnetized state is retained which constitutes the storage; retrieval is by sensing the direction of storage. Rewriting is done by erasing and magnetizing again.

Floppy Drive is possibly the simplest magnetic storage device. Data storage is in circular tracks in the floppy disc. The disc – when active – is continuously rotated. Read/write head moves over the track in close proximity to the disc surface. The head selects the track and the location in the tract for read/write operation. 1 M byte storage capacity is typical of floppy drives. Use of floppies as back up storage devices has plateaued due to competition from other devices.

Hard Disc Drives (HDD) function on the same principles as the floppy drive. The base is a hard disc. The disc with drive is assembled and sealed during manufacture. The gap between the disc surface and the moving read/write head is one order smaller than with the floppy drive. Correspondingly the storage density on the disc surface is higher. A hard disc drive can have capacities of 100 Gigabytes or more. Magnetic tapes form another source of data storage. A typical tape may have 8 data tracks each representing a byte; data storage is serial – byte-wise. Access is serial. Unidirectional access limits the range of application of tapes to archiving. Whenever data is read, it is read in bulk into an HDD.

5.6.2 Optical Storage Devices

All optical storage devices are in the form of discs – called Compact Discs (CD). Data storage is in successive circular tracks in the disc. The read/write head moves over the tracks for access, for reading as well as writing. In these respects the CD drive is similar to a floppy drive or an HDD. The un-programmed storage track is a finely polished mirror surface. Data is written at any selected location by burning the area with the help of a laser beam and making it non-reflective. The selected location being fully reflective or fully non-reflective is interpreted as storage of 0 or 1 respectively. Reading is done using a finely focused laser beam and directing it to the selected location. The reflected light is sensed. Its presence or absence is interpreted as a 1 or a 0 being stored.

CDs of the read only or read/write type are available. 700 MB storage capacity is common. At the present state of technology, CD forms the most economic medium for bulk storage and transfer of data. The availability of blue laser is likely to increase the CD storage capacity by one order and make it more economic too.

5.6.3 Flash Memory

Flash E2PROM based bulk storage units have come into vogue in the recent years. Compactness and high capacity (~ 1 GB) – though not cost – are in their favour. They are available as outfits that can be directly plugged into the serial ports of PCs. They function with the power for operation being tapped from the PC itself.

5.7 EXERCISES

1) A and B are two 16-bit registers. They are linked through an 8-bit bus. Design a scheme to transfer the data from A to B and B to A through suitable clock sequences. If a control line $C = 1$, data will be transferred from A to B. If a second control line $D=1$, data will be transferred from B to A.

2) What is the capacity of memory that can be addressed by a 10-bit address?

3) A memory IC has 10 address lines and 8 data lines. Data is stored in it with one byte in each address location. The IC is interfaced to an 8-bit bus on the address side as well as data side.

   Design a scheme to transfer the 10-bit address to the address register of the IC, through the 8-bit data bus. The address is available in a 10-bit source register S

   The IC is organized to store 16-bit data, each data being stored in two consecutive locations. Design a scheme to read data from a pair of locations and transfer it to a 16-bit destination.

   Device a scheme to take data from a given register G and write it into a selected pair of locations within the IC.

4) Two ICs of the type in 5.3 above are used together to form a 16-bit wide memory. One is used to store the upper byte and the other the lower byte. Design the memory scheme functioning around an 8-bit bus. The scheme should have the following facilities:

   Transfer a 10 bit address from a source register to the memory address register.

   Take a 16-bit data from a given source G and write it to a selected location in the memory.

   Read data from a selected location and transfer it to a destination register D.

   Compare the scheme in 3 and here in terms of capacity, response time and hardware requirements.

CHAPTER 6

DATA TRANSFER UNIT

6.1 INTRODUCTION

The circuit blocks and the functional elements that have been built up so far can be combined judiciously to build digital systems. It used to be the practice a few decades ago. But an attractive alternative is to combine them in a manner which will offer the facility to the designer to craft his own system; such circuits are called 'programmable circuits'. In fact over the last three decades programmable systems have evolved into a few clear cut structures. One of the popular and widely used of these structures is the 'Microprocessor' [16]. The microprocessor structure is built here in stages.

6.2 AN ELEMENTARY PROCESSOR

All the building blocks required to assemble a Microprocessor have been discussed. The microprocessor can be built in a few more definite steps. Different aspects of working with the microprocessor will also be examined. Figure 6.1 shows a set of functional elements linked together to facilitate a variety of data transfers; we call the whole assembly a 'Data Transfer Unit (DTU).

6.2.1 RAM File Registers

A RAM is essentially a series of registers each assigned with an address. The RAM file here is basically a RAM - eight bit wide. Each register location in the RAM file has an associated address. Thus a RAM file with a seven bit address will have 128 address locations; each of the locations has a byte wide register. Each such location can be accessed and data stored there, read from there, or existing data there replaced by a new data byte. The file address is stored in the RAM address register – seven bit wide in the specific case here. The address is decoded by the RAM address decoder and the decoded address used to access the file location. The RAM file unit has five control signals associated with it; these are designated as RAWE, RAMACLK, WE, RCLK, and RE.

- RAM Address Write Enable (RAWE): The RAM address register has to store a 7-bit address; it is seven bits wide. When RAWE signal is activated, it connects the 7-bit incoming address to the input lines of the RAM address register.
- RAM Address Clock (RAMACLK). The address given at the set of seven bit RAM address lines is latched into RAM address register, by giving a pulse on the RAMACLK line. As such RAMACLK is used to latch the memory address into RAM address register.
- Write Enable (WE): The RAM has an 8-bit input port: The WE line – when activated - connects the external 8-bit data bus to its input lines. The data to be written is ready and available within the

RAM file. The address decoder would have selected the desired file address location. Hence, the data to be written is available at the input side of the memory location.

- Write Clock (RCLK): When WE is enabled, data to be written is made available at the input side of the selected memory location. The same is written at the location by giving a pulse input through RCLK.
- Read Enable (RE): When activated, RE connects the eight bit output port of the register file to the 8-bit data bus. The data in the location accessed by the memory address decoder is made available on the data bus.

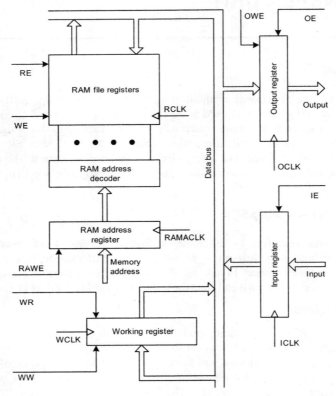

Figure 6.1 DTU block diagram

Using the above set of control signals, 'Memory Read' is carried out as follows:-

- Make the memory address available on the memory address bus. Give a pulse input to RAMACLK line. The memory address is loaded into RAM address register. It is decoded and the desired location selected. The eight bit output lines of the register file are connected to the 8-bit external bus.
- Activate RE. Data at the location selected by RAMACLK is directly made available on the external bus.
- No separate read clock signal is necessary.

'Memory Write' is carried out as follows:-

- The location where data is to be written is accessed by loading the concerned memory address in the RAM address register; it is done on the same lines as with memory read above.
- Activate WE: The external data bus is connected to the 8-bit input bus of the register file. The data on the bus is made available on the input bus. Location of interest is already selected; the data to be written is available at the input side of the location.
- Apply a pulse on the RCLK line. With this, the data to be written is loaded at the selected location.

6.2.2 Input Port

Input port is a single 8-bit wide register. Its input lines are available outside the DTU to input any data into the block. The input port has two control signals associated with it – IE and ICLK. When ICLK is given a pulse input, the data byte on the input lines is latched into the port register. It is frozen and available to be read into the DTU. When IE is activated, the 8 data lines of the port are connected to the data bus of the DTU and the data latched in the port is available on the data bus.

6.2.3 Output Port

The output port is a single 8-bit wide register. Its output lines are available outside the DTU to output data from the DTU block. The output port has three control lines associated with it – OE, OCLK and OWE. When OWE is activated, the DTU data bus is connected to the input lines of the port. Data on the bus is available at the port input leads. Once this is done, OCLK is given a pulse. It latches the data at its input into the port register. When OE is made active, the data is read from the port. Note that such port reading is done by the external device which wants the data.

6.2.4 Working Register

Working register is an 8-bit wide register directly connected to the data bus of the DTU. Three control signals are associated it – WCLK, WW and WR. The data on the data bus can be loaded into it. WW is activated to connect the bus to the input side. WCLK is given a pulse to latch the data into the register. To read the data stored in working register onto the DTU bus, WR is activated. All the operations with the DTU are centred on the working register. Working register plays a pivotal role in DTU functioning.

6.3 DTU OPERATION

The DTU as a whole has 14 control signals. Five of these are clock signals used to load data into the respective registers. Instead of retaining all these five clock signals, one can use a single clock signal for the whole unit. To avoid an element being given the clock input for an unrelated operation the clock signal can be suitably gated; Figure 6.2 shows the scheme of generating the individual clocking signals from a single master clock.

Data can be transferred from one of the elements to another in a structured manner, by activating relevant enable inputs and giving clocking pulses. For example, referring to Figure 6.1, data from the input port can be transferred to the working register by enabling IE and WW and giving a clocking pulse. When IE is enabled, data in the input port is put on the data bus. With WW enabled, the same data is connected to the input lines of the working register. Further with WW active, the next clocking signal pulse is steered to WCLK and the data is loaded into the working register. This completes reading the input port and loading the read byte into the working register. Similarly data in the working register can be transferred into a selected location of the register file. As a prelude to this, the address is to be clocked and latched into the RAM address register. Here too one actuates the necessary enable signals and follows it up with a clocking pulse. Such transfer operations possible with the DTU are given in Table 6.1. The control signals to be activated in each case are also given in the table. Each of the operations that the DTU can carry out is called an 'Instruction' in microprocessor parlance.

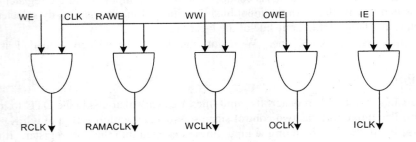

Figure 6.2 Generation of individual clock signals from a master clock

Observations:
- All instructions here – except those with serial numbers 2, 6, and 7 – involve the working register
- Each instruction requires the clock input to be activated once
- Instructions can be carried out in sequence to achieve more involved operations. Thus to transfer data from one RAM file location Ls to another Ld, proceed as follows :-
- Write address of Ls into RAM address register [Execute instruction No.2]
- Transfer data from RAM location Ls into working register [Execute instruction No.5]
- Write address Ld into RAM address register [Execute instruction No.2]
- Transfer data in working register into the selected RAM file location Ld [Execute instruction No.3]

6.4 ENHANCED DTU (EDTU)

In actual practice the instructions in the table may have to be carried out in a desired sequence – called a 'program'. The sequence (program) may be different for different applications. To put this into practice, the program to be executed may be loaded into a memory – called 'Program Memory (PM)'. Each location in the program memory can be fetched and the instruction loaded there transferred into another register, which has custody of the instruction to be executed: This register is called the 'Instruction Register (IR)'. The content of the instruction register can be directly connected to the relevant DTU control lines. Accessing program memory at a location, transferring the instruction stored there into the instruction register and executing this instruction is done sequentially until all the instructions stored are executed.

The scheme requires program memory locations to be accesses successively and sequentially. Addition of a counter – called the 'Program Counter (PC)' – for the purpose adds a touch of elegance to the DTU. At any instant the PC stores the address of the program memory location to be accessed. After the instruction to be executed is transferred from there to the instruction register, the PC is incremented to point to the next instruction to be executed. The PC, PM and IR have been added to the DTU and shown in Figure 6.3. The clock signals to the three units – namely PC, PM, and IR too – are also shown in the figure; the other control signals to these are present but have been left out in the figure. Once an instruction sequence (program) is prepared and loaded into the PM, the EDTU is ready for operation: The operational sequence is as follows:
- The PC is initialized to 0. It accesses (points to) the 0th location of PM.

- At the next clock pulse, the instruction stored at location 0 in PM is loaded into IR. This phase is called the 'instruction fetch' phase of the EDTU operation.
- At the next clock cycle, the instruction stored in IR is executed. This is called the 'execution phase' of the EDTU operation.
- When one instruction is being executed in the execution phase, the PC is incremented and points to the location from where the next instruction is to be fetched.

Each instruction here is a binary word of seven bits only. Practical processor may have a much larger number of functional elements inside; each will have its own control input lines. Combining these to form control words – instructions – is cumbersome and unwieldy. Common practice is to represent all such instructions in a compact and coded form. For example if 5 bits are used to represent an instruction, they together can represent a total of $2^5 = 32$ instructions. The IR can be 5 bits wide. Any bit combination loaded into it can be decoded using a decoder. The decoder output lines (32 lines maximum) can be used to activate the relevant enable lines. The decoder here is referred to as an Instruction decoder.

The DTU can be improved substantially in its versatility by adding an ALU to it. The unit with the addition of Instruction Decoder and ALU is shown in Figure 6.4 in block diagram form; the control signals have been left out and only the functional blocks are retained in the drawing. The ALU added, is eight bits wide. It can do two types of operations; firstly it can access a byte in one of the RAM locations on the working register and complement it, increment it or decrement it. Secondly, it can access a byte in one of the RAM locations and a second byte from the working register, and do an operation on the 2 bytes together. The operation can be an algebraic one as addition or subtraction; it can also be a logic operation like AND, OR, or XOR. The result can be stored in the working register or back in the RAM location. For all such two byte operations, one operand is from the working register. The other is from a RAM location.

Table 6.1 List of operations carried out by the DTU

S. No	RAWE	RE	WE	IE	OWE	WW	WR	Activity
	Status of control lines							
01	0	0	0	1	0	1	0	Transfer data from input port to working register
02	1	0	0	0	0	0	0	Write address into RAM address register (get ready to select an address location in the register file)
03	0	0	1	0	0	0	1	Transfer data from working register into the selected RAM location
04	0	0	0	0	1	0	1	Transfer data from working register to the output port
05	0	1	0	0	0	1	0	Transfer data from selected RAM location into the working register
06	0	1	0	0	1	0	0	Transfer data from selected RAM location into the output port
07	0	0	1	1	0	0	0	Transfer data from input port to the input register

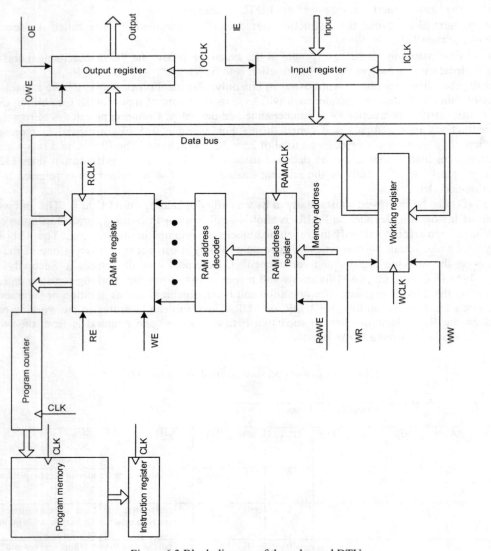

Figure 6.3 Block diagram of the enhanced DTU

6.5 OPCODE AND PROGRAM

The addition of ALU and instruction decoder makes the DTU versatile in many respects. Firstly, the operations one can carry out have increased in numbers and in capacity. One can add many more instructions – each as a bit combination. Each instruction represents a binary sequence; it stands for an operation that can be carried out. The bit combination representing an instruction is called an instruction 'Opcode' (derived from 'operation code'). The whole set of instructions – represented by respective Opcodes – is called the 'Instruction Set' of the processor.

The unit is used to carry out a desired sequence of operations. Typical such sequences are as follows:-

- Read a desired number of bytes from the input port and store them in the RAM in successive locations.
- Take a set of bytes stored in successive locations of RAM and output them sequentially at the output port.
- Add a set of bytes stored in successive locations in the RAM and output the sum.
- Take a byte from one location in the RAM and do EXOR operation on 8 bytes stored in succeeding locations stored in the RAM.

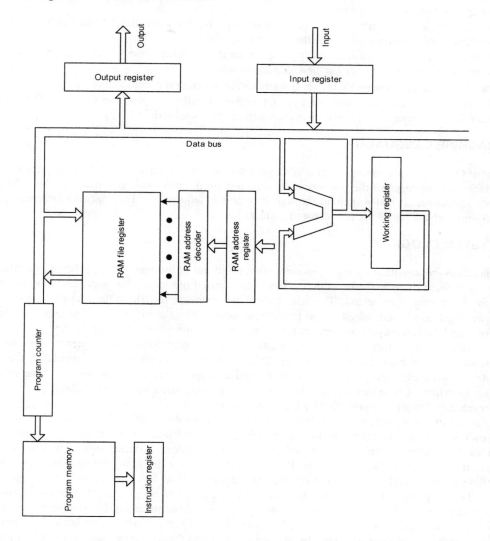

Figure 6.4 Enhanced DTU with the addition of ALU and instruction decoder

Each sequence of operations one carries out can be represented by a corresponding instruction sequence. An instruction sequence that can be loaded into the instruction register one by one and executed to achieve a definite and well defined objective is called a 'Program'. Each sequence above can be represented by its own program of instruction sequences. In practice a program to be executed is loaded into the program memory described earlier. The program execution sequence proceeds in a step by step and cyclic manner as follows:-

- PC is initialized and points to the location where the first instruction to be executed is stored
- The instruction is fetched and loaded into the instruction register; this constitutes the fetch cycle phase of instruction execution
- The instruction in the instruction register is decoded by the instruction decoder and executed; this constitutes the execution phase of instruction execution.
- PC is incremented. The instruction fetch operation during fetch cycle and execute operation during execute cycle are carried out for the second instruction.
- The PC is again incremented, the third instruction fetched and executed and so on.
- The instruction fetch and execute cycles together is called a 'Machine Cycle'. Thus program execution is a sequence of machine cycles carried out as desired.

6.6 MACHINE LANGUAGE

The set of Opcodes representing all possible operation that can be carried out by the processor is called the 'Machine Language' of the processor. It is as though the processor as a machine has a language, and its vocabulary. It can articulate through its Opcode sequences. The machine language and the instruction set characterize the processor to a great extent.

6.7 SYSTEM CLOCK

The operations represented by each Opcode, are carried out at the instance of a few sequential clock pulses. The fetch cycle and execution cycle too are carried out at regular intervals. The same holds good for the machine cycle too. The whole machine operates in a rhythmic manner with a 'clock' giving out regularly timed pulses – akin to the heart beats. With the PIC series of processors, the clock waveform and the other cyclic sequences take place in the manner shown in Figure 6.5.

The basic system clock is a periodic signal of square waveform. Its frequency may be in the 10MHz range for a typical microcontroller. Often it is built around a crystal oscillator; the crystal oscillator ensures excellent frequency stability and can guarantee precisely timed operations. For applications which are not time critical, the crystal can be dispensed with and oscillator configured in a simpler manner. Details are provided by the manufacturer.

Referring to Figure 6.5, the clock has 4 phases – designated $\Phi1$, $\Phi2$, $\Phi3$ and $\Phi4$. Four successive clock pulses together represent an interval in which an instruction can be fetched or executed. It is called the 'instruction cycle'. With the scheme shown in Figure 6.5 during one instruction cycle one instruction is fetched and loaded into the IR. It is executed in the following instruction cycle. As such a machine cycle – fetch and execute operations together – lasts for 2 instruction cycles or 8 clock periods. From Figure 6.4 one can see that the hardware segment responsible for an instruction fetch and an instruction load (PC, Program memory and IR) are different from that required to execute an instruction (Instruction decode, ALU, W, Register File and ports). This allows both the segments to function simultaneously but sequentially. Thus an instruction from memory location P is being fetched and loaded into IR during instruction cycle N; simultaneously the instruction fetched and loaded (from location P-1) in the previous instruction cycle – (N-1) – is being executed. The overlap here speeds up the processor by 100%; it is as though we have a pipeline of instructions at different phases of execution. In fact the scheme is called 'pipe lining'. Here, the pipeline has 2 instructions. Instruction N in the pipeline is fetched and loaded into IR in instruction cycle N. It is executed in instruction cycle N+1. The two operations together form the machine cycle. The execute function of machine cycle N and the fetch function of the machine cycle N+1 overlap and take place in the instruction cycle N+1.

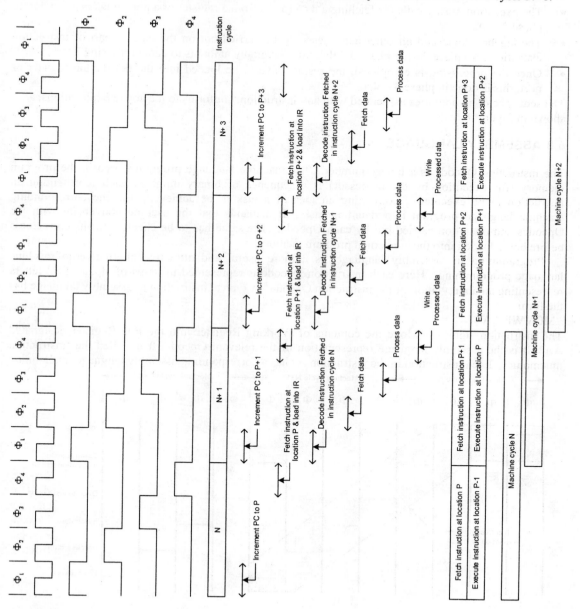

Figure 6.5 Processor operation – timing and sequence

Two activities take place in the fetch cycle.   The PC is incremented in phase $1 - \Phi1$. The instruction is fetched and loaded into IR in phase $4 - \Phi4$. $\Phi2$ and $\Phi1$ are idle phases. The execution cycle broadly calls for 4 activities:-

- The instruction available in the IR is decoded in phase $1 - \Phi1$.
- The execution mostly calls for fetching a data (byte) from a register or a port; it takes place during phase $2 - \Phi2$
- The algebra, logic and all other activities - to be carried out on the data as part of instruction execution take place during phase $3 - \Phi3$; this essentially amounts to data processing
- Once the processing is completed, the results have to be loaded into the destination register or port; this is done in phase $4 - \Phi4$.

The sequence of all activities associated with any instruction, conforms to the above explanations, as shown in Figurer 6.6.

6.8 ASSEMBLY LANGUAGE

The instruction Opcode is a binary number. The machine language program stored in the program memory (for execution by the processor) is a sequence of binary numbers each representing an instruction to be executed.   Converting an idea of a task to be achieved, into the corresponding machine language program is a daunting task.   It demands that the user be conversant with the Opcodes and operation carried out by each Opcode.   The same has to be checked for its correctness before being loaded into the concerned program memory.

Programming in 'assembly language' is a more useful and attractive alternative to machine language programming.   Here each instruction Opcode is represented in terms of alphabetical letters representing an abbreviation of the instruction Opcode.   A typical instruction in assembly language has the form:

'MOVWF AA'

The instruction signifies 'Move the contents of Working Register into the File Register designated AA'. It is the assembly language representation of the relevant Opcode. It is called the 'instruction mnemonic'. Each instruction in the instruction set has a corresponding assembly language mnemonic.

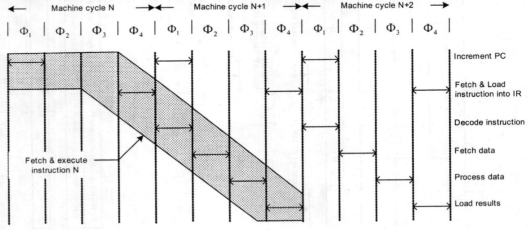

Figure 6.6 Processor clock and operational sequence

The programmer has to familiarize himself with the assembly language and prepare the required program in terms of the mnemonics. The task is much simpler than that of preparing the machine language program. The processor manufacturer supplies a PC based software which accepts the assembly language program file and translates it into equivalent machine language program. The latter can be directly downloaded into the program memory of the target processor. The translator routine here is called the assembler. One rarely does programming directly in machine language. The lowest level programming done is in assembly language. Each instruction in assembly language directly represents a corresponding operation by the processor. Being aware of the hardware structure of the processor, the programmer can visualize the ramifications of execution of each such instruction. Assembly language program represents the lowest level program – one closest to the machine (processor).

CHAPTER 7

PIC® 16CXXX SERIES - PROCESSOR

7.1 INTRODUCTION

PIC® 16CXXX represents an 8 bit series of microcontrollers widely used for embedded applications. The μC has 2 sections; one section is the processor proper - a slightly modified and enhanced version of the enhanced DTU discussed earlier. It carries out all the processor operations according to the stored program. The second section has all the peripherals attached to the processor. The peripherals enhance the processor flexibility substantially and help in realizing a compact hardware design [12, 20]. They are dealt with in detail later. The focus here is on processor per se.

7.2 ARCHITECTURE OF PIC 16®CXXX

Figure 7.1 shows the processor in block diagram form [11]. Many of the peripherals have been lumped together and shown as one block – to retain focus on the processor. The processor here differs from that of Figure 6.4 in a few respects; some of these differences are only quantitative while others add conspicuously to the processor power.

Each instruction opcode is 14 bits wide. Hence the PM and IR too are 14 bit wide. Depending on the device, the PM can have capacities up to 8K words (of 14 bits width each). In turn the PC is 13 bits wide (2^{13}= 8K). Being an 8 bit processor, the working register and each location of the register file are 8-bit wide. The ALU is also of 8 bits; it can do logic or algebra on one or two bytes as desired. Separation of data bus and program memory bus allow instruction fetching and execution to be carried out separately and without any clash: Hence they are carried out simultaneously but sequentially.

7.2.1 Ram File

The RAM is made up of a set of registers called 'File Registers'. Each register in the register file is one byte wide. In the most versatile μC in the PIC series, the register file can be 512 bytes in size. It is arranged in four banks each of 128 bytes as shown in Figure 7.2. The register banks and their address ranges are as follows:
- Bank0 with file address range 0x00h to 0x7Fh
- Bank1 with file address range 0x80h to 0xFFh
- Bank2 with file address range 0x100h to 0x17Fh
- Bank3 with file address range 0x180h to 0x1FFh

Further in each bank the set of 32 registers at the beginning of the bank is dedicated to specific functions connected with the processor or the peripherals. They are used to define assignments, do control and reflect the value of a variable or parameter connected with the peripheral. Being dedicated registers these are not to be used for any other purpose. These registers are called 'Special Function Registers (SFR)'. The rest of the registers are available for storing intermediate results of the program, data specific to the program etc. This is the 'scratch pad' area of the RAM. When a RAM location is to be specified in an instruction, the bank in which it is located as well as the address of the register within the bank is to be specified. Addressing is dealt with in detail below.

When the µC is active and functioning, only one of its banks is active and directly accessed for operations. Seven bits are enough to access any of the registers in an active bank ($2^7 = 128$). Two bits in one of the SFRs – namely bits b_6 and b_5 in the SFR called the 'STATUS register' (See Section 7.4 below) – specify the active bank; these two bits together can address all the four banks. The active bank is changed by writing a fresh byte into the STATUS register with the values of its bits b_6 and b_5 changed suitably.

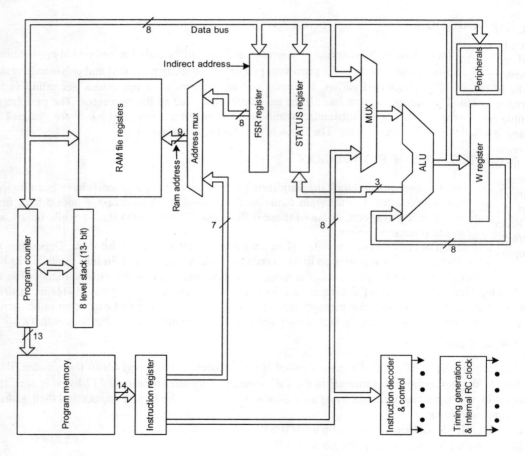

Figure 7.1 Microprocessor of the PIC16®CXXX controllers in block diagram form (The numbers by the side of the buses indicate the bus width)

7.2.2 Addressing

The processor can be supplied a byte to be processed in different ways. The way a byte is made available to it is called 'Addressing mode'. The byte can be what is stored in the working register. It can be supplied through the input port by reading the byte from the port. Apart from these, data bytes can be supplied by three modes of addressing.

7.2.2.1 Direct Addressing

An instruction in 'direct addressing' mode carries the byte address as part of itself. Consider an instruction in assembly language.

CLRF

The corresponding 14 bit instruction word is as shown in Figure 7.3. The 7-bit combination 'fff fff' in the instruction in Figure 7.3 directly specifies the address of the location to be cleared in the active bank. One can specify any value in the range 00h to 07Ah and clear the corresponding location in the currently active bank. Hence for all register based instructions seven bits (b_0 to b_6) in the instruction word are kept apart to specify the address in the active bank of RAM area. The file register banks and addressing a location in them, in direct mode is shown in Figure 7.4. The bits RP1 and RP2 (bits b_6 and b_5 in the STATUS register) specify the active bank. If a location in a different bank is to be cleared in the above manner, the values of RP1 and RP2 are to be changed beforehand. As another example consider the assembly language instruction

ANDWF f, 1

The corresponding 14 bit instruction opcode is shown in Figure 7.5.

The bit combination 0fff fff b specifies the address location from where the byte is to be fetched; subsequently the fetched byte is ANDed with the byte in W register (AND operation is carried out separately for each pair of bits in the 2 bytes) and the result stored back in location 0fff ffff b. Here again, the source and destination addresses are directly specified in the instruction.

Direct addressing is used when the operand value is not known beforehand or when it can take different values at different times or stages in a program; operation on an input byte is a typical example of this category.

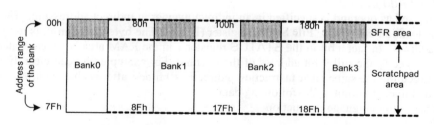

Figure 7.2 RAM file banks – arrangement and allocations

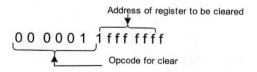

Figure 7.3 Details of instruction 'CLRF'

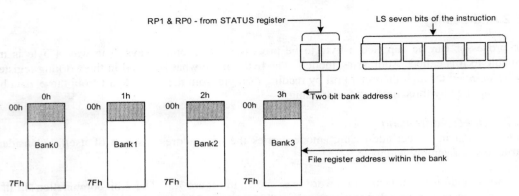

Figure 7.4 Formation of File register address in direct addressing mode

7.2.2.2 Immediate Addressing

In immediate addressing, the data byte which forms the operand is supplied as part of the instruction word. Consider the assembly language instruction.

ANDLW K

The instruction opcode has the form shown in Figure 7.6. The 8-bit number 0kkkk kkkk b represents one of the operands. The other is the content of the working register itself. The instruction is executed by ANDing the bits of the 8-bit word supplied with the respective bits of the working register. The set of 8 result bits is stored back in the working register. Here the operand concerned is available in the immediate vicinity of the instruction – hence the addressing mode is referred as 'immediate addressing'. Immediate addressing is used whenever the operand value is known beforehand. It finds application in different situations. For example, when a particular bit in the accumulator – say b_7 – is to be retained and all other bits are to be made zero, the accumulator is to be ANDed with 01000 0000 b(070h). Since the operation to be done is known and decided beforehand, immediate addressing may be resorted to.

7.2.2.3 Indirect Addressing

The byte to be used as operand is specified in an indirect manner in indirect addressing. The processor RAM has one location called the 'File Select Register (FSR)'. The byte stored in it, is used to specify the operand address. In fact a bit in the **STATUS** register – in the RAM area is concatenated with this byte in FSR to specify the full 9 bit address of the operand. For example one can clear 10 successive locations by executing the same clear instruction indirectly 10 times; after each execution, FSR may be incremented and made to point to the following data.

Consider the assembly language instruction

INCF .INDF, 0

The corresponding opcode is shown in Figure 7.7.

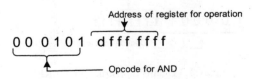

Figure 7.5 Details of instruction 'ANDWF f, 1'

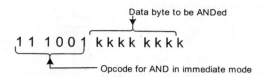

Figure 7.6 Details of instruction 'ANDLW k'

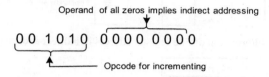

Figure 7.7 Details of instruction 'INCF INDF, 0'

The seven LS bits in the opcode being set to zero, signifies that the operand is at the location whose address is in FSR. This implies 'indirect addressing' mode. The operand to be incremented is at a location decided by FSR content. When the instruction is executed, the operand is fetched, incremented and stored back in the same location. The FSR has eight bits. Indirection requires the full 9 bits to be specified. The 9th bit (MS bit) is provided by bit b_7 of **STATUS** register within the processor; it is called the IRP bit [Register bank select bit for indirect addressing]. Indirect addressing requires the IRP bit in **STATUS** register as well as the FSR register to be loaded in advance. Figure 7.8 shows the formation of the indirect address using the IRP bit and the FSR content. The fact that FSR content can be changed behind the back of the instruction execution operation, adds considerably to the power and flexibility of the processor. Indirect addressing is useful in a variety of situations.

7.2.3 Program Memory

The devices in the PIC®16CXX family have the program memory arranged in pages of 2k words each. The low end devices have only one page of such program memory; the high end ones have 4 pages of program memory. Figure 7.9 shows the program memory organization. The addressing range required for the 4 pages of program memory – total size of 8k – is 13 bits wide. Hence the PC too is 13 bits wide. The LS byte of PC is the SFR – PCL. One can access PCL for reading as well as writing. The more significant bits of the PC – namely < b12:b6> – are in the register PCH. PCH is not directly accessible for reading or writing. PCLATH – a 5-bit wide SFR – is closely related to the PCH. Execution of certain instructions results in the bits of PCLATH redefining PCH contents. All such cases result in the PC content being changed and hence the program control being transferred to a different location in the program memory. Instructions which affect the PC content are discussed in Section 7.3.13 and in Chapter 9.

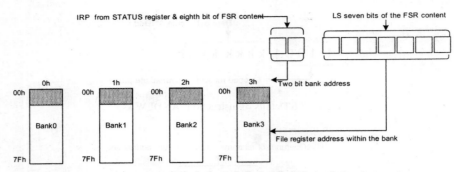

Figure 7.8 Formation of File register address in indirect addressing mode

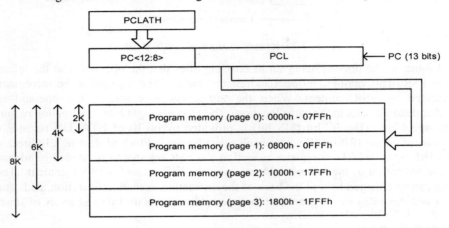

Figure 7.9 Program memory organization

7.2.4 STATUS Register

The PIC®16CXXX Processor has instructions to carry out algebraic operations on bytes. Multi-byte algebra is carried out doing it on byte pairs sequentially. Such an operation requires carry or borrow bit from one byte pair to be used with the next pair. To facilitate such multi-byte algebra and other similar operations, 3 status bits are provided. Depending on the result of algebraic and logic operations, their values get set or reset. These can be used for decisions in subsequent operations. The status bits are in the STATUS register – an SFR. The bit assignments of the STATUS register are shown in Figure 7.10. Roles of IRP, RP1, and RP0 have been discussed earlier. Discussion of the roles of the other two bits – $\overline{TO}$ and $\overline{PD}$ – is differed to the later chapters.

7.2.4.1 Zero Bit
The Zero bit is designated as 'Z'. It gets set if the result of execution of an algebraic or logic instruction is zero – that is, all the bits of the resulting byte are zero.

7.2.4.2 Carry Bit
The carry bit is designated as 'C'. It is affected only if an addition or subtraction instruction is executed. C is set if the result of executing an Add instruction causes carry bit output from the most

significant bit position; else it is reset. Subtraction is done in 2's complement form; here C has the significance of a borrow bit. If the result is positive, C is set. If it is negative, C is reset and the result is in 2's complement form.

7.2.4.3 Digit Carry

The digit carry bit is designated as DC. If is useful when doing algebra with BCDs. DC is affected by execution of Add or Subtract instructions. It is set if a Carry occurred by the addition of lower order nibbles; else it is reset. BCD digits can be corrected by checking the value of the BCD bit. Three bits – namely b_2, b_1 and b_0 in the STATUS register in the processor (Figure 7.10) are dedicated to Z, DC and C bits respectively. One can access the status register at its address and read any of these bits.

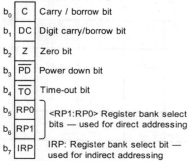

Figure 7.10 STATUS register and its bit assignments

7.2.5 Bank Selection

The STATUS register has 2 bits – b_5 and b_6 – dedicated to specify the active register bank. If the bit pair b6b5 in the STATUS Register is assigned values

$$b_6 b_5 = 00$$

bank0 is selected and active. All instructions in direct addressing mode are executed with the registers in this bank. If the bit pair is assigned the value

$$b_6 b_5 = 01$$

bank1 is selected and execution of all instructions in direct addressing mode take place with bank2; same holds good for bank2 and bank3 also.

The active register bank can be changed by rewriting into the bit pair b6b5 in the STATUS register. Bank changing can be fruitfully utilized in many applications. Let us consider one example. The program may carry out a task by running a program segment. The results and any intermediate values can be stored in the scratchpad area of the active bank – say bank0. At this stage a bank switching can be effected and bank1 made active. Next stage of the program can use the scratchpad registers in bank1. All the while the values stored in the registers in bank0 remain undisturbed. Continuing in the same manner one can switch to bank2 and bank3 also. Fast interrupt servicing (See Section 9.3) is another case where the bank switching facility is helpful.

As explained earlier, access to the RAM area in the indirect addressing mode is slightly different. The bits – b_6 to b_0 – in the FSR specify the address location in the selected bank. The bit IRP – bit b_7 in the STATUS register together with the bit b_7 of the FSR specify the bank for the location to be addressed (See Figure 7.11).

The banks provide two more levels of flexibility. Firstly the key Special Function Registers associated with the processor functions are common in all the banks. Any bank switching does not affect these. As a result the **STATUS** register, **FSR**, **PCL**, and **PCLATH** remain unaffected. In turn peripheral status, assignment etc., do not change with switching of banks.

Sixteen registers at the top end of the general purpose register area are common for all the four banks forming a common scratch pad area (See Figure 7.11). This facility can be useful in many situations. One can carry out execution of a program segment with bank0 and generate specific data out of this. The data can be stored in the common area. Bank switching can be effected at this stage and bank1 made active. The values generated earlier using bank0, are available directly for the task using bank1.

Another use for the common scratchpad area is in 'context saving' when embarking on context switching (See also Section 9.3).

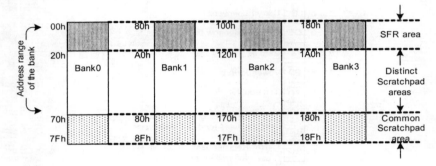

Figure 7.11 RAM file registers –Banks and allocation

7.3 INSTRUCTION SET

The PIC®16CXXX series uses a processor core of the 'Reduced Instruction Set Computer (RISC)' type. The instruction set available is compact and small; it has only 35 instructions. The set is characterized by the following:

- All instructions are single word instructions
- Most instructions are executed in one instruction cycle (See Section 6.5). There are a few instructions for branching or skipping an instruction. Their execution may extend to the following instruction cycle. Only these extend for 2 instruction cycles.
- An operation carried out by an instruction, cannot be carried out by another instruction or instruction combinations within the same time or with the same resources.
- Any operation not possible to be executed with an instruction within the instruction set, can be carried out by a sequence of instructions within the available instruction set.

Many other microcontrollers in use have a 'Complex Instruction Set Computer (CISC)' structure and a much larger instruction set. Comparison of RISC and CISC architectures [3, 14] is outside the scope of this book.

7.3.1 Instruction Format

Each instruction is 14 bits wide. Each has an Opcode and one or two operands. The instruction format is of 3 types.

7.3.1.1 Byte Oriented Instructions

There are 14 instructions of this category. The instruction format is as shown in Figure 7.12. The 5 MS bits represent the Opcode. The 7 LS bits specify the address of an SFR which is the source address. The intermediate bit at b7 position specifies the destination address. If d = 0, the result after execution of the instruction, is stored in the working register. If d = 1, the result after execution of the instruction is stored back at the SFR specified by the bit combination fff ffff.

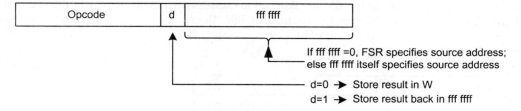

Figure 7.12 Instruction format of byte oriented instructions

If fff ffff   0, direct mode of addressing is implied. The full address of the SFR is formed by concatenating the register bank address – (RP1:RP0) of STATUS register – to it, to form the full 9 bit address. If fff ffff = 0, indirect mode of addressing is implied. The 8-bit FSR content concatenated with the IRP bit in the STATUS register specifies the address location for the data.

7.3.1.2 Bit Oriented Instructions

There are 4 instructions of this category. The instruction format is shown in Figure 7.13. Bits b_6 to b_0 specify a SFR address. Bits b_9-b_7 specify the bit address within the selected SFR. Bits b_{13}-b_{10} represent the opcode. If fff, ffff ≠ 0, direct mode of addressing is implied; fff ffff itself represents the source and destination byte address; the concerned bank is specified by (RP1:RP0) bit pair in STATUS register; If fff ffff = 0, indirect addressing is implied; FSR content concatenated with the IRP bit of STATUS register specifies the source and destination address.

7.3.1.3 Literal Based Operations

There are 9 instructions of this category. The instruction formats are shown in Figure 7.14. For algebra and logic operations [Figure 7.14(a)] b_{13}-b_8 specifies the opcode and b_7-b_0 specifies a byte value. The 8-bit k value forms a byte operand; the instruction is in immediate mode. Result of the operation is stored in the working register. Two additional instructions [Figure 7.14(b)] specify a 10 bit K value and use a 3 bit opcode – these are CALL and GOTO. In both cases k is directly loaded into the PC and the processor branches off to a new location in program memory. The instructions are discussed in detail later.

7.3.2 Other Instructions

There are 8 more instructions which do not fall into any of the above categories. Two of these (CLRF and MOVWF) specify the address of a RAM location as the destination. If the address is zero, content of FSR is taken as the address (operation in indirect mode). The other 6 instructions are fully specified by the instruction opcode and do not carry any additional operand information. The full instruction set is reproduced in a compact form in Appendix A. Each instruction is dealt with in details in the Mid-Range Manual [12]. We discuss only the salient features of each category.

7.3.3 Conventions in Assembly Language

Explanation of some symbols and conventions used in assembly language is in order here:

- (Ab) signifies content of location identified with the name 'Ab'. Ab may be a register location or a memory location.
- ((Ab)) signifies content of location (Ab) i.e., Ab points to a location; its content is treated as an address and the byte (word) therein is the operand. Figure 7.15 shows a register file; two segments are shown. The registers in the file are assigned addresses in ascending order. In the segment in Figure 7.15 (a) location Ab stores the hex number 38h. Hence

$$(Ab) \quad = \quad 38h.$$

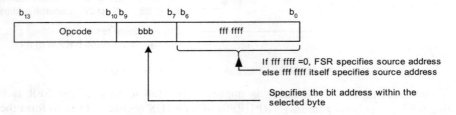

Figure 7.13 Instruction format of bit oriented instructions

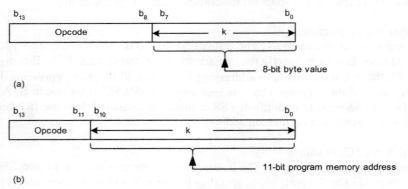

Figure 7.14 Instruction formats of literal based operations

- The segment in Figure 7.15 (b) shows addresses in the neighbourhood of address 38h. We can see that

$$(38h) \quad = \quad C4h$$

- From the two relations, we can see that

$((Ab)) = (38h) = C4h$

'$\rightarrow$' is the symbol for assignment. The statement

$$A + B \quad \rightarrow \quad C$$

implies that the values of variables A and B are added and the sum is assigned to variable c. The statement

$$(Ab) + B \quad \rightarrow \quad C$$

implies that Ab is an address. The data stored there is (Ab); it is added with the value of variable B and the result assigned to variable C.

- The symbol pair $< >$ implies the values of bits enclosed within.

- A<b3> represents the value of bit b3 in the bit stream A
- A< b3:b2> implies the value of bit combination b3b2 in the bit stream A
- 0→ A< b3> implies that zero value is moved into bit b3 in the bit stream A

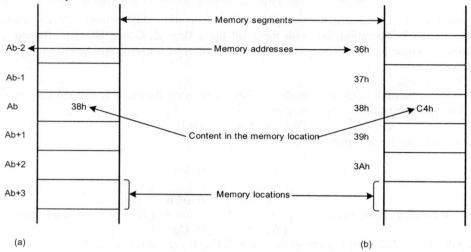

Figure 7.15 Two segments of memory with details of addresses and contents

7.3.4 Logic and Algebraic Instructions Involving Literal

Consider the assembly language instruction

ANDLW   k

When executed, working register content is ANDed with the literal K and the result stored back in the working register itself. Z flag is set or reset depending on the result. The operation can be represented by the statement.

$$(W) \text{ AND } K \quad \rightarrow \quad (W)$$

Before execution of the instruction, if

$$(W) \quad = \quad 0x38h$$
$$= \quad 0X0011\ 1000b$$

And

$$K \quad = \quad 0xE3h$$
$$= \quad 0X1110\ 011b,$$

after execution of the instruction, we have

$$(W) \quad = \quad 0x20h$$
$$= \quad 0x0010\ 0000b$$
$$(Z) \quad = \quad 0.$$

Alternatively if

$$(W) \quad = \quad 0x38H$$
$$= \quad 0x00111000b$$

and

$$k \quad = \quad 0xC3H$$
$$= \quad 0x11000011b$$

before execution of the instruction, after execution of the instruction, we have

$$(W) \quad = \quad 0x00H$$
$$= \quad 0x0000\ 0000b$$

and

$$(Z) \quad = \quad 1$$

Instructions IORLW k and XORLW k are also similar in operation. Instructions ADDLW k and SUBLW k are also similar in operation, but with them, all the 3 flags Z, C and DC are affected. Different possibilities and combinations exist; only a few are considered for illustration here.

The instruction

ADDLW k

adds the value of literal k to the byte stored in working register; the result is also stored in the working register. The operation can be represented as

$$(W) + k \quad \rightarrow \quad (W)$$

All flags are affected by the operation

With

$$(W) \quad = \quad 0x89h$$

and

$$k \quad = \quad 0xBCh$$

Execution of the instruction, results in the sum byte being stored in W. As a result we have

$$(W) \quad = \quad 0x45h$$

The carry flag C and DC carry bit are set to 1 each. Z flag bit remains reset at 0.

As another example, consider the following case:

$$(W) \quad = \quad 0x70h$$
$$k \quad = \quad 0x90H$$

Execution of ADDLW K results in

$$(W) \quad = \quad 0x00h$$

The C bit and Z bit are set. DC bit remains reset.

The SUBLW instruction functions in a similar manner; subtraction is carried out in 2's complement form. The Z, C and DC bits are affected accordingly.

7.3.5 Logic and Algebraic Instructions Involving Registers

Consider execution of the instruction

ANDWF f, d

The operation carried out can be represented as

$$(W)\ AND\ (f) \quad \rightarrow \quad (W)$$

if

$$d \quad = \quad 0$$

and

$$f \quad = \quad 0.$$

If

$$d \quad = \quad 1$$

and

$$f \quad = \quad 0$$

we have the operation

$$(W)\ AND\ (f) \quad \rightarrow \quad (f).$$

On the other hand if

$$d \quad = \quad 1$$

but
$$f \quad = \quad 0$$
the operation is
$$(W) \text{ AND } ((FSR)) \quad \rightarrow \quad ((FSR)).$$
If
$$d \quad = \quad 0$$
and
$$f \quad = \quad 0$$
the operation is
$$(W) \text{ AND } ((FSR)) \quad \rightarrow \quad (W)$$

The value of bit d and the value of f both have a say in deciding destination. Table 7.1 summarizes the different possibilities depending on the values of f and d.
Only the Z bit is affected by the operation.
Let

$$
\begin{aligned}
(W) &= 070h \\
f &= 0x38h \\[6pt]
(38h) &= 0x05h \\
d &= 1
\end{aligned}
$$

The contents of various locations after execution of the instruction are

$$
\begin{aligned}
(W) &= 0x70h \\
f &= 0x38h \\
(38h) &= 0x00h \\
(Z) &= 1
\end{aligned}
$$

Alternatively let

$$
\begin{aligned}
(W) &= 0x70h \\
f &= 0x00h \\
(FSR) &= 0x38h \\
(38h) &= 0x05H \\
d &= 1
\end{aligned}
$$

The contents of various locations after execution of the instruction are

$$
\begin{aligned}
(W) &= 0x70h \\
f &= 0000h \\
(38) &= 0x00h \\
(Z) &= 1
\end{aligned}
$$

Instructions IORWF f, d; XORWF f, d; ADDWF f, d; and SUBWF f, d are similar to the above in all respects. In the case of ADDWF f, d and SUBWF f, d the C and DC bits (apart from Z bit) are also affected by the operation.

Table 7.1 Instruction execution possibilities depending upon the values assigned to f and d

f	d	Operand and destination
0	0	Content of location with address 'f' is the source operand; destination is the working register.
0	1	The source and destination address is 'f' itself.
= 0	0	Source address is in FSR; destination is the working register.
= 0	1	The source and destination address is available in FSR.

7.3.6 Logic and Algebraic Operations with a Single Byte

The content of any desired location can be complemented by executing the instruction
COMF f, d.
The alternatives for the operation conform to Table 7.1. The operation carried out by execution of the instruction is

$$\overline{(f)} \quad \rightarrow \quad (W)$$

if

$$d = 0$$

and

$$f = 0;$$

that is the content of location with address f is complemented bitwise and the result loaded into W.
But

$$\overline{(f)} \quad \rightarrow \quad (f)$$

if

$$d = 1$$

and

$$f = 0.$$

Further if

$$d = 0$$

and

$$f = 0,$$

$$\overline{((FSR))} \quad \rightarrow \quad (W).$$

But if

$$d = 1$$

and

$$f = 0$$

$$\overline{((FSR))} \quad \rightarrow \quad ((FSR)).$$

Z flag is affected by the execution of the instruction. C and DC flags are not affected. As an example let

$$(W) = 0x38h$$
$$(f) = 0x94h$$

and

$$d = 1$$

before execution of the instruction.
After execution of the instruction, contents of the different locations are

$$(W) = 0x38h$$
$$(f) = 0x6Bh$$
$$Z = 0$$

Note that working register is not involved in the operation and it remains unaffected.
Let

$$(W) = 0x38h$$
$$(f) = 0x00h$$
$$(FSR) = 0x85h$$

$$(85) \quad = \quad 0xFFh$$
$$d \quad = \quad 1$$

After execution of COMF f, d; contents of various locations are

$$(W) \quad = \quad 0x38h$$
$$(f) \quad = \quad 0x00h$$
$$(FSR) \quad = \quad 0x85h$$
$$(85) \quad = \quad 0x00h$$
$$Z \quad = \quad 1.$$

DECF f,d and INCF f,d are instructions to decrement or increment the destination content by 1. Their execution is similar to the COMF f,d explained above. Execution of INC f, d can cause a bit overflow when the content is incremented from the value 0xFFh. But the **C** bit remains unaffected. Similarly when the operand is incremented from 0XXFh, there is a carry from lower nibble to the higher nibble. But the **DC** bit remains unaffected. Execution of DEC f, d also affects only the **Z** bit. **C** and **DC** bits remain unaffected. The operation is similar to the INC f, d instruction.

7.3.7 Instructions for Data Transfer

Consider the instruction
MOVLW k
The operation carried out can be represented as

$$k \quad \rightarrow \quad (W).$$

The operand k – a byte – is moved into the working register. The flags are not affected by execution of the instruction. The instruction
MOVF f, d
moves the contents of f to a destination. If d = 0, W is the destination. If d is 1, f itself forms the destination. The movement is in indirect mode if d = 1 and f = 0. The operation carried out can be represented as

$$(f) \quad \rightarrow \quad (W).$$

if

$$D \quad = \quad 0$$

and

$$(f) \quad = \quad 0.$$

And

$$(f) \quad \rightarrow \quad (f)$$

if

$$D \quad = \quad 1$$

and

$$(f) \quad = \quad 0.$$

Further if

$$D \quad = \quad 0$$

and

$$(f) \quad = \quad 0$$
$$((FSR)) \quad \rightarrow \quad (W).$$

Again if

$$D \quad = \quad 1$$

and

$$(f) \quad = \quad 0$$

$$((FSR)) \quad \rightarrow \quad ((FSR))$$

All the execution possibilities conform to Table 7.1. Z flag is affected by the execution of the instruction; in this respect, this instruction is different from the above two. The operations

$$(f) \quad \rightarrow \quad (f)$$

and

$$((FSR)) \quad \rightarrow \quad ((FSR))$$

can be used to test operand value for zero. They do not serve any other purpose.

7.3.8 Clearing a Destination

There are 3 instructions to clear designated locations; clearing the content of a selected location, makes all the bits there zero. Clearing, resetting, and setting to zero mean the same.
The instruction
CLR W
implies the operations

$$0 \quad \rightarrow \quad (W)$$
$$1 \quad \rightarrow \quad (Z)$$

Similarly the instruction
CLR F
implies the operations

$$0 \quad \rightarrow \quad (f)$$
$$1 \quad \rightarrow \quad (Z)$$

if

$$f \quad = \quad 0$$

Note that with both instructions, Z flag is set.
The processor has a Watch Dog. It has a Watch Dog Timer associated with it. The instruction
CLR WDT
clears the watch dog timer contents. Watch Dog facility is discussed in a later chapter.

7.3.9 NO Operation

The instruction
NOP
does no specific operation. It is apparently harmless. But its presence serves two purposes. Some routines call for measured time delays of short duration, to be introduced intentionally. Execution of NOP lasts for one instruction cycle and serves this precise purpose. In an application, the assembly language program prepared (during prototyping and debugging stages), may be altered for the final version of the product. The altered version may require deletion of specific instructions in the program. The preferred practice is to fill up such deleted locations with NOP rather than deleting and shifting all subsequent instructions. This is the second useful purpose of executing a NOP instruction.

7.3.10 SLEEP

When SLEEP instruction is executed, the processor automatically shuts down the oscillator clock and the whole unit enters a low power mode. However many of the registers and peripherals retain the values stored in them. The system can be 'woken up from sleep' and brought back to full operation through one of a set of designated operations. The mode is useful for battery operated devices where the energy drain is to be kept to the minimum. Details of SLEEP mode are discussed later.

7.3.11 Rotate Instructions

There are two rotate instructions; one rotates the binary string of 8 bits of the operand plus the carry flag bit, right through one bit. Content of bit bo is shifted into C flag, b1 to bo and so on as shown in Figure 7.16(a).
The operations executed can be expressed as:

$$(b0) \rightarrow (C)$$
$$(C) \rightarrow (b7)$$
$$(b7) \rightarrow (b6)$$
$$.\quad\quad.$$
$$.\quad\quad.$$
$$.\quad\quad.$$
$$(b2) \rightarrow (b1)$$

$$(b1) \rightarrow (b0)$$

The instruction mnemonic is
RRF   f, d
If

$$d = 0$$

and

$$f = 0,$$

content of f register is rotated right through carry bit once and the result is stored in the working register.
If $d = 0$ and $f = 0$ content of register whose address is in FSR, is rotated through carry and the result stored in the working register.
If $d = 1$ and $f = 0$, content of f register is rotated right through carry bit once and the result stored back in f register.
If $d = 1$ and $f = 0$, content of register whose address is in FSR, is rotated right through carry bit once and the result stored back there itself.
The instruction
RLF f, d
carries out the rotation of the operand bits along with carry bit, left by one bit (See Figure 7.16(b)). Depending on the values of f and d, four possibilities arise – these are similar to the four cases discussed above with RRF instruction. All the execution possibilities are in line with those in Table 7.1. The source and destination address is 'f' itself.

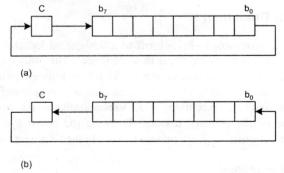

(a)

(b)

Figure 7.16 Execution of 'rotate' instructions (a) Rotate right (b) Rotate left

7.3.12 Swapping the Nibbles

The instruction
SWAP f, d
swaps the upper and lower nibbles of the operand. None of the flag bits is affected.
The operation can be expressed as

$$(f < 3:0>) \rightarrow (f<7:4>)$$
$$(f <7:4>) \rightarrow (f<3:0>)$$

Four variants of the instruction exist depending on the content of the f and d.
By way of illustration let

$$(W) = 0x39h$$
$$(f) = 0x47h$$
$$d = 1$$

before instruction execution. After the instruction is executed, the W and (f) contents will be

$$(W) = 0x39h$$
$$(f) = 0x74h$$

Note that W remains unaltered. Alternatively if we execute the instruction with

$$(W) = 0x39h$$
$$(f) = 0x47h$$

and

$$d = 0$$

we get

$$(W) = 0x74h$$

and

$$(f) = 0x47h$$

If we start with

$$(W) = 0x39h$$
$$(f) = 0$$
$$d = 0$$
$$(FSR) = 0x47h$$
$$((FSR)) = 0x86h,$$

execution of the instruction, alters W content. The new content of W is given by

$$(W) = 0x68h$$

Contents of other registers remain unaltered.

7.3.13 Instructions for Branching

Many programs require the processor to branch off to a designated location in the program memory rather than monotonously executing instructions in a sequence. The branch off may be conditional or unconditional.   PIC®16series has an instruction to do unconditional branching. With conditional instructions the branching may have to be effected if a condition is to be satisfied. The capacity to take such logical decisions in program execution, adds substantially to the processors' capability and power. In fact this is one of the primary reasons for using µC based systems so widely. It has 4 additional conditional instructions which are pivotal to such branching operation.

7.3.13.1 Unconditional Branching
The instruction
GOTO k

loads the LS bits of PC with literal k. Note that here k is 11 bits wide. The PC itself is 13 bits wide. Along with the above operation, the 2 MS bits of the PC are loaded with bits 3 and 4 of the PCLATH. The PCLATH is a Special Function Register in the RAM register file. It is repeated in all the 4 memory banks. It has address 02h. Once the PC is loaded with a new 13 bit word, the subsequent instruction is fetched from this new address. As the PC content is incremented, program execution continues from the new location. Prior to execution of the GOTO instruction, the bits PCLATH<b4:b3> are to be loaded with appropriate values.

The unconditional GOTO operation is illustrated in Figure 7.17. PC contains the program memory address 0299h where GOTO 1245H is stored as an instruction. The literal string K supplies 11 bits 0100100 0101. The bits b4 and b3 of PCL are 0 and 1 respectively. These two sets of bits are concatenated and loaded into PC. The new PC content is 1245H. Program execution continues with the instruction fetched from there. Since the instruction is fetched from a new location, the processor misses an execution cycle. In short GOTO is executed in 2 instruction cycles. This is a factor to be reckoned with where execution times are to be precisely estimated.

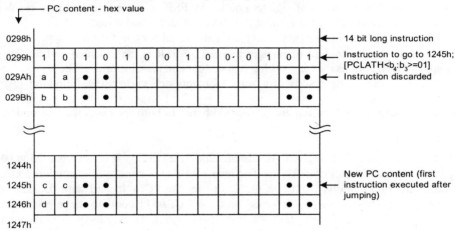

Figure 7.17 Details of execution of GOTO instruction

7.3.13.2 Conditional Skip Instructions

The instruction·

BTFSC f, b

is a conditional skip instruction. As with other instructions the field f is 7 bits wide and specifies an address in the RAM area; the field b is 3 bits wide and specifies a bit in the register specified by f. The instruction execution tests the specified bit; if it is clear, the following instruction is skipped and execution continued with the subsequent instruction. If the tested bit is not clear, execution sequence continues undisturbed; here execution is in one cycle. But if the next instruction is skipped, the following instruction which was already fetched and is already in the execution pipeline, is discarded. A NOP is executed in the following cycle. Thus total execution extends over two cycles. If f = 0, the bit in the location specified by FSR is tested and decision for skipping (or otherwise) taken based on that. In most programs, the instruction that follows BTFSC instruction will be a GOTO. The program will vector to the address specified thereon. GOTO is also a 2 cycle instruction. Hence whether instruction skipping occurs or not, one extra cycle time is required for the program sequence.

BTFSS f, b

is another instruction which tests a bit and skips the following instruction. Here skipping takes place if the specified bit is set. In all other respects this instruction execution is identical to BTFSC f, b instruction.

The instruction

INCFSZ f, d

increments the operand register. If incrementing results in the operand value becoming zero, execution of the following instruction is skipped. Program execution continues with the subsequent instruction. Since the instruction to be skipped would have been fetched already in the previous instruction cycle, it is discarded and a NOP is executed in the following instruction cycle. Thus the skipping, if it occurs, lasts for 2 instruction cycles. As mentioned earlier, the instruction following an INCFSZ instruction will be GOTO in many of the cases. Its execution extends over 2 consecutive machine cycles. In short either route of execution will result in an additional cycle being consumed for execution. Depending on the value of d and f, operation has 4 alternatives (Table 7.1).

- If d = 0 and (f)   0, (f) is incremented and the result placed in W
- If d = 0 and (f) = 0, the content of register pointed by FSR is incremented and the result placed in W. In both of the above cases, the content of source register remains undisturbed.
- If d = 1 and (f)   0, (f) is incremented and the result placed back in f itself
- If d =1 and (f) = 0, the content of register pointed by FSR is incremented and the result placed back there itself. In both cases the working register content remains undisturbed.

In all the cases if incrementing results in a zero, as mentioned earlier, the following instruction is skipped.

Two examples of instruction execution are considered here. In both the cases the instruction executed is

INCFSZ Ab, 1.

In the case shown in Figure 7.18, register Ab contains the byte 0xFEh. It is incremented to 0XFFh and the new value stored back in register Ab itself. Since zero flag Z is not set, execution continues with the very next instruction namely, GOTO 0X392h. In turn the PC is loaded with 0X392h; execution continues from there (It is implicitly assumed that the PCLATH contents have not been changed before embarking on the execution of the GOTO instruction here.). During the pass in the next iteration cycle, shown in Figure 7.19, register Ab contains the byte 0xFFh. It is incremented and overflows to 0X00h. The new value is stored back in register Ab itself. The zero flag Z is set; the processor skips the very next instruction; execution continues with the instruction following it at location 0x029Bh – namely, LMNO.

The instruction

DECFSZ f, d

decrements the operand. If the result is zero, execution of following instruction is skipped. Else, execution continues undisturbed. In all other respects the instruction is identical to the instruction INCFSZ f, d described above; hence it is not elaborated further here.

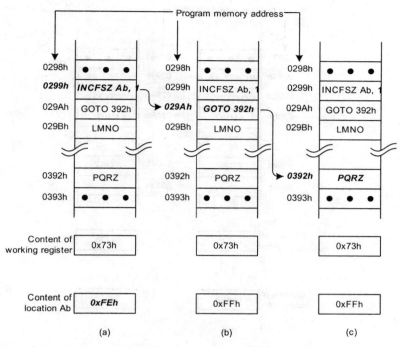

Figure 7.18 Illustration of execution of instruction INCFSZ Ab, 1: (a) Contents of different locations before the execution of instruction INCFSZ Ab, 1 (b) Contents of different locations after the execution of instruction INCFSZ Ab, 1 (c) Contents of different locations after the execution of the following instruction GOTO 0x392h. In each case contents of locations active during execution of the current instruction are shown in bold italics. The arrows show the progress of processor execution.

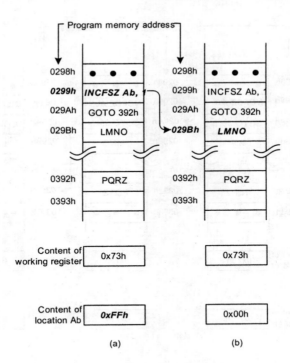

Figure 7.19 Illustration of execution of instruction INCFSZ Ab, 1: This shows the execution in the iteration cycle following that in Figure 7.18 above. (a) Contents of different locations before the execution of instruction INCFSZ Ab, 1 (b) Contents of different locations after the execution of instruction INCFSZ Ab, 1: In each case contents of locations active during execution of the current instruction are shown in bold italics. The arrows show the progress of processor execution.

CHAPTER 8

FIRMWARE DEVELOPMENT

8.1 INTRODUCTION

The program downloaded into the μC is firmware. It may be of a few hundred words in the low-end systems but may extend to 20,000 words in systems at the high end. The program in Assembly Language cannot be prepared in one go, especially in the case of high-end systems. A structured approach for program development has evolved in the last few decades [8]. Three broad activities are involved here – Assembly language programming, testing and debugging, and Programming of the μC. Out of these programming of the μC is done using a Programmer. It has the hardware necessary to download the program as a binary file into the μC. It is done by conforming to the programming signals specified for the μC concerned.

Assembly Language Programming (ALP), testing and debugging are carried out with the help of software tools available for the purpose. All such software tools are available in a common platform called the 'Integrated Development Environment'– often referred as the 'IDE'. The IDE is supplied by the μC manufacturers themselves as well as third parties. They vary in detail from supplier to supplier. All of them possess some commonly used features which are discussed here.

8.2 PROGRAMMING IN ASSEMBLY LANGUAGE

Consider the group of assembly language statements in Figure 8.1. Two numbers 0x34h and 0x12h are obtained from the program memory itself. Out of these, 0x34h is first loaded into register location 005A; Subsequently the number 0x12h is moved into the accumulator: It is added to 0x34h and result stored back at 005A itself. The group of statements here is an assembly language program. Any μC based system will have a similar program running within. The program itself is in the machine language of the μC concerned. It may typically run into a few thousand lines of machine code. It was mentioned in the previous chapter that the assembly language has been developed to facilitate preparation of such machine language programs. Structured development of the program for a given application involves the following well defined steps.

- Identify the tasks to be carried out by the μC. Prepare the flowchart representing the sequence of tasks.
- With the flowchart as basis, prepare the assembly language program on paper. Often this may call for repeated attempts and modifications to the flowchart. The interactive process leads to an acceptable flowchart and its Assembly Language Program (ALP).
- The ALP has to be tested repeatedly, made acceptable as a running program, and as a program which does what is precisely expected of it.

We shall go through the details of ALP preparation in its final form.

```
MOVLW 0X34h;
MOVWF 0X5Ah;
MOVLW 0X12h;
ADDWF 0X5A, 1;
```

Figure 8.1 A segment of an assembly language program

8.3 FLOWCHART

A flowchart is a chart based representation of a program [19]. The chart is in the form of an activity flow. As an example the flowchart to multiply two numbers N and M and obtain the product P is shown in Figure 8.2. The steps involved in the multiplication process are as follows:-

- Read the numbers N and M
- Declare a number S and assign the value 0 to it.
- Add N to S and decrement M by 1.
- Check the value of M;
- As long as M is non-zero, repeat the above two steps in the same sequence
- When $M= 0$, assign the content of S to P– the desired product.

We have identified the steps involved in the multiplication process as a set of primitive operations. Each of the operations has been assigned a symbol and the operations are linked together showing the flow of program execution. The symbols used are shown separately in Figure 8.3.

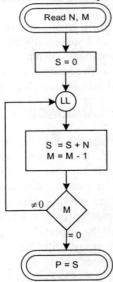

Figure 8.2 A flowchart to multiply two numbers

Figure 8.3 Symbols used in flowcharts

Each rectangle in a flowchart represents a simple algebraic or logical operation. Each diamond represents a logical decision box. Each line and associated arrow represents program flow; it points to the following step. Each small circle represents a 'branch in' point. The activity represented by the diamond calls for a logical decision to branch out to one or another part of the program. In this sense a diamond represents a decision block. In contrast the activities represented by each of the other blocks are assigned tasks. The branch out paths may be assigned labels which identify them. Thus in the flowchart of Figure 8.2 if M is non-zero, the program branches out to 'LL'. If M is zero, the program continues in its main path without branching out. Such continuation does not call for a branching out; hence it has not been assigned a label. Some flowcharts may have more than two branches to choose from. In such cases all the branch out directions are assigned labels to identify them (as is the case with C language). Such situations do not arise at the Assembly Language level. The double ellipse-like enclosure is an entry point to the program or an exit point from it. Any program can be represented as a flowchart by repeated and judicious use of these symbols relating the program activities.

In a practical situation, the flowchart may have many more logical / algebraic operations and decision boxes. As the number of decision boxes increases beyond four or five, the flowchart can become intractable. Whenever possible, it is preferable to have a modular approach; there flowcharts are prepared separately for clearly identified and demarcated segments of the overall program; these are combined to form the overall flowchart.

8.4 PROGRAM CODING

Consider the simple task of clearing a set of N registers in the scratch pad memory. A flowchart to achieve the task is given in Figure 8.4. Conventions used for assignments conform to those in Section 7.3.3. The program for the task is shown in Figure 8.5 (The program requires some alterations to make it work. These are discussed in the following section). The steps involved in the program are as follows:

- Load the address of the first memory location to be cleared into FSR.
- Load the number (of successive registers to be cleared) into register F1.
- Clear the first register – address location pointed by FSR.
- Decrement the content of location F1.
- Increment the content of FSR.
- Repeat the above three steps until F1 content is zero. When F1 content is zero, the task is complete.

The following observations are in order here:

- The program has 8 instructions. It occupies 8 memory locations.
- If N = 1, it takes 8 instruction cycles for completion of execution; if N = 2, it takes 12 instruction cycles for completion of execution since the group of 4 instructions.

```
LL    clrf INDF;
      incf FSR, 1;
      decfsz F1, 1;
      GOTO LL;
```

is executed again. In general the execution lasts for $5+4*N$ instruction cycles.

- The working register W and FSR are used in the routine. If they stored data of use for the program segment to follow the 'registers-clearing' task here, such data has to be saved prior to the routine here; after the routine the same data has to be retrieved and brought back to W and FSR respectively.

- Change in the value of N does not alter the program length. It affects only the execution time.

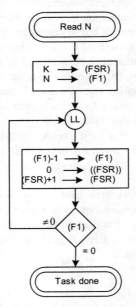

Figure 8.4 Flowchart to clear N successive locations

```
      ;Routine_1 to clear successive memory locations
        movlw k;
      movwf FSR; address of first register to be cleared  in FSR
      movlw  n:
      movwf F1;No. of registers to be cleared in F1
LL: clrf INDF;
      incf FSR, 1;
      decfsz F1, 1;
      GOTO LL;
        ; NEXT ROUTINE
```

Figure 8.5 Program conforming to the flowchart of Figure 8.4

8.5 ASSEMBLER

The assembler accepts a file of ALP as input. It has three roles to play:
- Translation of the program into a machine code sequence: The same will form the input to the programmer. The translation can be carried out by the assembler on receipt of a specified command. The embedded systems designer need not concern himself with its internal details.
- Provision of a set of directives to aid program development: The commands in frequent use are discussed here.
- Preparation of a list which shows the assembled program, the corresponding machine language program and the associated details: The list depicts the step by step program flow and facilitates tracking of the μC operation.

8.5.1 Assembler Directives

A number of Assembler Directives are available. Some of them are common to assemblers of many µCs; others may be specific to one µC or family of µCs. A set of commonly used assembler directive are discussed here. The aim is essentially to get a working knowledge of the assembler and be able to do coding. Designers get mastery over the set of directives with some practice of programming.

8.5.1.1 org and end Directives

The role of the directive – org – can be understood with the help of an ALP. A simple ALP is shown in Figure 8.6. A number – 0x04 – is loaded at the scratch pad register location – 0x30. It is repeatedly decremented until 0; When 0, the number 0x04 is loaded into the register once again. The sequence of activities continues ad infinitum. The set of instructions representing the program is repeated in Figure 8.7. The column to the left has the program memory address of each of the instructions. The instruction

GOTO 0x02

makes the µC go to location 0x02 and execute the instruction there – namely

Decfsz 0x30, 0x1

Similarly the instruction

GOTO 0x01

makes the µC go to location 0x01 and continue execution there from. The directive at the beginning of the program

Org 0x00

tells the assembler that the code segment that follows should commence at location 0x00. The line

End

at the end of the program tells the assembler that the program has 'ended here'. Assembly will continue until the assembler encounters the end directive and not beyond. The assembler will ignore all statements beyond the end directive. One can see that any ALP will have only one end statement in it.

```
;SIMPLE_1
 ;PROGRAM TO REPEATEDLY COUNT DOWN FROM A NUMBER TO 0.
 ORG      00H
         MOVLW     0X4
         MOVWF     0X30
         DECFSZ    0X30, 0X1
         GOTO 0x2
         GOTO 0x1
 END
```

Figure 8.6 A program to Illustrate the use of org and end directives

```
 ;SIMPLE_1
  ;PROGRAM TO REPEATEDLY COUNT DOWN FROM A NUMBER TO 0.
  ORG 00H
00          MOVLW     0X4
01          MOVWF     0X30
02          DECFSZ    0X30, 0X1
03          GOTO 0x2
04          GOTO 0x1
      END
```

Figure 8.7 Program in Figure 8.6 with the addition of address locations of the instructions (shown at the left)

A well organized program will be in a set of clearly identified program segments – each segment positioned in a different program memory area. A statement of the type 'org ABh' precedes each such segment. The assembler locates the segment that follows, from program memory location ABh; it continues until another org statement is encountered or until the end of the program. The program in Figure 8.6 has been modified and shown in Figure8.8. The latter has two org statements. After reset the μC starts execution at location 0x00; the first statement

```
Org 0x00
```

signifies as much. The μC jumps to location 0x10 by way of execution of the instruction

```
goto 0x10
```

and continues execution from there. The statement

```
org 0x10
```

directs the assembler to locate the program segment that follows from location 0x10; the program memory area up to 0x10 has been left out in the process. One can accommodate additional program segments in the same manner.

```
            ;SIMPLE_a
          ;PROGRAM TO REPEATEDLY COUNT DOWN FROM A NUMBER TO 0.
          ORG  0x00
   00          GOTO 0x10
          ORG  0x10
   10          MOVLW    0X4
   11          MOVWF    0X30
   12          DECFSZ   0X30, 0X1
   13          GOTO 0x12
   14          GOTO 0x11
          END
```

Figure 8.8 The program in Figure 8.6 modified to illustrate use of org directive to split a program into two segments: The numbers at the left represent respective program memory addresses.

8.5.1.2 Labels

As the program length increases, keeping track of the address of each instruction becomes difficult. Labels come to our rescue. A label is a name attached to a location. One can relate the label to the task at hand which helps in the coding activity. LED, switch, panel, show, rega, regx, . . . , are typical examples of labels. The program of Figure 8.6 has been modified with the use of labels and shown in Figure 8.9. Two Labels 'AA' and 'BB' have been used here. Consider the ALP line

```
BB movwf 0x30
```

The assembler encounters the symbol set 'BB' starting at the first column; Any such symbol set starting at the first column is interpreted by the assembler as a label. Thus 'BB' is interpreted as a label and identified as representing the address of the program memory – 0x01. Further wherever the same label appears in the program, it is taken as the memory address – 0x01 and replaced by it. Similarly 'AA'is interpreted as the label for the program memory address 0x02. Use of labels in longer programs makes programming job easier. It allows the programmer to link labels to physical quantities or events directly and makes the program more readable, understandable, and tractable. They are also useful with branching instructions.

```
;SIMPLE_2
 ;PROGRAM TO REPEATEDLY COUNT DOWN FROM A NUMBER TO 0.
 ;LABELS ARE USED TO IDENTIFY REGISTERS
 ORG     00H
         MOVLW   0X4
BB       MOVWF   0X30
AA       DECFSZ  0X30, 0X1
         GOTO    AA
         GOTO    BB
 END
```

Figure 8.9 Program of Figure 8.6 modified to show the use of Labels

8.5.1.3 *Defining Assembler Constants and Variables*

A number used in an ALP can be assigned a label apt for it. The value can be assigned for it separately. The assembler offers different options for doing this. Consider the statement
num equ 0x4
The assembler interprets num as a number and assigns the numerical value 04h to it; it offers different options to do this. Where ever the label 'num' appears in the program, the assembler replaces it by the value – 0x4 during assembly. As an example consider the ALP in Figure 8.10 which is essentially the same as that in Figure 8.6. The numerical constant 0x4 has been replaced by the label 'num' wherever it appears within the ALP; num has been assigned the value 04h separately through the above statement. The label num itself can appear repeatedly anywhere in the program. The single statement
num equ 0x4
is enough to assign the value 0x4 to it at all the locations. The facility helps the programmer in different ways. During testing and debugging a value can be assigned and test carried out fast enough – essentially to test the logic of the program. The exact value to be used within the routine can replace the test value later. A code segment used in an ALP can use one value of num. The same segment can be used at another location within the ALP with a different value assigned to num (In fact the 'set' directive is to be used here).

The ALP in Figure 8.6 has been altered slightly and given in Figure 8.11. A label reg_1 has been assigned to the scratch pad register location 030h through the statement
reg_1 equ 0x30
The assembler will replace reg_1 by 0x30 throughout the ALP. The difference between the two uses of equ is only conceptual. In one case a constant has been defined and assigned a value. In the second case a RAM address has been defined; the assembler does not discriminate between the two uses.

The pair of directive cblock and endc can be us together to assign a RAM area to a set of variables with assigned names. Consider the ALP in Figure 8.12. A RAM block has been defined; the statement
cblock 0x25
assigns addresses to the identified block starting with the RAM location 0x25. Since only one variable – reg_1 – has been specified within the block, it is assigned to the addres 0x25. In practice the cblock-endc set can be used to assign as many variable addresses as desired. The program in Figure 8.13 is an illustration of such a use (See Section 8.6 for its working). The block can be used in an ALP repeatedly.

```
;SIMPLE_3
 ;PROGRAM TO REPEATEDLY COUNT DOWN FROM A NUMBER TO 0.
 ;A CONSTANT USED IN THE PROGRAM IS ASSIGNED A NAME 'num'
num equ 0x4
 ORG    00H
        MOVLW  num
BB      MOVWF  0X30
AA      DECFSZ 0X30, 0X1
        GOTO   AA
        GOTO   BB
 END
```

Figure 8.10 A modification of the program in Figure 8.6 to illustrate the use of assigning a value to a label

```
;SIMPLE_6
 ;PROGRAM TO REPEATEDLY COUNT DOWN FROM A NUMBER TO 0.
 ;THE REGISTER HAS BEEN ASSIGNED A NAME & SAME USED WHEN
REFERRING TO THE REGISTER
num       equ 0x4
reg_1     equ 0x30
 ORG      00H
          goto   start
 org      0x12
start
          MOVLW  num
BB        MOVWF  reg_1
AA        DECFSZ reg_1, 0X1
          GOTO   AA
          GOTO   BB
 END
```

Figure 8.11 A modification of the program in Figure 8.6 to illustrate the use of assigning values to a labels

8.5.1.4 include *directive*

The SFRs and specific bits in the SFRs are assigned specific labels in the PIC series. Programmers remember these and prefer to have their ALP using these as corresponding labels. It eliminates the need to remember SFR addresses and corresponding bit addresses and using them directly. The #include directive facilitates this. Each μC has its associated header file. It has all the labels associated with the μC and the addresses assigned to them. The ALP line

#include

automatically includes the header file within the program. μC specific labels can be used in the program as labels. The assembler will replace them with respective addresses. Figure 8.14 shows an ALP which illustrates the use of #include directive. It is a modification of the ALP in Figure 8.11. An SFR in Bank1 is used to store the number. Before accessing it, Bank1 is selected by setting the IRP1 and IRP0 bits in the STATUS register. 'IRP1', 'IRP0' and 'STATUS' are used as labels in the program. Thanks to the '#include p16f877a' directive the assembler will automatically insert the concerned addresses in place of the labels. The '#include' directive can be used to include other source files also in the ALP.

```
;SIMPLE_7
 ;PROGRAM TO REPEATEDLY COUNT DOWN FROM A NUMBER TO 0.
 ;A DATA/REGISTER FILE BLOCK FOR THE USER HAS BEEN IDENTIFIED & USED
num        equ 0x4
           cblock  0x25
           reg_1
           endc
 ORG       00H
           goto    start
 org       0x12
start
           MOVLW   num
BB         MOVWF   reg_1
AA         DECFSZ  reg_1, 0X1
           GOTO    AA
           GOTO    BB
 END
```

Figure 8.12 Modification of the program in Figure 8.6 to illustrate the use of cblock – endc pair of directives

```
 ;SIMPLE_9
 ;PROGRAM TO REPEATEDLY COUNT DOWN FROM A NUMBER TO 0.
 ;THE NUMBER CAN BE 24 BITS LONG & STORED IN A SET OF 3 REGISTERS
numl      equ 0x4
numh      equ 0x2
numn      equ 0x3
     cblock  0x25
     regl, regh, regn
     endc
     ORG      00H
     goto     start
     org 0x12
start
           clrf    regh
           clrf    regl
           clrf    regn
BB         movlw   numn
           movwf   regn
CC         movlw   numh
           movwf   regh
DD         movlw   numl
           movwf   regl
AA         decfsz  regl, 0X1
           goto    AA
           decfsz  regh, 0X1
           goto    DD
           decfsz  regn, 0X1
           goto    CC
           goto    BB
     END
```

Figure 8.13 A program to illustrate the use of cblock – endc pair of directives

8.5.1.5 banksel *DIRECTIVE*

Switching amongst RAM banks to select specific RAM locations and SFRs is facilitated by the banksel directive. For example reg_1 in the program in Figure 8.14 is in BANK1. The statement

banksel reg_1

in the ALP ensures that Bank1 is selected and reg_1 accessed. Bank1 remains selected for the program segment that follows. One can switch back to bank0 (if necessary) using another banksel statement. The ALP in Figure 8.15 illustrates the use of banksel directive.

8.5.1.6 Numeric Constants

The Assembler offers the facility to define numeric constants in different forms. The default mode is hexadecimal. Possible number definitions are given in Table 8.1. Whenever necessary the default radix can be changed to binary, octal or decimal using the 'radix' directive.

```
;SIMPLE_6_C
;PROGRAM TO ILLUSTRATE THE USE OF '#INCLUDE'
DIRECTIVE
;THE REGISTER HAS BEEN ASSIGNED A NAME & SAME USED
WHEN      ;REFERRING TO THE REGISTER
#include p16f877a.inc
num       equ 0x4
reg_1     equ 0xA0
 ORG      00H
          goto    start
 org      0x12
start
          BCF  STATUS,RP1;SELECT BANK1
          BSF  STATUS,RP0
          MOVLW   num
BB        MOVWF   reg_1
AA        DECFSZ  reg_1, 0X1
          GOTO    AA
          GOTO    BB
 END
```

Figure 8.14 Modification of the program in Figure 8.11 to illustrate the use of '#include' directive

Table 8.1 Radix specifications

Type	Syntax	Example
Decimal	D'digits'	D'100'
	.'digits'	.'100'
Hexdecimal	H'hex_digits'	H'a5'
	0xhexdigits	0xa5
Octal	O'octal_digits'	)'763'
Binary	B'binary_digits'	B'11001010'
ASCII	A'character'	A'c'
	'character'	'c'

```
;SIMPLE_6_D
 ;PROGRAM TO ILLUSTRATE THE USE OF 'banksel' DIRECTIVE
 ;THE REGISTER HAS BEEN ASSIGNED A NAME & SAME USED WHEN
;REFERRING TO THE REGISTER
#include p16f877a.inc
num        equ 0x4
reg_1      equ 0xA0
 ORG       00H
           goto    start
 org       0x12
start
           banksel reg_1
           ;BCF     STATUS,RP1;SELECT BANK1
           ;BSF     STATUS,RP0
           MOVLW   num
BB         MOVWF   reg_1
AA         DECFSZ  reg_1, 0X1
           GOTO    AA
           GOTO    BB
 END
```

Figure 8.15 The program in Figure 8.13 Modified to illustrate the use of 'banksel' directive

8.6 SIMULATION

Simulation is done with two purposes – to see whether the program runs through the prepared sequence and to see that it does what it is expected to do. Facilities are provided for both the functions. In general the simulator runs the program with a simulated clock. It involves incrementing the PC and executing the instruction at the location concerned. Further at every stage the status of different registers within the µC are monitored. This is done in every mode of simulation provided.

8.6.1 Single Stepping

In Single-Stepping mode at a time, only one instruction is executed by the simulator on a command. The command is given by pressing a designated key or an icon within the simulation window. After execution of an instruction the simulator captures the status / contents of all the registers within the µC and stores the same. These can be viewed in convenient dedicated windows to confirm whether the captured values are as expected. Single stepping is done repeatedly until the program execution is completed. It helps in early stages of learning to work with the IDE and also to pay more attention to selected segments of a program.

8.6.2 Animation

Animation is the process of executing the program at a reduced pace. It is executed at intervals of a few seconds. Simultaneously the status of different registers in the µC is displayed. The mode helps in visually monitoring the execution. In the process anomalous behavior can be detected.

8.6.3 Breakpoint

One can carry out execution of a selected segment of a program and halt the simulator at a specified breakpoint. The breakpoint is specified in terms of a specified program line. One can locate the breakpoint at key locations of the program and carry out simulation. More often breakpoints are inserted in programs being simulated in RUN mode. Breakpoints can be inserted and removed from test programs through commands available in the simulator.

8.6.4 RUN mode

The simulator runs the program in simulation time in the RUN mode. The RUN mode is invoked at the final stage in program testing and debugging. The program is treated as a black box and run to confirm that its behavior is on predictable lines.

8.6.5 'Stepping in' and 'Stepping out'

A program may have loops and recursive loops within it. Once a program loop is tested and debugged, it can be treated as a black box during subsequent simulation. Such an approach helps in focusing attention to the loop and the main program separately and makes the simulation more structured. Simulators have facilities to segregate simulation of the loops and main programs. One can 'step into' a loop from the main program. With that one can test the loop. The second facility is to execute the program segment within the loop and then halt. With this a loop can be debugged and for the rest of the simulation treated as a black box; subsequent attention can be on the segment of the main program that follows. Such bypassing of program loops can be done repeatedly for each loop.

8.6.6 Displays for Monitoring

IDEs provide different windows for displaying different register contents. These can be simulated during simulation to monitor the register contents. The formats are suited for easy viewing and directly relating to the program flow.

8.6.6.1 Register File
The register file as a whole can be viewed in a matrix form. Alternatively specific registers can be selected and their labels and contents viewed.

8.6.6.2 Program Memory
The Program memory content can be viewed as a sequence. Often it will be displayed along with the corresponding instruction in assembly language.

8.6.6.3 Timing
Simulation time can be viewed in different form – cumulative time from start of simulation, elapsed time interval between execution of specified instructions and so on.

8.6.6.4 Trace
Trace is the facility which keeps a record of the progress of simulation activity. Contents of different registers, time stamp up to the current instruction and so on are recorded and displayed. Normally trace is stored In a FIFO. Records for a specified number of instructions up to the last one executed, are maintained.

8.7 ADDITIONAL ROUTINES

The simple task of clearing a set of sequential locations can be achieved in different ways; One routine may be easy to code but lengthy: Another may be compact in size but may call for skill in realizing: Yet another may use minimum resources within the μC: Yet another may be executed in the shortest possible time. Such choice possibilities and their pros and cons hold good with the alternatives for other tasks as well. A few examples are examined here by way of illustration.

8.7.1 Alternate Routine to Clear RAM Area

A slightly modified version of the routine considered in Figure 8.4 and in Figure 8.5 is shown in Figure 8.16. Instead of clearing the locations directly, the routine loads the number oxo (number zero in hex form) into each of the locations concerned. In all other respects the routine remains the same as the earlier one. It occupies 9 program memory locations. Execution extends for 5+5*N instruction cycles. The CLR group of instructions affect the Z flag (Z flag is set when the instruction is executed). The MOVWF instruction used here does not affect the zero flag.

When the memory locations to be cleared are only a few in numbers, one can do so directly by executing the instructions

CLRF REG1;

CLRF REG2;

CLRF REG3;

....

The clearing operation here is direct; it is preferable to be used when the locations to be cleared are not sequential in the register file.

8.7.1.1 Numeric Constants

The Assembler offers the facility to define numeric constants in different forms. The default mode is hexadecimal. Possible number definitions are given in Table 8.1. Whenever necessary the default radix can be changed to binary, octal or decimal using the 'radix' directive.

```
; ROUTINE 2: routine to clear successive memory locations
        #include p16f877a.inc
N equ   0xa
F1 equ  0x21
    org     0x00
        movlw   0x40;
        movwf   FSR; address of first register to be cleared  in FSR
        movlw   N;
        movwf   F1;No. of registers to be cleared in F1
        movlw   0x0;
next:   movwf   INDF;
        incf    FSR, 1; FSR = FSR + 1
        decfsz F1, 1;
        GOTO next;
        goto $
        ;NEXT ROUTINE
    end
```

Figure 8.16 A modified version of the routine in Figure 8.5

8.7.2 Time Delay in Software

Figure 8.17 shows the flowchart for generating a delay in software. A bit in a selected location is cleared. A number N_Dly is loaded into a scratchpad register B. Subsequently, the number is decremented continuously until it becomes zero. Once the number at B becomes zero, the bit is set to 1. The bit value remains at zero for the definite period of time of execution of the routine. The routine is shown in Figure 8.18. The total time delay is 3* N_Dly + 3 instruction cycles times.

During the period of execution of the routine, the µC is tied up and cannot carry out any other activity. It appears to be a wasteful use of the µC resource; it can be justified only if the µC does not have any other task to perform during the time.

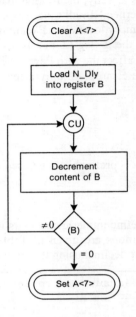

Figure 8.17 Flowchart to generate a desired time delay

```
;Routine_3   time delay routine - clear a bit
;keep it zero for a specified time & then make it 1
A          equ 0x30
AB         equ 0x31
N_Dly      equ 0x20
 org           0x00
           bcf A, 7
           movlw N_Dly
           movwf AB
cu         decfsz AB,1
           GOTO cu
           bsf A, 7
           goto    $
 end
```

Figure 8.18 A routine to generate a time delay

Figure 8.19 shows a slightly modified form of the routine. Three NOP instructions have been inserted in the routine of Figure 8.18 to form this one. Every NOP instruction within the count down loop increases the duration of delay by N_Dly instruction cycles. Every NOP instruction outside the count down loop increases the duration of delay by one instruction cycle – it can be used to fine tune the delay duration. The total duration of the time delay here can be seen to be 5* N_Dly + 4 instruction cycle times.

Observations:

The time delay duration can be doubled by repeating the instructions in the figure en bloc. The procedure can be continued further for longer delays.

The timer peripheral discussed in Chapter X, generates delays of the type discussed here in a different manner. With its use, the µC is freed of the mundane task of count down; it is free to do other activities.

The flowchart of a more detailed time delay routine is shown in Figure 8.20. It uses a register pair <NH:NL> to form a 16 bit number. The number is decremented until it becomes zero, to generate the time delay. The decrementing is done in 2 parts. NL is continuously decremented. Every time it becomes 0, NH is decremented. Decrementing is stopped when NH becomes zero. The routine is given in Figure 8.21. The total delay duration is NH*(256*3+3) instruction cycle times (This is an approximate value; the actual duration is different by a few instruction cycle times). The routine of Figure 8.21 has been altered slightly and shown in Figure 8.22. Here NL is loaded with the original value every time NH is decremented (it does not count down from FF every time). The delay duration is approximately equal to NH*(3*NL+3) instruction cycle times. The program in Figure 8.13 provides a delay decided by a 24-bit number.

```
;Routine_4   time delay routine - clear a bit
;keep it zero for a specified time & then make it 1
A          equ 0x30
AB         equ 0x31
N_Dly      equ 0x20
 org           0x00
           bcf A, 7
           movlw N_Dly
           movwf AB
           nop
cu         nop
           nop
           decfsz AB,1
           GOTO cu
           bsf A, 7
           goto      $
 end
```

Figure 8.19 A routine to generate a longer time delay compared to that in Figure 8.18

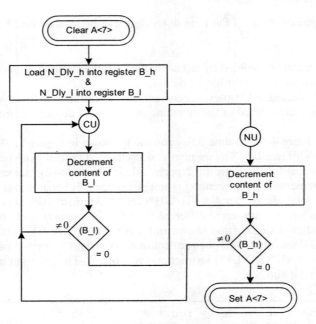

Figure 8.20 A routine to generate a time delay substantially longer than that in Figure 8.18

```
;Routine_5:Software delay based on a
16-bit register
N_Dly_h equ 0x30
N_Dly_l equ 0x40
 cblock     0x25
        a,N_l,N_h
 endc
     org 0x00
        bcf a, 7;
        movlw N_Dly_h;
        movwf N_h;
        movlw N_Dly_l;
nu      movwf N_l;
cu      decfsz N_l,1;
        goto cu;
        decfsz N_h,1;
        goto cu;
        bsf a, 7;
        goto $
 end
```

Figure 8.21 A routine to generate a longer time delay compared to that in Figure 8.18

```
;Routine_6: A PROGRAM TO GENERATE A LONG TIME DELAY
     CBLOCK  0X25
     a,b_h, b_l
     endc
N_Dly_h      equ 0x03
N_Dly_l      equ 0x04
     org 0x00
     bcf a, 7;
     movlw N_Dly_h;
     movwf b_h;
     movlw N_Dly_l;
nu   movwf b_l;
cu   decfsz b_l,1;
     goto cu;
     decfsz b_h,1;
     goto nu;
     bsf a, 7;
     goto $
end
```

Figure 8.22 A routine to generate a longer time delay compared to that in Figure 8.18

8.8 WORKING WITH THE SIMULATOR

The MPLAB® IDE is the tool used for editing, assembling, simulating, debugging, and downloading an assembly language program into the µC. The procedure for invoking and using the tool is briefly described here.

- Double click the MPLAB® icon on the desktop to open the IDE. In the IDE window, we get the following menu displayed in the same sequence:

File Edit View Project Debugger Programmer Tools Configure Window   Help

- An untitled workspace window and output window are also shown. The output window is used for displaying warnings, messages and compilation results (errors etc). It expands as follows:
- To configure a required device, click on Configure menu. In Configure menu, we get the following:

        Select device…
        Configuration bits…
        External memory…..
        ID Memory …
        Settings …

- Click on Select device to select the required device. A popup window is displayed from which a desired device can be selected. For example, key in

Select PIC 16 F877

- The corresponding programmer support and debug tool support available are highlighted in the window. Click on OK to confirm the selection.
- The selected device name will appear in the third column of the status bar shown at the bottom of the main window.
- To select a tool for simulating and debugging the code, click on Debugger menu. It expands as follows:

        Select tool   ▶

Clear Memory   ▶
- Click on Select tool; The expansion is as below:

None
1 MPLAB ICD2
2 MPLAB ICE 4000
3 MPLAB SIM
4 MPLAB ICE 2000

- Only supported tools are highlighted. Choose:

 3 MPLAB SIM

which will enable simulation options.

- The Debugger Menu provides different options to simulate a program.   Now, we are ready to prepare a program and simulate it.

The professional way of doing a code is to use a Project, add source files to it, configure compiler/assembler, linker etc., and build it to form a hex file which could be downloaded into a chip. But for the restricted purpose of checking a part or a Module of the Code (or small code), the Quick Build Mode is preferable.

8.8.1 Quick Build Mode

- Click on File Menu: Choose New from the options displayed in the pull down menu. An untitled editor window will appear, in which we can key in the assembly language program to be tested. This forms the source file. The text appears in grey colour.
- Click on "File" option. We get a pull down menu.
- Choose "Save as". In the popup window, select a directory of your choice (Default directory may be the installation directory of MPLAB®. Avoid saving user files into that directory to avoid crashing of the software).
- Give a suitable file name with extension ".asm" say "test-pulse.asm". Click on "Save". The source file has been created and saved. It is ready for compilation (assembling).
- labels, mnemonics, literals and assembler directives are highlighted in different colours and formats (bold) in the editor window.   For example, the mnemonics are highlighted in blue colour and bold format; the assembler directives are highlighted in blue colour and normal format and so on.
- For assembling a program available as a source file, click on "Project:" menu. From the pull down menu, choose "Quick build test-pulse.asm". The file name to be assembled, (one in the active editor window) is shown in the option along with the word "Quick build"
- The Program will be checked for syntax errors and assembled. If the program has no errors, the message "BUILD SUCCEEDED" will be displayed in the last line of the output window. Otherwise, the errors are indicated in the same window.
- If there are errors, the source file can be opened and edited. Once again compilation can be attempted. The procedure has to be repeated iteratively until all the errors in the source file have been removed and compilation is successfully completed.
- While writing the code, some disciplined format has to be followed. Violation of these will generate warnings. Warnings will not in any way affect the simulation or operation of the code; but they are intended for improving readability. For example, the following can generate warnings
- Writing an opcode or assembler directive in column 1 (starting from the left most end of the line). A tab in the beginning of a line would be a remedy.
- Labels are supposed to be in column 1. If they are away, a warning will be generated.

- The output window displays some other messages also; an example is a situation where we use a register in bank1 or bank2 – that is a bank different from what is selected (default bank selected is bank0, and it will be shown in the 9th column of the status bar at the bottom of the window).
- Debugging can be done, if the code is compiled successfully. For debugging, a step by step execution of the code will help (especially if the length of the code is small enough).
- If the program has been compiled successfully, an arrow in green colour will appear on the editor window pointing to the first line of the code. It indicates that the code is ready for simulation.
- For simulation, we would like to view the file registers of the µC, where the scratch pad memory is present. The variables (or registers) used in the program will be within the scratchpad.
- To view the scratchpad registers, click on "view" menu, and choose the option "4 File Registers" from the drop down menu.
- Choose "Symbolic" button present at the bottom of the File R Window to get the symbols displayed. The whole of File register memory is displayed in the window, where we can find all the special function registers, and user defined registers at the corresponding addresses.
- The File Register Window and the Code Editor Window can be tiled side by side to view both simultaneously. For this, choose the option "Tile Vertically" from the "Window" Menu.

Some of the important registers to be monitored while debugging, are given in the status bar at the bottom of the Main Window. Program Counter (PC) is displayed in the 4th column, work register (W) is displayed in the 5th column, Status flags (Zero, Digit Carry and Carry) are displayed in the 6th column. The values of the corresponding registers are updated during simulation. For example in the case of Status flags, an upper case letter indicates that the corresponding flag is set and a lower case letter indicates that it is cleared

- 'Z' indicates that zero flag is set and 'z' indicates that it is cleared.

8.8.2 Simulation

As mentioned earlier, with comparatively smaller programs simulation in Single Stepping mode is preferable. For single stepping, the option "Step into" can be chosen from the "Debugger" drop down menu. It can be also be done by clicking on the icon for it shown in the tool bar (See Table 8.2); alternately the shortcut key "F7" can be used for the purpose.

- Initially the processor can be reset by pressing the shortcut key "F6" or clicking on the icon for it (See Table 8.2).
- Pressing "F7" successively will proceed through the program, executing the mnemonic pointed by the "green arrow" displayed in the editor window. The "green arrow" is a symbolic representation of the program counter; as 'F7' is pressed the green arrow advances to the next instruction which logically follows the currently executed instruction.
- After executing every instruction, the file registers (as well as 'W') which are affected (whose content is changed by the instruction) are highlighted in the File Register Window. The corresponding line in the F.R. Window, changes its colour into red, indicating that the data of the register is modified by the currently executed instruction.
- After every execution step one can check the code for logical errors, by verifying the expected data and verifying the final result. In case of an error the code may be suitably modified, compiled again, and debugged for logical errors.
- The last line of the code must contain 'end' which is a compiler directive, specifying the physical end of the program.
- The logical end of the program is specified by the instruction "GOTO $" where $ represents the current PC value. This assembler instruction is exactly equivalent to a HALT instruction, which prevents the processor from executing junk instructions from the unused part of the memory.

Table 8.2 Icons, Function keys, and the functions they represent

Icon	Function	Function key
▶	Run	F9
❚❚	Halt	F5
▶▶	Animate	
{↓}	Step into	F7
{}↓	Step Over	F8
{}↷	Step Out of	
📄↓	Reset	F6

8.9 EXERCISES

1) A set of numbers is available in 8 consecutive locations. Write and test the programs for the following tasks:-
   Get the sum of the numbers.
   Get the sum of all the even numbers
   Count the number of numbers greater than 5.
   Count the numbers divisible by 4.

2) A set of numbers is available in 8 consecutive locations. Output all odd numbers on Port B. Output all even numbers on Port C.

3) A 16-bit number is stored as two bytes in 2 consecutive locations. Interchange b0 and b15, b1 and b14, . . . and b7 and b8.

4) A 16-bit number is stored as two bytes in 2 consecutive locations. Rotate the whole number by one bit in the right direction.

5) Write a program to multiply 2 8-bit numbers. Use it as a subroutine in a program to multiply 2 16-bit numbers.

6) Write a program to divide one byte by another. The quotient and the remainder are to be stored in two separate locations.

7) Write a program to divide one 16-bit number by another 16-bit number. The quotient and the remainder are to be stored as two separate 16-bit words.

CHAPTER 9

SUBROUTINES AND INTERRUPTS

9.1 INTRODUCTION

One of the novel features of µCs is their capacity to do varied tasks concurrently. They can keep track of each task consistently and without any mix-up. Concurrent operation is through subroutines and interrupt services. This feature is examined in detail in the present chapter.

9.2 SUBROUTINES

All programs require specific tasks to be done repeatedly at different stages of the program. Multiplication of bytes and adding multi-byte numbers are typical examples. Such tasks are carried out by the same type of program segment. Instead of repeating such program segments at all such locations, they are written and stored separately; each such block of instructions which carries out a specific and well defined task is called a 'Subroutine'. Subroutines are stored in a separate area of memory. Each subroutine is 'called' by the running program whenever desired.

Well organized programmers use subroutines effectively. They segment the programming task into logically separate modules; each such module is written as a subroutine program. The main program proceeds by calling the subroutines in the desired order and manner. Calling a subroutine, executing it and returning to the main routine involve four steps.

- Halt the main program and provide for returning to the same point. This requires the PC content to be removed. When coming back, program execution has to be restored at next location. Hence, the address of the following program memory location is to be saved in a suitable location.
- Transfer control to the subroutine: It is done by loading the PC with the starting address of the subroutine
- Execute the subroutine
- After completion of subroutine execution, revert to the main routine and restore execution at the interrupted location. The address that was saved earlier is loaded back into the PC for this purpose.

Saving the PC content and restoring it are different from similar activities with the other registers. A separate memory module called 'Stack' is provided for the purpose. Whenever PC content is saved, it is 'pushed' into the stack. Whenever one returns from subroutine, the stack is 'popped'; that is its contents are brought back to the PC. In the PIC®16 series the stack is not accessible to the programmer. The stack is directly tied to the PC. Saving into the stack (from PC) and retrieving from stack (to PC) are unique, different from other similar operations and important to the processor functioning – the terms 'push' and 'pop' refer to this set only.

Figure 9.1 illustrates the subroutine call operation. When PC contains address aa-1, corresponding program memory location has instruction MOVWF. After its execution, the processor encounters the subroutine call instruction 'CALL bb' at PC location aa. The first step in the calling process is to

PUSH return address aa+1 into the stack. The second step is to load PC with address bb so that subroutine execution can continue from there. When subroutine execution is complete, that is when PC has program memory address cc in it, RETURN instruction is encountered. At this stage the processor accesses the stack; and the return address aa+1 is popped back into the PC. The µC re-enters main routine and continues execution of the main program. The arrows with serial numbers near each indicate the sequential order in which the processor activities take place. A few observations are in order here.

- The stack content should remain undisturbed during the execution of the subroutine. In PIC® 16 series, the stack is directly tied to the PC and not accessible to the programmer for read/write operations.
- Registers used in the main program and those used in the subroutine are not to be mixed up. The programmer has to ensure that any data saved in a register in the main program is undisturbed during execution of the subroutine.
- GOTO instructions are to be used with care. When executed in a subroutine; the programmer has to ensure that program control is brought back to the subroutine; eventually one has to RETURN to main routine to the calling area itself. Similarly care has to be exercised when using GOTO in main program, to avoid entering the subroutine area inadvertently.
- The CALL statement has to specify subroutine entry address. The RETURN statement need not specify any address.
- To execute a CALL instruction, the literal supplies only the Lower 11 bits of the called address. The MS bits are taken from PCLATH<4:3>. But the stack is as wide as PC (13 bits). So full PC content is saved and popped.

9.2.1 Subroutine Nesting

Consider the task of multiplying two numbers, each 2 bytes wide. The main routine may require it be used in different locations and hence have a subroutine for the task. Let the numbers be <p:q> and <r:s> where p, q, r and s are bytes, concatenated to form the two larger numbers. The multiplication here requires the partial products. p*r, p*s, q*r and q*s to be obtained and added with suitable shifts. One can have a subroutine to multiply bytes and another to add 2-byte numbers. These two can be called inside the subroutine doing the multi-byte multiplication. Such calling of one subroutine inside another is called 'subroutine nesting'.

Subroutine nesting is accommodated by extending the concept of pushing into stack and popping from stack and using it repeatedly. Figure 9.2 illustrates the case of one subroutine calling another within itself with one level of nesting. The processor is active executing instructions in the main program area until it comes to program memory location aa. Here it encounters instruction "CALL bb". The CALL bb instruction results in PUSHing return address aa+1 to the stack; the arrow with serial number 1 shows this activity in the figure. PC is loaded with program memory address bb – identified by the arrow 2 in the figure. Execution continues with instruction at bb in subroutine 1 – shown by arrow 3. When subroutine 1 is being executed, PC encounters a second CALL instruction – CALL ee at program memory location dd. Once again the execution is interrupted; return address dd+1 is pushed into stack – as shown by arrow 4. (The stack already has the return address aa+1 stored in it; the new return address stays above the already stored return address aa+1). PC is loaded with ee, the starting address of subroutine 2 – identified by arrow 5 in the figure. The processor embarks on the execution of Subroutine 2 – as shown by arrow 6. When execution of subroutine 2 is complete, RETURN instruction is encountered. The return address at the top of the stack – namely dd+1 – is popped up to the PC; arrow 7 shows this. PC reverts to subroutine 1 and resumes its execution – shown by arrow 8. At the completion of execution of subroutine 1, RETURN instruction (the last instruction in the subroutine) is encountered once again; the processor pops the stack content again. It being aa+1, the same is loaded into PC. Main program execution resumes at aa+1 – as shown by arrow 9 in the figure. Figure 9.3 clarifies the sequence of operations in subroutine nesting further; the focus is on pushing and popping. The following observations are in order here:

- Repeatedly the return addresses are pushed into stack one after the other. The stack contents are popped in reverse order. Always the address on top of the stack is the return address for the subroutine to be reactivated next.
- PIC®16CXXX series has an 8 level stack. At most 8 return addresses can be stored in a sequence at any point of time. It accommodates as many levels of nesting too.
- Stack is a 'first-in last-out' register file (Compare pipeline which is a 'first-in first-out' file)
- Allocation of general purpose registers is to be done carefully here, as explained earlier. The caution regarding use of GOTO instruction is equally valid.
- As in the case of single subroutine servicing, every CALL supplies only the LS 11 bits of called address. The other bits b12 and b11 are supplied by PCL<4:3>.

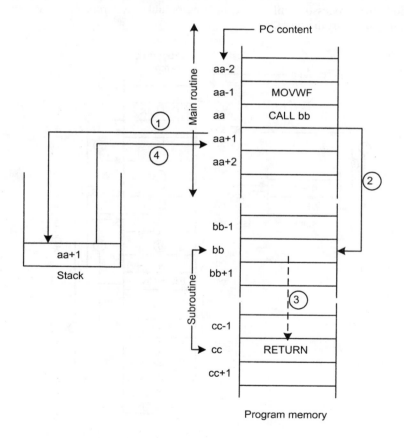

Figure 9.1 Sequence of operations in subroutine CALL& RETURN

9.3 INTERRUPTS

In many applications the processor will be executing an assigned task regularly. Over and above that, it will have some special tasks to be executed on a priority basis. Such special tasks can be from different sources and each such source may require a separate short time assignment to be executed. Once it is completed, the processor can resume its regular activities. In all these cases the regular operation of the processor is interrupted and a priority activity carried out; once the same is completed, the regular operation is resumed. Many examples of such situations can be cited. A few are given below by way of examples.

- A μC may be monitoring the AC power supply to a device on a continuing basis. If the supply is interrupted, it has to turn on an emergency back up supply and bring it into full swing as soon as possible. It will halt the monitoring and resort to the emergency supply start up routine; once the same is fully activated, the processor can resume the monitoring function.

- A μC may be required to measure temperature and pressure in a process at regular intervals. The processor will activate a timer and be 'dormant'. When the timer gives out a signal at the end of a preset period, the processor will 'wake up', carry out measurement activities; once same is completed, it will restart the timer.

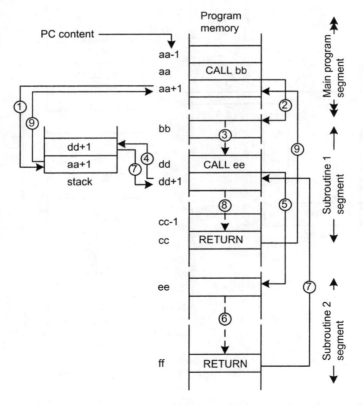

Figure 9.2 Sequence of operations in subroutine nesting – with one level of nesting

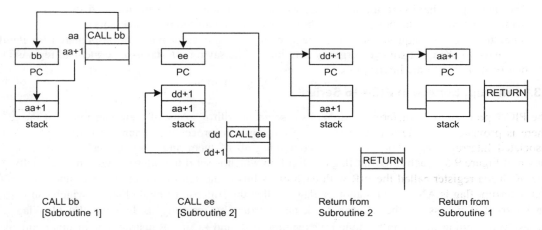

Figure 9.3 Sequence of operations in subroutine nesting – with one level of nesting; the CALL and RETURN operation are in focus

- A μC may monitor the bearing temperature and noise from a rotating machine. If the bearing temperature exceeds a set limit, it may set one type of alarm. If the noise exceeds a limit or deviates from a set pattern, it may set a second type of alarm. If both the abnormalities occur together, it may set a third type of alarm and also set an emergency operational sequence in motion. This is a case of the processor being called upon to handle multiple interrupts.

Situations of the above type demand a 3 step additional action for the interruption to be accommodated.

- Halt the activity in process or interrupt the ongoing activity
- Take up the new activity and complete it
- Once the new/short term activity is completed, revert to the regular activity
- Interrupting the ongoing activity and restoring it at the end of the interrupt service is done with the help of stack, as was with subroutines. A hypothetical and simplified case of interrupt service is considered first. Details specific to the PIC16® series will be dealt with subsequently.

The processor discussed thus far goes through a simple operational sequence repeatedly – fetch an instruction, execute it, increment the PC, fetch the next instruction, execute it and so on. Figure 9.4 shows the modification to the processor operation to accommodate the interrupt service. The processor can have a single bit register called the 'Interrupt Flag' dedicated to the interrupt service. The source seeking interrupt can set this flag when it demands attention from the processor. After execution of every instruction, the processor can check the flag status using its own built–in logic. If the flag remains reset, the processor continues regular operation. If it is set, the processor embarks upon servicing the interrupt. It completes execution of current instruction in the main program and saves the return address in the stack. The PC is loaded with a new address which is the starting address of the 'Interrupt Service Routine (ISR)'. Once the ISR is completed, the processor reverts to the main program itself by popping the stack and loading the PC with the address of the instruction to be executed next in the main program. The modification to the operating cycle of the processor to accommodate interrupt service is shown enclosed within a dotted box in Figure 9.4. The interrupt service as indicated here is similar to the execution of a subroutine but differs from it in some respects:

- Subroutine call occurs at specific and pre-determined locations in the main routine. In contrast the interrupt can ask for service without any prior notice; it can come at any point of the processor's working. It can be from a source completely independent of the processor.

- The interrupt is hardware initiated whereas the subroutine is software initiated (Such clear cut demarcation may not be possible in all the cases with some of today's processors).
- Since the interrupt request can come unexpectedly, the processor may not be prepared for it. It may have to carry out some emergency activities – like saving the status of scratchpad registers – before attending to the Interrupt.

9.3.1 Interrupt Service in PIC®16 Series

The PIC® processor has different peripherals associated with it. Any of them can interrupt the μC. There is provision for external devices too, to interrupt the processor. Each such interrupt has an associated Interrupt flag. The process of combining all of them and interrupting the processor is shown in Figure 9.5. Each interrupt flag is a flip flop which is set if the interrupt goes high (is active). The μC has a register called the PIR with each of its bits dedicated to one of the interrupts. Further each interrupt flag is ANDed with another flag – called the 'Interrupt Enable Flag' – which can mask the interrupt if necessary; the μC can mask an interrupt by resetting its Interrupt Enable Flag or unmask it by setting it. It can be done by executing BSF and BCμC F instructions or other suitable instructions. Outputs of all the AND gates – representing the different interrupts – are ORed together to form a common interrupt. The common interrupt here is further ANDed with a global mask bit – called the 'Global Interrupt Enable (GIE)' Flag to generate the final interrupt to the μC. In summary three conditions are to be satisfied for a source to interrupt the μC:

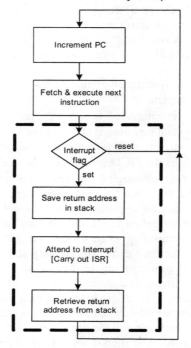

Figure 9.4 Flowchart of processor operation to accommodate interrupt service

- The interrupt flag has to be set by the Interrupt ReQuest (IRQ) going high; this represents the basic IRQ to the μC,
- The interrupt has to be enabled by the μC by setting the concerned Interrupt Enable Flag, and

- The GIE flag has to be set – again by the µC.

The µC hardware is structured such that the interrupt service starts at location 0x04h of program memory – by loading 0x04h into the PC at the start of the interrupt service. The µC is said to 'vector' to location 0x04h for interrupt service. The program segment executed by the µC by way of interrupt service is widely referred as 'Interrupt Service Routine (ISR)'. The µC has a variety of options to service the interrupts. Some of these are discussed here to bring out the flexibility at the disposal of the designer.

9.3.1.1 Single Interrupt

All except one interrupt can be masked – by setting the IE flag of the selected one but resetting all other IE flags. The GIE too has to be enabled. The ISR can be loaded starting at location 0x04h itself. Figure 9.6 shows the sequence of activities when the processor embarks on an interrupt response. Interrupt input may occur any time. It will be latched into the Interrupt Register within a specified time. The processor senses the interrupt flag at the end of every Φ_1 in the instruction cycle. Since interrupt enable flag is high, the interrupt is enabled. For the specific case shown, the interrupt request has become active during the instruction cycle T_n but after the clock cycle Φ_1. It is latched into the interrupt flag soon after; but it is sensed during the instruction cycle T_{n+1} . Concurrently instruction fetched in the previous cycle (at program memory location P) is executed. The processor saves address P+1 in stack; it discards the instruction fetched during instruction cycle T_{n+1} and executes 2 dummy cycles. During the second dummy cycle 0X004H is loaded into PC and execution continues from there. Prior to that – during instruction cycle T_n – GIE is disabled by the processor to avoid succumbing to the same interrupt source (or other interrupt sources). Typical applications will have a GOTO statement at location 0004H. The destination of GOTO instruction will have the detailed routine to decide on interrupt response details as well as to carry out the response. Execution continues from there. The processor takes 3 to 4 instruction cycles to respond to the interrupt request.

9.3.1.2 Interrupt Status Saving

Often interrupt service requests come unexpectedly. The processor may be in the midst of carrying out important activities and the data it has been processing cannot be sacrificed. All that data may have to be saved before servicing the request. Mostly saving the working register and status bytes may suffice. All such saving has to be done in software before embarking on the task dictated by the interrupt. When a number of registers is to be saved, one can resort to switching memory banks. The processor can switch to an alternate memory bank and leave the memory bank used by the main routine undisturbed. After completing interrupt service when it returns to the main routine, the original bank may be switched back into active mode. There may be situations where some data generated in the main program have to be used in the interrupt service routine or vice versa. In such cases, the common memory area with addresses 70h to 7Fh (which repeated in other banks with address range F0h to FFh, 170h to 17 Fh, and 1F0h to 1FFh) can be used to store such data. It eliminates the need to switch banks back and forth to access data during interrupt service.

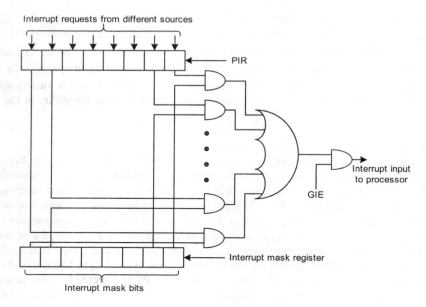

Figure 9.5 Generation of interrupt to the processor from all the interrupts and their respective interrupt masks

9.3.1.3 Returning from Interrupt

The instruction RETFIE is intended to return to main routine after completing service to the interrupt source. When executed the address saved on top of the stack is brought back to the PC. From the following cycle, execution of main routine is resumed. Further the GIE which was disabled before embarking on the ISR is enabled. This will allow any pending interrupt to be serviced in the same manner. Two observations are in order here.

- The stack is used to serve return addresses of sub-routines as well as interrupts. Total depth of stack remains 8. This has to be kept in view when doing repeated nesting of subroutines and servicing interrupts – all with the same processor.
- Memory bank switching facility adds a lot of flexibility to the system and the designer can exploit it fruitfully. However if not done with care, data from different segments can get mixed up.

9.3.2 Servicing Two Interrupts

Some applications can have two interrupt sources used. An interrupt to the µC can come from either of them. The interrupt source can be one of the peripherals or an external interrupt. Let us call these sources 'A' and 'B'. A number of issues come up regarding service to such interrupts as A and B; all these are to be addressed at the design stage itself.

9.3.2.1 Source Based Service

The first task in the ISR is to identify the source of interrupt and then branch to the corresponding service routine. The µC can read the PIR register (Figure 9.5) to identify the interrupt source. If the interrupt is from source A, the corresponding interrupt flag bit in PIR will be set; similar is the case with interrupt source B. The interrupt flags are checked in sequence. If the interrupt is from source A,

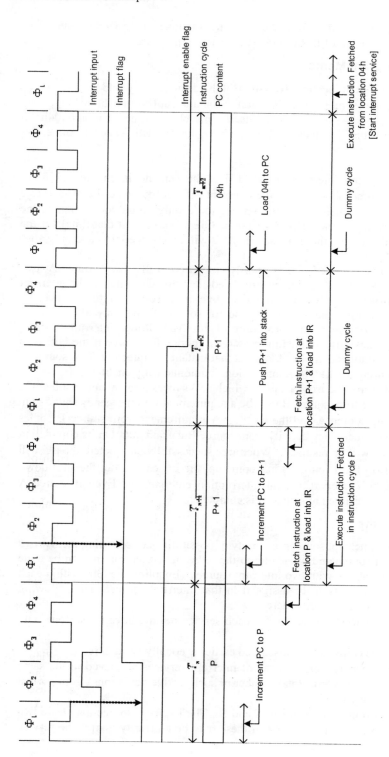

Figure 9.6 Interrupt request and response – Sequence of activities

the μC branches to service routine of A; if it is from source B, μC branches to service routine of B. The flowchart for branching and the portion of ISR to branch are shown in Figure 9.7.

9.3.3 Positioning of ISR in Program Memory

The μC has the provision to start the ISR at location 0x004h. But the designer may find it inconvenient to put the full ISR – completing the identification of the interrupt source, branching, Service Routine A, and Service Routine B – at this location itself. In such cases the location 0x04h can have a simple GOTO ISR instruction. The location ISR can have the detailed Service Routine.

9.3.4 Interrupt Prioritising

Different possibilities of prioritizing the interrupts exist [13]. The scheme in Figure 9.7 gives (essentially) equal priority to both sources A and B. If the interrupt is from source A, the μC services the IRQ from A. Since GIE is kept disabled during the service to A, only on its completion of the service to source A, source B is given a chance of service (since GIE is enabled at that time). If the interrupt from source B became active before that from source A, B will be serviced first and only after that source A will have its turn.

If the interrupt from source B is to be assigned higher priority compared to that from source A, two things are to be done in the ISR. Firstly the ISR is to be structured such that the status of the interrupt flag from source B is checked first and only after that the same from source A to be checked. Secondly the GIE is to be enabled within the service routine for source A and only after that the service task proper for source A, carried out. The procedure of this type allows 'interrupt nesting' (similar to that of nesting of subroutines). In contrast in the service routine for source B the GIE is to be kept disabled until the service activity is completed; only at that instant the interrupt from source A is to be enabled. It is done by executing the RETFIE instruction; it automatically enables GIE.

In a practical situation interrupt A may be from a source which seeks measurement and display of a variable at regular intervals of 10s: that from source B may be a requirement for emergency shut down. If the μC is in the midst of servicing source A and the emergency requirement from source B comes up, task A will be suspended and the emergency activity from source B carried out. On its completion the measurement and display service will be resumed. When it is successfully completed, the μC will revert to the main routine. If an IRQ from source B has come up, the μC carries out the shut down task; GIE is kept disabled during this interval and no interruption entertained. Even if source A requests service, it has to wait until the service to source B is completed.

9.3.5 Servicing Multiple Interrupts

If the application has more than two interrupt sources, a variety of possibilities of servicing them arises [20]. Identification of the interrupt source and branching to the relevant service routine can be done by extending the procedure followed above with two interrupt sources. Enabling the GIE within each service routine can be based on the priority to be assigned to that interrupt source. The following observations regarding interrupt service are in order here:

- Scratch pad register assignment and status saving for each service routine have to be done with care to avoid any mix up.
- Caution regarding the use of GOTO instruction discussed earlier is equally valid here.
- The return addresses of subroutines as well as those of interrupt services are saved in the same stack. Since the stack depth is only 8 words, total number of such 'active deviations' from normal working has to be limited to 8 at any time.
- Consider the application with two interrupt sources A and B. The interrupt source may have two components to the task it has to carry out; the first of these may be a priority item. The service

routine can complete that, enable GIE (to allow IRQ from source B), and then take up the (non-priority) second part of the task. Similar procedures can be adopted with the other cases as well in a more generalized manner.

- If the number of interrupts exceeds 4 or 5 and different priorities are involved, the ISR may become too complex to be programmed and debugged. A worthwhile alternative is to assign a different processor to one or more of the tasks concerned and simplify the programming task as well as the ISR itself. Such task segregation may turn out to be a less costly alternative too!

9.3.6 Interrupt Response Time

Consider a PIC μC working with a 10 MHz clock. Each instruction being of four clock periods is of 400ns duration. The μC takes 3 to 4 instruction cycles to complete execution of the active instruction, save return address in the stack and vector to location 0x04h. Saving of the μC status – if present – may last for another 3 to 5 instruction cycles. Thus the μC may take about 10μs to embark on the interrupt service proper. For many applications of μCs a response time of this order is no cause for worry. But with the μCs becoming faster and smarter by the day, and the designers becoming equally adept at it, more and more challenging tasks may be assigned to the μC. There may be situations where 10μs is too long a time to wait! In all such cases a μC dedicated to the task, is the preferred alternative. It will also free the main μC for other more mundane tasks – it may do these much better too!

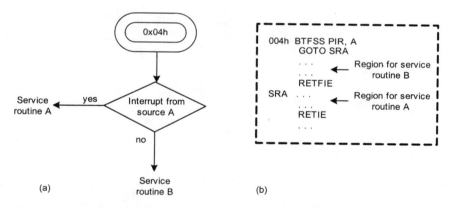

Figure 9.7 Branching to interrupt services depending upon the interrupt source (a) Flowchart (b) Program segment

9.4 EXERCISES

1) Form a 24-bit register by cascading three registers. Do a time delay routine with a 24-bit number.
2) Have a subroutine to count down from an 8-bit number to zero. Write a program to create a 16-bit delay. The delay routine will call the subroutine as man times as the number loaded in the eight MS bits of the 16-bit value.

CHAPTER 10

PORTS AND TIMERS

10.1 INTRODUCTION

Every μC has a set of peripherals. Presence of such frequently used peripherals is a distinct characteristic of the PIC® series of μCs. They are offered with enough flexibility. The designer can program each μC to suit the need. Further the company offers the μCs with a variety of combinations of the peripherals. A clear understanding of each peripheral and its working go a long way in realizing compact designs [20]. The peripherals on offer are discussed here and the following chapter. Some common applications are also discussed to bring out the features.

10.2 IO PORTS

IO Ports are the most commonly used peripherals. The IO pins can be assigned to IO or to one of the other peripheral functions. The assignment is programmable. Here we concentrate on the IO function. The peripheral functions are discussed later. Each IO port can have 4 to 8 lines. Some of the assignments are for the whole group while others are done separately to each bit line. One IO line of port B is shown in Figure 10.1 along with all the associated latches, control logic, and control lines. L_1 is the data latch; it is negative edge-triggered and WR-port is the clock signal for it. Another latch – called the 'TRIS latch' designated as L_2 –decides whether the pin is in output mode or input mode. L_2 is also a negative edge-triggered latch. If a '0' is latched into it, tri-state buffer B_2 is in ON state; output of L_1 is connected to pin P. Buffer B_1 has output at 1 state. The pull up transistor Q (PMOS) is in OFF state. The Pin P is in output mode and carries the output of data latch L_1. When latch L_2 is latched to 1 state, buffer B_2 is OFF and pin P is input mode (whether PMOS transistor Q is in OFF state or active). Let us examine the output and input operations more closely.

10.2.1 Output Operation

There are three distinct phases in the output operation. The relevant signal waveforms are shown in Figure 10.2 and the phases identified as F_0, F_1 and F_2.

- F_0 is the phase before the port pin is configured to be in output mode. WR TRIS control line is in 1 state and output of latch L_2 at 1 state. Buffer B_2 is in OFF state.
- During F_1 the port pin is configured to be in output mode. Data bus line carries a 0 bit; WR TRIS line goes low. At its negative going edge, the '0' in the data bus line is latched into L_2. Buffer B_2 turns ON and connects output of latch L_1 to pin P. With these the pin is configured to be in output mode and it carries the bit value earlier latched into output latch L_1.
- Once configured in output mode, data can be written into the port line. Whenever WR port control line goes low from a high state (negative going edge), the data on the data line is latched into L_1;

through buffer B_2, it is available on Pin P. Two instances of data output – designated A and B – are shown in the figure.

The whole port has 8 data latches and 8 TRIS latches associated with it similar to the ones shown in Figure10.1; there are 8 data lines giving input to both the sets of latches – The data latch outputs are steered to the output pins when the set of output buffers B_2 are ON. All of them are turned on simultaneously by the single clock signal WR TRIS. Subsequently a similar clock signal WR port latches the 8 bit wide data to be output, into the output latches and the output pins. The full bus and port latches are shown in a simplified form in Figure 10.3 which shows bus signals and control signals separately giving due importance to their width.

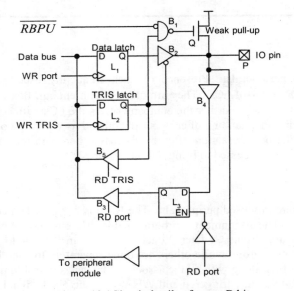

Figure 10.1 Circuit details of a port B bit

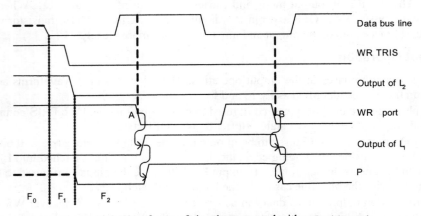

Figure 10.2 Waveforms of signals connected with output to port

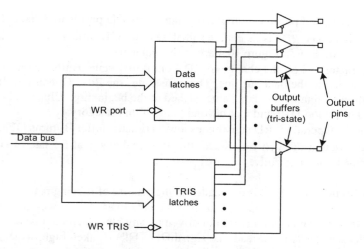

Figure 10.3 Scheme of output and its control for a byte wide port

A few observations are in order here:

- Once a 0 is latched into L_2 at the negative edge of **WR TRIS**, the 0 remains latched. It can change only at the next negative edge of **WR TRIS**. Until that instant, the port continues to be in output mode.

- Before '0' is latched into L_2, it is assumed that a '1' was latched into it. In turn the buffer B_2 is in off state and the I/O pin is isolated – neither in input nor in output mode (Its value can be erratic and not dependable).

- Once in output mode, data can be repeatedly written into the output port latches. These are done at the negative going edges of WR port line. The data values will be faithfully reproduced at the output pins. In Figure 10.2 the data line shown carries a '1' at the instant A when the **WR** port line is going to 0 state; hence the 1 value in the data line is latched into L_1 and made available at port pin P. Whatever be the subsequent changes in data line, the latch L_1 and pin P do not respond to them until the next negative edge on **WR** port. At instant B the **WR PORT** line goes to 0 state once again; the data line carries a 0; it is latched to L_1 and the port pin P.

- When the set of **TRIS** latches (eight numbers of L_2) are being reset, all eight lines of data bus carry 0 and the same is latched into them. When data is being output, the data lines carry corresponding bit values and these are latched into the set of L_1's. Due to the propagation delay, latch output transition is to follow the negative going clock transition with a delay; this delay sets an upper limit to the speed of operation of the device.

10.2.2 Input Operation

There are three distinct phases associated with the input operation. The signal waveforms are as in figure 10.4; the three phases are identified as F_0, F_1 and F_2. The pull-up transistor Q is assumed to be in the OFF state for the discussion here.

- F0 is the state prior to configuring the port to be in input mode. To configure it in input mode (See Figure 10.1), data bus line is made to carry a 1 bit. WR TRIS is at 1 state. During F_1 it goes to zero state. At its negative edge the '1' bit on the data line is latched into latch L_2. Tri-state buffer B_2 turns OFF isolating pin P from the output of L_1. With this the stage is set to do input operation.

- Throughout phase F_0 and phase F_1, the Read Control line RD port is in 0 state. Tri-state buffer B_3 is OFF. But latch L_3 has $\overline{RD}$ as its enable signal. Output of L_3 follows changes at its input pin P. The periods A–D and E–F in Figure 10.4 are of this category.
- Phase F_2 is characterized by the change in RD from 0 to 1 state. Buffer B_3 turns ON. Output of latch L_3 is connected to the data bus line. This constitutes the Read Operation. In fact throughout the read operation L_3 retains the data value latched at the beginning. Changes in input at pin P do not affect it. The period D–E and that beyond F are of this category.
- At the end of Read Operation, RD signal goes low. Tri-state buffer B_3 turns OFF. Data bus line is disengaged from output of latch L_3. Since $\overline{RD}$ (the enable signal for latch L_3) is high, latch L_3 follows input transitions faithfully.

Observations:

- Buffer B_4 ensures that the IO pin is not loaded in the process of it being read
- When the port is being read, output of L_3 drives the data lines. The processor (through its internal logic) ensures that the data bus lines are not driven by any other element inside.
- When the TRIS register is to be read, the signal RD TRIS is taken high; tri-state buffer B_5 turns ON and makes the output of latch L_2 available on data bus line to be read into the μC. It holds good of all TRIS latches of the port
- The set of latches L_1 constitute the output data latches; data loaded in them can be read by keeping the port configured in output mode and doing a port read operation.

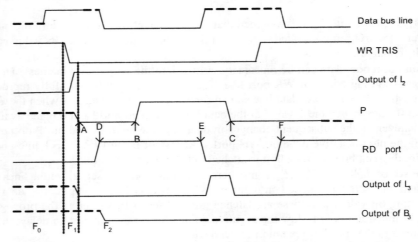

Figure 10.4 Waveforms of signals connected with the input operation of port

10.2.3 Role of Pull Up

In some applications, the wire connected to pin P in the input mode may not be connected anywhere else; this can happen under certain operating conditions. Such a 'floating' line can pick up stray electrical signals and cause damage to (the CMOS) IC. To avoid this, whenever the pin is left in input mode, a resistance may be connected from the pin to supply line. It is called a 'pull up'. The chip has the facility for such a pull up. A PMOS transistor kept in active state forms the pull up here. It is kept active by driving its gate to 0 state – the option is exercised by taking $\overline{RBPU}$ to 0 state – The port is to be in input mode when exercising the option; this is done by using output of Latch L_2 as a gating signal for the pull-up. Note that if the pull-up is connected in output mode, it will cause unnecessary loading on buffer B2.

10.2.4 Variations & Alternate Functions

The PIC®16 Series have ports A, B, C, D, E, F, and G – these are designated as PA, PB, PC, PD, PE, PF, PG, and GPIO. Different combinations of these are used in individual µCs. Features of individual ports and their interface are briefly outlined here.

- Every port has its port register. When in output mode, data is written into the port register. It is made available on the output pins depending on the configuration.
- Pins of Ports A, B, C, D, and E can be configured as input pins or output pins. Each pin can be configured separately. Each of these ports has an associated data direction register – 'TRIS' register. A latch of the TRIS register is selected and set to '1' state to commit the corresponding pin to be in input mode. It is reset to '0' state to commit the pin to be in output mode.
- Pins of ports A, B, and C are assigned alternate peripheral functions. Peripheral function and IO Port function are programmable.
- The pins of Port B and some pins of Port G have the facility to have an internal (weak) pull up. If necessary it can be activated. Note that such activation is for the whole group and not individually selectable. It is done by resetting [OPTION<7>] bit. The pull-up resistor assignment matters only for input mode.
- An 'interrupt on change' facility is provided on four pins of Port B - <PB7:PB4> - and also on the pins of Port G. If the value of the bit concerned is different on two successive port-read operations, a processor interrupt is automatically generated. If a pin of this group is assigned to output function, it is excluded from the facility.
- Ports B, C, D, F, and G are 8-bit wide. Ports A and GPIO are 6-bits wide. Devices which have the parallel slave port facility (discussed below), have only a 3-bit wide port E.
- Devices that have the LCD drive facility have ports F and G implemented. These can be used for LCD drive or as other input ports.
- The port configurations vary slightly from device to device. Specific data sheets/manuals deal with such details. With some ports, the port is configured in input or output mode as a whole. With other ports, the pin can be programmed to be in input our output mode individually and selectively. To improve functionality, additional peripheral functions are assigned to some ports; the port is connected to the port pins. In most applications port function is decided at design state and assigned accordingly. Details of alternate function and their assignments are discussed in relevant sections.

10.3 PARALLEL SLAVE PORT

When the µC has a port configured as a data input port, a read operation enables data bytes to be accepted by the µC. It is done at the convenience of the µC and the device supplying data has to wait for it. An alternative approach is for the µC to accept data and give out data as demanded by the external device – the device can be another µC based system accumulating data, or information as a set of bytes, and making the same available. Port D can be configured as such a port to function under the control of the external device; with such an assignment it is called the 'Parallel Slave Port (PSP)'. Operation of the port is through three control signals – $\overline{RD}$, $\overline{WR}$, and $\overline{CS}$ (see figure 10.5). Port configuration and assignment of the control signals are done through a set of SFR bits as follows:
- Load control word 00010111b into TRISE register. $b_4 = 1$ commits Port D to slave operation.
- Bits <b2:b0> configure <RE2:RE0>as input pins.

- The pins RE2, RE1 and RE0 have multiple assignment possibilities – as control lines for PD when it is configured as a slave port, as independent digital IO lines or as analog input lines. ADCON1 <PCFG3: PCFG0> are to be assigned values as specified in data sheets to commit RE2, RE1 and RE0 as control lines of Port D
- When configured with TRIE <PSP MODE>=1, TRIE<b7:b5> shall function as status bits linked to the parallel port operation.

10.3.1 Chip Selection ($\overline{CS}$)

A device (μC) may have more than one external peripheral device interfaced to it. The μC selects such a peripheral for communication, through appropriate chip select lines. For example if a μC has only two peripherals connected to it, access to either can be through one control line – CS – committed for the purpose. The first device can be selected (for reading and writing) by making CS = 1. The second can be selected (for reading and writing) by making CS = 0. The first one is said to be selected by CS and the second by $\overline{CS}$. If CS = 0, the parallel slave port is selected as can be seen from the figure. It enables Read and Write Operation of the port.

10.3.2 Write Operation

When the μC port (D) is selected with $\overline{CS}$ = 0, data can be written into the port by the external device under the control of $\overline{WR}$. The sequence of operations is as follows; related waveforms are shown in Figure 10.6.

- Select port with $\overline{CS}$ = 0
- Make data available on data lines <b7:b0>
- Take $\overline{WR}$ to low (0) state
- Data is latched into the port latches

After $\overline{WR}$ goes high, the Input Buffer Full (IBF) flag is set; it happens at the beginning of Q_4 period following $\overline{WR}$ going high. The μC can read this flag bit (if necessary) to ascertain whether valid data is available at the port. When μC reads the port, IBF flag automatically gets reset. The port register reading can be by polling. When IBF is set PSPIF is also set. The μC can read the port in interrupt mode. For this the PSPIE flag is to be set. On interrupt generation μC can identify the source of interrupt and read the port as part of the service routine. The device supplying data through the port has no mechanism to know whether the data has been read by the μC. Without knowing it, a second byte may be supplied by it. In such a case the μC has an overflow flag (IBOV) which gets set and indicates to the μC that there has been an overflow. The choice between polling IBF bit to read the byte and interrupt driven read depends on the application. The general guideline is to go for polled mode if the firmware overhead is not high. On the other hand interrupt driven byte reading is attractive if the hardware overheads are not high.

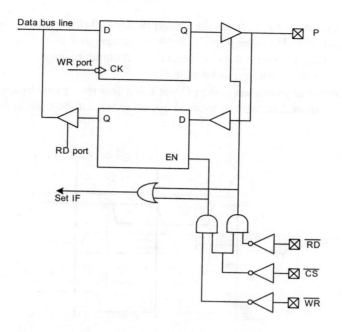

Figure 10.5 Circuit details of a parallel port line in block diagram form

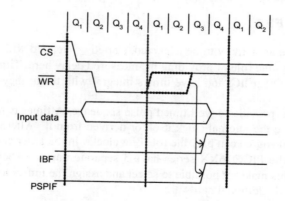

Figure 10.6 Waveforms related to data input at parallel slave port

10.3.3 Read Operation

The external device is ready to read the port if $\overline{CS} = 0$, that is the port has been selected for reading the data therein. The step by step sequence of operations is as follows. Figure 10.7 shows the related waveforms:

- The port is selected with $\overline{CS} = 0$
- RD is exercised low by the device

- The negative going edge of $\overline{RD}$, resets the Output Buffer Full (OBF) flag. The µC can detect this in its polling schedule and load the port register with the next byte to be output.
- The µC can make data to be output, available in the port output register.
- The device reads the port data; $\overline{RD}$ is taken high
- The PSPIF is set; it is an indication to the µC that it can gear up for next byte to be output
- As with write operation data loading for read operation too can be done in interrupt mode or by polling.

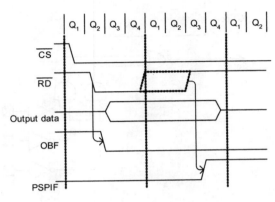

Figure 10.7 Waveforms related to data output at parallel slave port

10.4 TIMERS / COUNTERS

Many applications require an activity to be triggered, started or stopped with a definite delay from an earlier event. A timer plays the role of providing the measured delay here. Timers in the PIC® series of µCs are discussed here. Other µCs too have timers integral with them; they may vary in operational details only.

The clock within a µC provides a well timed pulse sequence. A timer is basically a counter which has a clock input; it can be the internal clock itself or derived from it. Alternatively an external clock or pulses from an external source can play the role of a clock. In the latter case the timer is adapted to function as a counter. The PIC®16XX series have 3 separate timers. Their functioning modes are configured differently. This makes it possible to select and assign the timers to applications of different requirements with minimal additional hardware.

10.4.1 Timer0

Figure 10.8 shows Timer0 in block diagram form. The timer is built around an 8-bit Timer Register. If the input at P is activated, the timer register counts up with every pulse there. The timer register can be accessed as an SFR (preferably when the input at P is not activated) to read its content as well as to write a number into it. The input pulses can be from an external source given at input pin – T0CKI; alternately it can be derived from the internal clock. All the six control bits – T0SE, T0CS, PS2, PS1, PS0, and PSA – are in the Option register (an SFR). The roles of the control bits are as follows:

- T0SE selects the falling (1) or rising (0) edge transition on the input line T0CLI for counting.
- T0CS selects between the internal clock (0) and the external pulse (1) for counting. It is done through the mux M_1.

- If the internal clock is selected, the machine cycle forms the basic period for counting.
- The bit **PSA** decides to route the pulse to be counted directly to the timer register (1) or through a programmable prescaler (ripple counter); the mux M2 does the selection.

The prescaler is an 8-bit ripple counter as shown in Figure 10.9. The pulse rate at the output of b0 is half of that at its input; the pulse rate at the output of b1 is $(1/4)^{th}$ of that at the input to the prescaler and so on. The three bits – PS2, PS1, and PS0 – form the input to a 3-to-8 decoder; their outputs are used to select one of the 8-bit outputs of the prescaler counter and assign to mux M_2. If output of bo is selected as the clock to the timer register, the timer register increments for every 2 pulses at R; if output of b1 is selected as the clock to the timer register, the timer register increments for every 4 pulses at R and so on. Prescale values possible are – 2, 4, 8, . . .and 256. Another bit in the **OPTION** register – **INTEDG** – decides the nature of transition from the timer register which causes the interrupt – **T0IF** – to go high (details explained below).

10.4.1.1 Operation

First consider the operation from the internal clock without the use of the prescaler (PSA = 1). Left to itself, the timer will count up from 00h to 0FFh continuously, the count increasing by 1 for every instruction cycle. Whenever the counter overflows from 0FFh to 000h, the timer interrupt flag – T0IF – is set. A bit in the Option register – **INTEDG** – selects the rising edge (**INTEDG** = 1) or the falling edge (**TNTEDG** = 0) of the interrupt signal to set the interrupt flag. The application program can utilize the flag as desired. The timer register can be accessed when desired and an 8 bit number written to it. Following this (in fact 2 instruction cycles later) the timer will continue to count up from the number loaded. Again as soon as the count reaches 0FFh, the interrupt flag is set and counter overflows to 0h. Thus if a number *N*h is loaded into the timer register, (FF-N-2) machine cycles later the interrupt flag is set. This allows any time delay from 0h to FFh machine cycle periods – to be generated using the timer. The delay value cannot be reduced below 7 machine cycle. With the use of the prescaler the delay can go up to much larger values.

10.4.1.2 Operation with External Clock

The clock signal to the timer can be assigned to an external source (with **T0CS** = 1). The timer register will count up for every pulse cycle from the external source. If the source is periodic, function is similar to that of the internal timer. More often the external clock input may be triggered by an 'event' occurring repeatedly. In such a case the unit functions as a counter. By loading a desired number *N* into the timer register, one can trigger interrupt as soon as the corresponding number of counts – namely (FF-*N*)h is reached. Operation as a counter in this manner expands the role of the unit considerably. Use of the prescaler increases the range of count in steps – there being 8 scales as explained above.

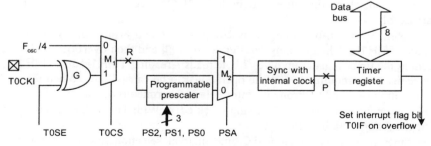

Figure 10.8 Timer0 in block diagram form

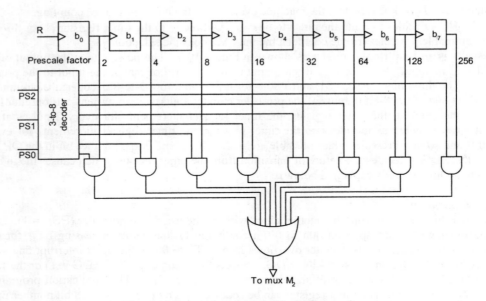

Figure 10.9 prescaler circuit in block diagram form

10.4.1.3 Synchronizing To External Clock

When functioning with internal clock without prescaler, timer operates at the highest speed possible; corresponding minimum clock period is one instruction cycle. Since the instruction cycle itself is composed of 4 clock periods, the minimum discernible time period remains 4 clock periods. Further the timer state transition is at the end of Q_3 of the instruction cycle. When external clock is used with the timer, it is synchronized to the internal clock. When transitions in the external clock period occur at intervals – large compared to 4 T_{osc}; operation is smooth and satisfactory. But when such transitions occur at periods closer to T_{osc}, caution should be exercised in the timer operation; for reliable operation the signal from the pulse source must stay at 1-state as well as at 0-state for at least 4 T_{osc} each.

10.4.2 Timer 1

In principle Timer 1 functions are similar to those of Timer 0. The timer increments with every pulse input. When the timer count overflows to 0, an interrupt is generated - its clocking can be from the internal processor clock or from an external one. The timer has a prescaler too. The bits for programming and control of operation are in the T1CON register (an SFR). The operational differences are as follows:

- The Timer Register is 16-bit wide. It is composed of two 8-bit registers (both are SFRs) – TMR1H and TMR1L. The counting goes up to 0FFFFh and then overflows to 0h. Being 16-bit wide, reading the timer register involves 2 byte reads – reading TMR1H and TMR1L. Refreshing the timer register too requires two Write operations to be carried out. When the timer is active, reading or writing can lead to erroneous information or erroneous operation.
- The timer incrementing is done at the rising clock edge. With Timer 0, one can select the trigger edge to be the rising or falling one.
- Since the timer register is 16 bit wide, time interval selection has a wider range 00000h to 0FFFFh. In contrast for timer 0, the range is restricted to 000h to 0FFh only.
- The prescaler is a 2 bit ripple counter; it provides only 4 scale values.

- Timer clock signal can be from the internal clock, with the timer incrementing every instruction cycle (similar to that with Timer0). Alternatively, it can be from an external source. The external source itself can be of two types. It can be an oscillator run completely independent of the master clock of the processor; or it can be an external asynchronous input – here the unit functions as a counter.
- When the clock source is external, operation can be synchronized to the internal clock as with Timer0. Alternatively the operation can be in asynchronous mode; incrementing is at the rising edge of the clock input to the timer register.
- Separate gating signal is provided to the timer clock. It can allow the clocking pulses through or prevent them. It permits selective enabling and disabling of timer operation.
- The timer has a separate reset signal. When exercised, it resets the timer register to 0000h value.
- Operation of the timer can be enabled or disabled by exercising the **TMR1ON** bit in **T1CON** register. It cannot be done with Timer0.
- Timer1 can function even when the µC is in **SLEEP** mode. For this the µC has to be assigned to the external clock. Alternately it can be generated through a dedicated crystal oscillator. The facility can be used to carry out a specific activity at regular intervals. Normally the µC can be in **SLEEP** mode with the timer functioning through the clock dedicated for it. At designated intervals the timer can interrupt the µC. The µC wakes up, carries out the assigned task through an ISR, and reverts to **SLEEP** mode. This manner of operation is suited for battery operated devices.

10.4.3 Timer 2

Timer 2 is shown in block diagram form in Figure 10.10. The timer register (an SFR) is 8 bit wide, counting up from 00h. Internal clock is the only timing pulse source. The timer has a 2-bit prescaler – possible prescale values being 1:1, 1:4 and 1:16. With the prescaler set to 1:1, the timer register increments once every instruction cycle. For prescale settings of 1:4 and 1:16, it increments once every 4 and every 16 instruction cycles respectively. All the programming and control bits of Timer2 are in the **T2CON** register (an SFR).

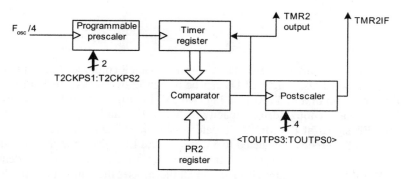

Figure 10.10 Timer 2 in block diagram form

The timer has an 8 bit Period Register – PR2 (an SFR) – and a comparator associated with it. The period register can be loaded with any number from 00h to 0FFh. When the count in the timer register matches that in the period register, the comparator output changes state. It resets the timer register and counting starts afresh. Thus if the period register has a number N loaded in it, timer output generates one pulse after every N instruction cycle periods. The operation is automatic and one need not load the period register after every reset. To this extent the operation is simplified. The timer output can be used as the basic clock source for the serial communications module. When desired it can be put to other uses as well. The timer can be enabled or disabled through a dedicated gating signal – TMR2ON.

10.4.3.1 Postscaler

The timer output – represented by TMR2 bit – is given as clock to a 4-bit Postscaler. The postscale factor can be set to any value from 01h to 0Fh. In effect the postscaler counts the number of times the timer output matches set value in the PR2 register; when the postscaler overflows, the interrupt flag – TMR2IF – is set.

10.5 DISPLAY AND KEYBOARD INTERFACE

Keys and displays are the most commonly used devices through which we interact with µC based systems. There are a variety of ways of interfacing these with the µC. We examine some commonly used configurations, since they constitute applications of the peripherals – ports and timers.

10.5.1 Displays

Many of the µC based systems have displays associated with them. The display size may be large with hundreds of elements (CRT monitor) from which a selection is to be made and displayed. Such displays function with a dedicated µC doing the control. Barring these, for comparatively smaller displays – involving only a few characters, the display can be directly interfaced to the µC. Typical and representative displays linked to the µCs are discussed here.

10.5.1.1 Led Based Displays

An LED that is turned on is functionally a forward biased diode. It has a forward voltage drop in the 1.6 V – 2.0 V range. It remains in the OFF state if the voltage across it is below 1V. Reverse voltage across the LED is avoided to prevent permanent damage to it. When in the ON state even a small increase in voltage across the LED results in the current through it increasing unduly causing damage. Hence LEDs have current limiting resistors in series with them. With the circuit of Figure 10.11(a) the current through the LED will be

$$I = \frac{5-1.7}{220} A$$
$$= 15mA$$

For a normal single LED, 15 mA is a fairly large current. Such a continuous current can be sustained only by fairly large sized LEDs. Single LEDs can be driven in two ways. In Figure 10.11(b) lead P (typically an output port pin of a µC) is driven high (to 5V) to turn it ON. If P is low, the LED is OFF. If the µC pin does not have enough drive capacity, A MOSFET can be used to drive the LED. Pin P can give the ON/OFF signal at the MOSFET gate as shown in Figure 10.11(c). Alternatively, the Anode of the LED can be directly connected to 5V supply and LED driven at the cathode side by a µC output port pin as shown in Figure 10.11(d). If the port pin drive capacity is not adequate, the drive can be through a MOSFET as in Figure 10.11(e). In all these cases the output port pin driving the LED can be of the latched type. The µC can write a '0' or '1' in the port bit as desired. Once this is done, the µC need not do anything else to maintain the status of the led. Single LEDs of different colours of the type discussed here, are used to indicate the status of a system, or that of a system variable etc. 'Communication is active', system is operating safely', 'Air conditioner is in the heat mode', 'Fuel level is low', . .are examples of such indications.

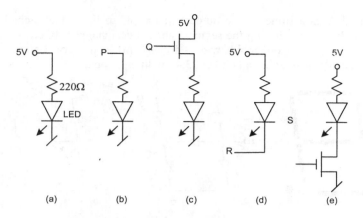

Figure 10.11 Different modes of driving a single LED

10.5.2 Alphanumeric Character Displays

Numeric and alphanumeric displays are done with LEDs arranged in segments. The widely used 7 segment format to display a digit and a dot (decimal point) to its side is shown in Figure 10.12 (a). Each segment as well as the dot has an LED for it. If the LED in place 'a' is lit, segment 'a' is lit. Different digits and the LEDs to be selected and lit to display each digit are shown in Table 10.1. The eight segment LEDs can be connected with the cathode common as in Figure 10.12(b). The common lead and the eight anode leads can be brought out of the 7-segment module. The module will have a total of 9 pins connected. Alternatively the module connection can be as in Figure 10.12(c) with a common anode and eight individual cathodes brought out separately. The way the digits appears when the LEDs are selected and lit, is shown in Figure 10.13.

Alphanumeric character displays use a 14 segment or 16 segment type display as in Figure 10.14(a) and Figure 10.14(b). Dot matrix displays are more versatile in the sense that they can display other types of characters too. We confine our discussion to the 7 segment display here. Once the principle is understood, it can be directly extended to the others. One possible interface circuit is shown in Figure 10.15. It uses a multiplexed drive scheme. The display has a 4-digit display formed with a set of 4 seven segment modules – M_1, M_2, M_3 and M_4. The modules are of the common cathode type. Each cathode has a MOSFET to drive it; the MOSFETs have 4 output leads from the μC connected to their gates. If lead Com1 is taken to logic high state, Q_1 is turned ON. When Q_1 is ON, the segments of module M_1 can be turned ON by driving respective anode leads to high state. Thus if the digit '1' is to be displayed, anodes of LED segments b and c (See Table 10.1) are to be taken to the high state; anodes of all the other segments will be at the low state. Similarly, Q_2, Q_3, or Q_4 too can be turned ON to display a digit in the M_2, M_3, or M_4 module. At any instant only one of the MOSFETs is turned ON; in turn only one of M_1, M_2, M_3, or M_4 modules is active. The 'a' segments of all the four modules are connected together and driven by a port lead. Resistor R_a is a current limiting resistor for the set of four a segments. Similarly all the four 'b' segments are tied together and driven by a second port lead again through a current limiting resistor – R_b. It is done with all the eight segments of the module set. One can see that one full eight bit port is used to drive the eight segments at the anodes. One nibble of a second port is used to drive the set of four common cathodes through respective drive transistors – Q_1, Q_2, Q_3, and Q_4. Consider a segment which is given a short current pulse at periodic intervals. If the current value is large enough and the pulse repetition rate is 30/s or more, the segment appears continuously lit; this is due to 'persistence of vision' that we – as human beings – have. This fact is exploited judiciously in multiplexed displays of the type here. Pulse Repetition frequency of 50/s or more is common. Let us assume all the outputs of the μC to be of the latched type. We have 2 registers – first one-byte wide driving the segment anodes designated a, b. c....h. Let this register be called 'R_{seg}'. Another register – one nibble wide – drives com1, com2, com3 and com4; let this register be called 'R_{com}'. Consider displaying a number 23.45 with the scheme.

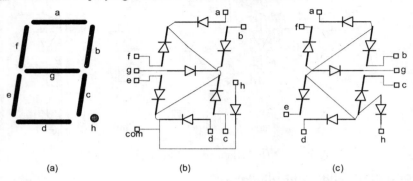

| (a) | (b) | (c) |

Figure 10.12 Seven segment display layout and the LED connections: (a) Layout of segments (b) LEDs with the cathodes being common (c) LEDs with the anodes being common

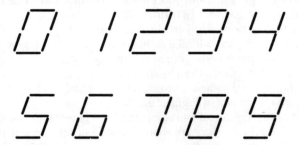

Figure 10.13 The way the digits 0 to 9 appear with 7-segment displays

We proceed as follows:

- Turn ON Q_1 and keep Q_2, Q_3 and Q_4 turned OFF. It is done by loading 1000 to nibble R_{com}. The status continues as long as the value loaded in R_{com} remains unchanged.
- Take segments a, b, d, e, and g to high state and all other four segments to low state. The binary number 11011010b is loaded into port register R_{seg} for this. One can see that the segments a, b, d, e, and g of module M_1 are lit; all other segments of M_1 and all segments of M_2, M_3, and M_4 remain OFF. The digit 2 is displayed by module M_1.
- Wait for 5 ms [1/4th of the period of 50/s frequency]. During this time, M_1 keeps displaying digit 2 at the position assigned to it. At the end of 5ms, turn OFF Q1 by loading 0000b into the Rcom nibble. The display is fully turned OFF.
- Load 11110011 into register R_{seg} and 0100 into register R_{com}. It makes M_2 active. The digit 3 is displayed at the position of M_2. The decimal point to the right of digit 3 is also lit and displayed. Sustain this status for 5ms. Turn OFF Q_2 by loading 0000 into R_{com}. This turns OFF the full display.
- Repeat above type of procedure for 5 ms to lit and display digit 4 at M_3 position. Repeat procedure for 5 ms to lit and display digit 5 at M_4 position.
- Repeat all the above steps in the same sequence cyclically.

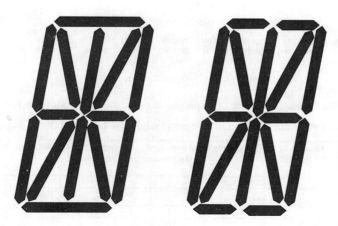

Figure 10.14 Multi-segment display for alphanumeric characters (a) 14 segment scheme (b) 16 segment scheme

The multiplexed and repetitive scheme of displaying the modules is further illustrated in Figure 10.16. Due to persistence of vision we see a continuous display of the number 23.45.

10.5.2.1 Implementation and Outline of Software

One timer in the µC can be dedicated for the display. The timer can have a repetitive period of 5 ms and interrupt the µC at the end of every 5 ms period. The µC can go through the refresh and update routine as its ISR. The flowcharts for the main program as well as the ISR are shown in Figure 10.17. It has a counter N (a scratch pad register in the µC) which repeatedly counts down from 4 to 1. Register R_1, R_2, R_3 and R_4 store the binary equivalents of segments to be lit in M_1, M_2, M_3 and M_4. A register stores the nibble used to turn ON Q_1, Q_2, Q_3 and Q_4 sequentially and periodically. The ISR functions as follows (after each interrupt):

- If $N = 4$, load R_1 value into port B. Load 1000 into the nibble of R_{com} – nibble of port C. Decrement N. Display digit at position of M_1.
- If $N= 3$, load R_2 value into Port B. Shift R_{com} nibble right by one bit. Decrement N [Display digit at position of M_2]
- If $N = 2$, load R_3 value into port B. Shift R_{com} nibble right by one bit. Decrement N [Display digit at position of M_3]

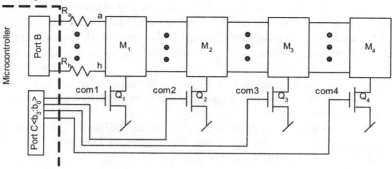

Figure 10.15 A scheme of interfacing four 7-segment modules to a µC

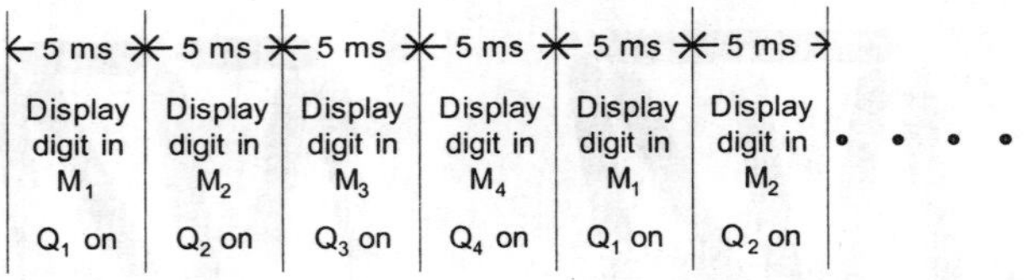

Figure 10.16 Sequential scheme of displaying digits in the four modules in Figure 10.14

Table 10.1 Digits and the segments of 7-segment module to be turned ON to display the digit

Digit to be displayed	Segments to be turned on – 1 represents an ON segment & 0 represents an OFF segment						
	a	b	c	d	e	f	g
0	1	1	1	1	1	1	0
1	0	1	1	0	0	0	0
2	1	1	0	1	1	0	1
3	1	1	1	1	0	0	1
4	0	1	1	0	0	1	1
5	1	0	1	1	0	1	1
6	1	0	1	1	1	1	1
7	1	1	1	0	0	0	0
8	1	1	1	1	1	1	1
9	1	1	1	1	0	1	1

- If $N = 1$, load R_4 value into port B. Shift R_{com} nibble right by one bit. Decrement N [Display digit at position M_4].
- If $N = 0$, reset N to value 4 and R_{com} nibble to value 1000. Go to step a).

Observations:

- At 10 MHz clock rate, ISR execution takes ~100 µs. Since it occurs at 5ms intervals, interrupt service takes 2% of the processor time which is acceptable.
- Whenever necessary the number to be displayed can be changed in the main routine. During that time the interrupt from the timer display is to be disabled; this step avoids display flicker.
- With LEDs, the increase in light output is more than proportionate to an increase in the working current. This in effect increases the brightness level of the display when it is worked in multiplexed mode.
- The bits of port B drive only one LED (or no LED) at any time. The lead may have this drive capacity. In contrast the common cathode transistors are called upon to drive all eight segments simultaneously (in the worst case). This explains the need to use the transistors at the common cathodes and not load the lines com1, com2, com3 and com4 directly with the full drive current
- There are other alternatives to implement the display; each has a corresponding hardware/software trade off.
- As the display size increases in number of digits, the software overhead too increases. Beyond a certain state the scheme may become too cumbersome to be implemented. A dedicated µC for the display can be a better alternative.

A display panel of the most versatile type is formed by having uniformly spaced pixels all over. The pixels are lit by a matrix of column and row wires. These are activated by scanning cyclically. Such display units have dedicated controllers for their operation.

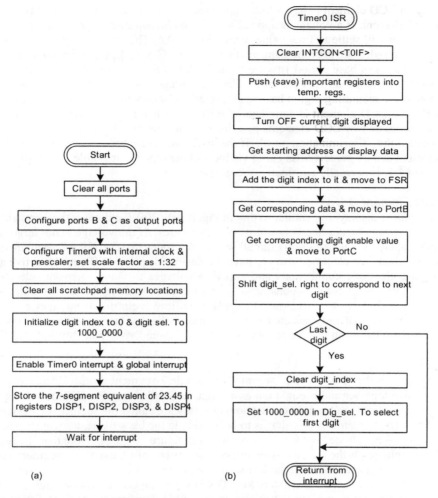

Figure 10.17 Flowcharts for the display with 7-segment modules (a) Main program (b) Interrupt subroutine

10.5.3 LCD Based Displays

LCD displays are constructed by combining a number of segments and providing electrodes on either side. By applying voltages to selected sets of such electrode pairs different shapes can be displayed. Displays of different sizes and shapes can be easily made and interfaced to µCs. Their power consumption is negligible; they are in wide use in battery operated µC based devices. Let us consider a single LCD display element as shown in Figure 10.18. It has 'liquid crystal' between two transparent electrodes. A screen type background is provided behind the combination. In the absence of any voltage between the electrodes the element is transparent. When a voltage is applied between the electrodes, the transparent element becomes opaque and visible. The LCD element is said to function in 'reflective' mode. An alternate scheme has the element opaque in the absence of voltage between electrodes. When a voltage is applied, the element becomes transparent. The element will have a uniformly distributed light source behind. When transparent, with light allowed to pass through, the element becomes visible. The LCD element is said to function in 'transmissive' mode.

Electrically an LCD element behaves like a lossy capacitor; the capacitance value is of the order of 1nF/cm2 area of element. If a voltage is applied across the element through a 10 kΩ source resistance, the voltage across it will settle to final value within 50 µs. Any DC voltage beyond 50 mV across the LCD element is avoided to prevent damage to it; only AC voltages can be applied across LCD elements. AC voltages of base period 1ms are typical of LCD drive waveforms. Current consumption by an element will be less than 1µA per cm2. The rms voltage across an element versus its brightness or contrast with the surrounding region has the characteristic shown in Figure 10.19. If the rms voltage is less than V_{th}, the LCD element is completely invisible. When rms voltages V_1 and V_2 are applied across the electrodes of two LCD elements in a display, their relative visibility is decided by the ratio (V_1/V_2). It is called the 'discrimination ratio'. If the discrimination ratio is above an accepted minimum value, one element can be perceived as being visible and the other as invisible. A discrimination ratio value larger than 1.7, is normally considered acceptable.

10.5.3.1 LCD Element Drives

Consider a 7-segment display unit similar to that in Figure 10.12(a). To display the segment 'a', a large enough AC voltage is to be applied between its two electrodes. If the voltage is not large enough, the segment is not displayed. One can bring out all the 16 electrodes (8 x 2) of the 7-segment display unit and selectively activate the electrode pairs to have the desired display. Different attractive alternatives of combining the electrode connections and applying voltages to them, to make selected elements visible, are available. Such combinations reduce the number of leads to the display unit and simplify interface. But in each case it is necessary to apply specific waveforms of voltages to the electrode combinations to ensure that only selected segments are on display but the others are not displayed. The electrode drive schemes in use are of two categories – 'Static Drive' and 'Multiplexed Drive'.

10.5.3.2 Static Drive

Figure 10.20(a) shows the 'Static Drive' scheme for a single 7-segment display. The electrodes in the rear are all bunched together and brought out as a single lead called the 'Back Plane'- designated as B. The electrodes in the front are brought out separately – they are called the segments – designated as S_a, S_b, ... , and S_h. The waveforms of voltages to be applied to the back plane, a segment – S_a to be kept ON, and a segment – S_b – to be kept OFF — are shown in figure 10.20(b). Note that if a segment is to be kept ON, the voltage on the electrodes on either side are out of phase. If the segment is to be kept OFF, the voltages on the electrodes on the 2 sides are in phase. A frame period of 20 ms (50 frames/s) is typical. The rms voltage across a segment kept ON is vV while that across the segment kept OFF is 0V. The discrimination ratio is infinity. Achieving perceptible discrimination is not difficult.

Static drive provides the simplest LCD interface. Wiring is most elaborate but software development is easy – only two types of waveforms are to be stored and reproduced. The electrodes and backplanes can be directly connected to the port pins of a µC. The voltage levels corresponding to the 0 and 1 states of logic used can be the two voltage values used for the electrode drives. The desired drive waveforms are generated by outputting the 0 and 1 values in the desired sequence. The interface to the LCD unit is achieved with no additional hardware. Display of N characters requires 8*N+1 leads to be driven separately – keeping the Back Plane common for all of them. If the number of segments per display unit is different, a corresponding number of drive lines are to be used. The PIC®16 series have some µCs with 4 common and 24 segment drive lines. They can drive three 7-segment display units.

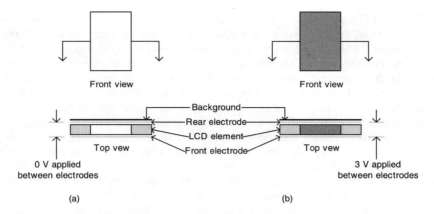

Figure 10.18 Simplified representation of a single element LCD display of the reflective type: (a) With 0V applied across the electrodes the element is invisible. (b) With a definite voltage (3V) applied across the electrodes the element is invisible.

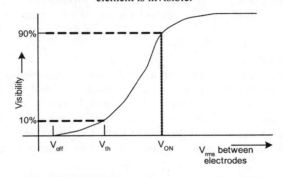

Figure 10.19 Contrast ~ Voltage characteristic of an LCD element

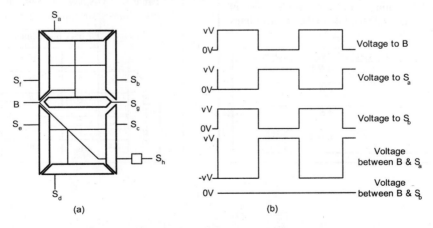

Figure 10.20 A single 7-segment LCD module & the direct drive for it: (a) Connections to the electrodes of the module; (b) Waveforms of voltages applied to the backplane & two of the electrodes (The backplane voltage values are taken the reference for the differential voltage waveforms).

10.5.3.3 Multiplexed Drives

The static drive for N segments requires N+1 drive lines to be interfaced to the µC and two drive waveforms to be output. The scheme is simple and useful if the display size is small and involves only 10 to 15 segments. Further, the µC should have sufficient number of pins to be committed for the purpose. Alternate drive schemes are to be used if the number of segments increases beyond this range. Electrodes behind and in front can be combined and connected in judicious ways to reduce hardware requirements. It also requires appropriate interface waveforms to be output from the µC. The waveforms should be such as to ensure that the segments selected have enough voltage (rms value) across to make them visible but those not selected should have lower voltages to keep them invisible; it requires a high enough discrimination ratio to be realized. Further the voltage waveforms have to be of AC type with zero dc value.

Different schemes are in vogue. Each uses the back plane segments bunched together into a set of common drive lines. A scheme that uses N such common lines for the back plane is called a '1/N mux' type. Simultaneously the electrodes in front too are combined into an appropriate number of segments. Further the drive waveform may use different voltage levels; with M levels, M-1 voltage intervals can be realized. $1/(M-1)$ is called the bias for the waveform. Values of N up to 4 and M up to 4 are common. Larger values bring down the discrimination ratios to unacceptable levels. They also require cumbersome voltage waveform schemes to be used. We illustrate the drive schemes through two specific examples.

10.5.3.4 ½ Mux- ½ Bias Drive

Figure 10.21 shows a connection scheme for a single 7-segment display unit. The back plane electrodes are segregated into two groups and have two leads – B1 and B2. Segments a, e, f, and g have their back plane electrodes connected together and brought out as one lead – B1. Similarly the back plane leads of the remaining 4 electrodes – b, c, d, and h are connected to a second lead – B2. The front side electrodes are formed into 4 groups; to form 4 segment lines designated as S1, S2, S3, and S4. The front side electrodes of segments are assigned to segment lines S1, S2, S3 and S4 as shown in Table 10.2.

Table 10.2 Assignments of front side electrodes of segments of the LCD module to the segment lines

Segment line		Front electrode segments connected
	S_1	a & b
	S_2	e & d
	S_3	f & c
	S_4	g & h

Table 10.3 Display segments and drive lines shown in matrix form

		back plane lines	
		B_1	B_2
Segment line	S_1	a	b
	S_2	e	d
	S_3	f	c
	S4	g	h

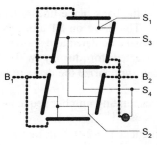

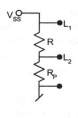

Figure 10.21 ½ Mux scheme of driving a 7-segment display unit

Figure 10.22 Generation of LCD segment drive voltages from the supply

The display segments and their respective drive lines are shown in matrix form in Table 10.3. A bias of 2 requires three voltage levels to be used for the drives. A scheme to generate the voltage levels is shown in Figure 10.22. V_{SS} is the supply voltage to the μC. With $R = R_p$ the three voltage levels are $V_{SS}V$, $V_{SS}/2V$, and 0V. Two of these are available at the leads L_1 and L_2 and the third is the ground level of the μC supply itself. Electrode leads of the LCD unit can be connected to the three voltage levels in succession at defined time intervals and any desired set of drive waveforms can be obtained. It is customary to use fixed waveforms for the back plane leads; Figure 10.23 shows one possible set of waveforms for B_1 and B_2 lines with voltage levels 0V, v1V, and v2V: Only one frame of the waveform is shown. The figure also shows a waveform for all the segment lines and the corresponding voltage waveforms across different segments. The illustration corresponds to the case when no segment is selected for display. We can simplify the discussion by designating the voltage levels by proportionate non-dimensional numbers – 0, 1, and 2. From the figure we get the of rms voltage value across every segment as

$$V_{rms} = [(0^2 + 0^2 + 1^2 + 1^2)/4]^{1/2}$$
$$= 0.707 \text{ V}$$

Assigning numbers 0, 1, and 2 to the three voltage levels, the voltage waveforms can be represented as corresponding number sequences. For the illustration under discussion, these are shown in table 10.4. Each cell in the table represents the voltage during one basic period of a frame. The rms values of voltages across the segments are also given in the table. From the table we can see that each of the segments has the same voltage level across it – 0.707V; as explained below, this is below the threshold level and none of the segments is displayed.

Another case is represented in Table 10.5. The back plane drives remain the same: The drives on S2, S3, and S4 too remain unaltered. The drive on S1 alone is changed. The rms voltage level across segment a can be obtained as

V_{rms}-a $\qquad = [(2^2 + 2^2 + 1^2 + 1^2)/4]^{1/2}$
$\qquad\qquad = 1.58 \text{ V}.$

Similarly, rms voltage across the other elements can be obtained as

V_{rms} $\qquad = [(0^2 + 0^2 + 1^2 + 1^2)/4]^{1/2}$
$\qquad\qquad = 0.707 \text{ V}$

Discrimination ratio $= 1.58/0.707$
$\qquad\qquad = 2.24$

The voltage levels used should be such that a discrimination ratio of 2.24 suffices. Referring to Figure 10.19, a V_{rms} of 0.707V has to be below V_{th} and a V_{rms} of 1.58V well above V_{th}. Resistance R_P in Figure 10.22 is adjusted to fine tune the voltage levels and achieve necessary contrast. With such an adjustment, only one segment – segment a– is selected for display. No other segment is displayed. The discrimination ratio achieved (shown in block letters in the table) is 2.236.

A third case is illustrated in Table 10.6. Here too the drives on the back plane lines – B_1 and B_2 – as well as the segment lines – S_2, S_3, and S_4 – remain unaltered. The drive on S1 alone is changed again. The rms values of the voltages and discrimination ratios are also calculated and shown. One can see that both segments – segment a and segment b – are selected for display with the same discrimination ratio level. From the three cases considered, one can summarize the drive requirements of segment lines S_1, S_2, S_3, and S_4 as follows:

- Keep the drives for back plane lines B_1 and B_2 fixed with corresponding number sequences as 0-2-1-1 (for B_1) and 1-1-0-2 (for B_2) respectively.
- For a segment line use the drive waveform of number sequence 0-2-0-2 if neither segment connected to it is to be displayed.
- For a segment line use the drive waveform of number sequence 2-0-2-0 if both segments connected to it, are to be displayed.
- For a segment line use the drive waveform of number sequence 2-0-0-2 if the segment common with B1 is to be displayed.
- For a segment line use the drive waveform of number sequence 0-2-2-0 if the segment common with B2 is to be displayed.

From the above one can see that any of the digits 0 to 9 can be selected for display (along with or without the associated decimal point) by a proper selection of drive waveforms for S_1, S_2, S_3, and S_4 from the four sets above. Displays of multiplex 7-segment display units can be driven by extending the idea. All of them can have the same set of two back plane drive lines. The segment lines are to be provided separately for each unit. To drive N number of 7-segment display units, 4N+2 drive lines are required in the ½ mux, ½ bias scheme – two for the backplane drives with fixed waveforms and 4N for the segment drives with four waveforms to select from. In all the cases the scheme gives a discrimination ratio of 2.236 which is quite adequate.

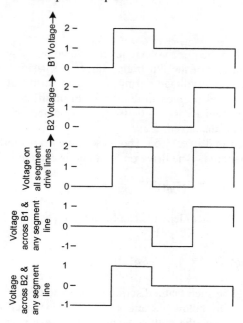

Figure 10.23 ½ mux, ½ bias drive: Backplane and segment line drive waveforms for one frame when no segment is selected (The backplane voltage values are taken as the reference for the differential voltage waveforms)

Table 10.4 Voltage levels & rms values of voltages for the case in Figure 10.19: ½ mux, ½ bias drive: no segment is selected

	Drive voltage levels in successive clock periods of a frame				rms value
B_1 line	0	2	1	1	
B_2 line	1	1	0	2	
All segment lines	0	2	0	2	
Voltage between B_1 & a segment line	0	0	-1	1	0.707
Voltage between B_2 & a segment line	-1	1	0	0	0.707

Table 10.5 Voltage levels & rms values of voltages for ½ mux, ½ bias drive: The only segment selected is common to B_1 & S_1. The discrimination ratio is shown within brackets (The backplane voltage values are taken as the reference for the differential voltage waveforms).

	Drive voltage levels in successive clock periods of a frame				rms value
B_1 line	0	2	1	1	
B_2 line	1	1	0	2	
Segment line S_1	2	0	0	2	
All other segment lines	0	2	0	2	
Voltage between B_1 & S_2 / S_3 /S_4	0	0	-1	1	0.707
Voltage between B_2 & S_2 / S_3 /S_4	1	-1	0	0	0.707
Voltage between B_2 & S_1	1	-1	0	0	0.707
Voltage between B_1 & S_1 [segment a]	2	-2	-1	1	1.581 [2.236]

Table 10.6 Voltage levels & rms values of voltages for ½ mux, ½ bias drive: the only segments selected are common to B_1 & S_1 and B_2 & S_1. The discrimination ratio is shown within brackets (The backplane voltage values are taken the reference for the differential voltage waveforms).

	Drive voltage levels in successive clock periods of a frame				rms value
B_1 line	0	2	1	1	
B_2 line	1	1	0	2	
Segment line S_1	2	0	2	0	
All other segment lines	0	2	0	2	
Voltage between B_1 & S_2 / S_3 /S_4	0	0	-1	1	0.707
Voltage between B_2 & S_2 / S_3 /S_4	-1	1	0	0	0.707
Voltage between B_2 & S_1[segment b]	1	-1	2	-2	1.581 [2.236]
Voltage between B_1 & S_1[segment a]	2	-2	1	-1	1.581 [2.236]

10.5.3.5 ¼ Mux 1/3 Bias Drive

Hardware interfacing can be made more compact by rearranging the connections to the 7 segment display units. Figure 10.24 shows one possible connection to two 7-segment display units. It uses 4 back plane drive lines – B_1, B_2, B_3 and B_4 – in a ¼ mux scheme. Four segment lines – designated S_1, S_2, S_3 and S_4 – are used for the front electrodes. The drive lines and the display elements they can activate in pairs, are shown in Table 10.7 in a matrix form. Four voltage levels are used corresponding to a 1/3 bias drive. The voltage levels can be generated using a potential divider similar to that in Figure 10.22. Adopting the convention used in the example above, the numbers 0, 1, 2, and 3 can be used to represent the voltage levels. The number sequences for back plane drives and for the segment line SG1 for one frame of the drive waveform are given in Table10.8. Corresponding rms voltage values for the different segments are also given in the table. The case corresponds to segment 'd_1' being selected for display with a discrimination ratio of 1.732. As with the previous example the waveform templates for different segment combinations to be displayed can be identified and used as required. For example a segment drive with a – '2-1-2-1-2-1-2-1' – sequence for the frame will ensure that no segment is selected for display.

Observations:

- As the number of back plane lines used to multiplex increases, the number of individual template waveforms to be used too increases; same holds good of the voltage levels too
- As the multiplex index increases, the discrimination ration reduces; Mux values up to 4 are in common use. Combinations of voltage levels up to 4 and mux lines up to 4 are in common use. If display size cannot be accommodated within this (corresponding to 16 segments), display groups are duplicated. For example four 7-segment units can be driven by 4 back plane lines and 8 segment lines (two sets of four segment lines).
- In the drive voltage waveforms, changes to the voltage used take place at regular intervals of a basic clock period 'T'. (The period is typically 1ms). If the number of mux lines is N, the back plane drive waveform used extends for 2 NTs.
- 2NTs is the basic frame period. The frame repetition rate is $1(2\ NT)$ Hz.
- With a frame period of 2 NTs, the back plane drive waveforms are all identical, each shifted with respect to others by multiples of $2T$s.
- Persistence of vision demands that the frame repetition rate be at least 30/s to avoid display flicker. With a basic clock period of 1ms, the frame duration can extend up to 32 ms. It corresponds to 16 mux lines.
- An alternate scheme which has frame duration of NTs is also in vogue.

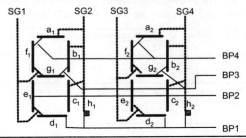

Figure 10.24 Display scheme of two 7-segment LCDs with 1/3 bias – ¼ Mux drive

Table 10.7 Matrix of drive lines & the segments
connected to them for the display scheme of Figure 10.20

		Segment lines & the segment electrodes connected to them			
		SG_1	SG_2	SG_3	SG_4
Back plane lines & the segments they drive	BP_1	d_1	h_1	d_2	h_2
	BP_2	e_1	c_1	e_2	c_2
	BP_3	g_1	b_1	g_2	b_2
	BP_4	f_1	a_1	f_2	a_2

Table 10.8 Voltage levels & rms values of voltages for 1/4 mux, 1/3 bias drive: only one specific set is shown; the only segment selected is d1. The discrimination ratio is shown within brackets (The backplane voltage values are taken as the reference for the differential voltage waveforms).

		Drive levels in successive clock periods of a frame								
Segment drive lines	BP$_1$	3	0	1	2	1	2	1	2	rms voltage
	BP$_2$	1	2	3	0	1	2	1	2	
	BP$_3$	1	2	1	2	3	0	1	2	
	BP$_4$	1	2	1	2	1	2	3	0	
	SG$_1$	0	3	2	1	2	1	2	1	
Segments	BP$_1$-SG$_1$ (d1)	3	-3	-1	1	-1	1	-1	1	1.732 [1.732]
	BP$_2$-SG$_1$	1	-1	1	-1	-1	1	-1	1	1
	BP$_3$-SG$_1$	1	-1	-1	1	1	-1	-1	1	1
	BP$_4$-SG$_1$	1	-1	-1	1	-1	1	1	-1	1

10.5.3.6 LCD Drives from PIC® Processor Series

Some of the PIC 16 series processors have the facility to interface LCD modules directly. The facility includes the following:

- Provide voltage bias levels – 3 or 4 as required
- Assign a selected set of port pins to the back plane drives (common lines); a single common line can be provided for static drives; 2, 3, or 4 lines can be provided for multiplexed drives.
- Assign a selected set of port pins to segment drives; the number can go up to 24.
- Once the above assignments are made a corresponding set of SFRs are directly linked to them.
- Select the frame clock source and clock rate by suitable pre-scaling.

The above activities are carried out as part of initialization. Subsequently the data registers (which are assigned to the segment drive lines and back plane lines) are loaded with appropriate bytes corresponding to the desired segment drive voltage sequence. The back plane drive is enabled. At the end of a frame, the registers concerned may be refreshed. Respective data books may be referred for details.

10.5.4 Keyboard Interface

A key is the simplest mechanical device that can be interfaced to a µC. Depending on the number of keys to be interfaced and the type of application, different approaches to interface are available. We discuss some of the commonly used ones here.

10.5.4.1 Key Types

Keys can be of two types – normally ON and normally OFF. A normally OFF key is in the OFF state. When the key is pressed, it goes to the ON state and reverts to OFF state when released. A normally ON key is in the ON state. When the key is pressed, it goes to the OFF state and reverts to ON state when released. With µC based systems, the normally OFF key is in wider use. Unless specifically mentioned, for our discussion here, the keys used are assumed to be of the normally OFF type. The simplest key will change state when pressed but revert to the normal state when released; it may have only a simple spring action to aid its operation. Alternatively, the key can be of a latching type. The key may be normally OFF. When pressed, it may get latched and remain at ON state even after it is free. It has to be pressed again to be released and to revert to OFF state. The keys we consider are of the non-latched type. The concept can be directly extended to the latched types also.

10.5.4.2 Key Debounce

A key is a mechanical device with spring action. Consider the normally OFF non-latched key. When pressed, it is to change to the ON state. It may take a few ms to settle to the ON state. In the interim period, the contact may be made and broken repeatedly due to the vibration of the moving mechanism. The behaviour reflects as the ON-OFF waveform in Figure 10.25(a). When released, it reverts to the OFF state after a similar period of vibration. Similar behavior for the normally ON key is shown in Figure 10.25(b). The duration of bounce and the high frequency of vibration associated with it depend on the key structure. The status of the key is not clearly defined during the bounce duration. The key can be 'debounced' (the effect of bounce nullified) in hardware or software.

Software debounce scheme requires the key status to be checked and reconfirmed after the bounce period. The processor may be made to poll the status of the key at intervals of 15ms (greater than the maximum expected bounce duration). The key is normally in OFF state. If pressed, at the next poll it ON status is returned. The status is confirmed with a 15 ms delay. If the two coincide, the status is taken as changed to ON state. µC will initiate any response measure only at this state. A key pressed by a finger may be released with a delay of about one second. The debounce scheme should refrain from sensing a key a second time within this period. It is to avoid the µC interpreting a single (longer) key closure as two successive key closures. Software debouncing requires the key to be polled at frequent intervals. It increases the µC overhead unduly if the number of keys used is more than 2 or 3. It is not in common use in such cases.

A simple hardware debounce scheme is shown in Figure 10.26. The RC time constant can be of the order of 5 ms to filter out bounce generated voltage spikes. When the key is pressed, the voltage at point X increases from 0V to 5V smoothly. The buffer B has threshold voltages for the up and down transitions; Y changes to high state (from the earlier low state) when the voltage at X crosses the threshold. The debounce circuit behaviour is similar for the transition from high to low state. With the scheme in figure, latching has to be done separately.

Debouncing can be done by employing a simple RS latch as shown in Figure 10.26. The latch is normally in the reset state. When the key is pressed, the latch changes to set state; output Q changes from low to high state. The instant of transition is decided by the first spike at input which is wide enough to make the latch set. Once the latch is set, it remains in set state even after the key is released. The µC has to reset it by applying a high state pulse at R; it is to be done with a reasonable time gap of about 1s after key closure. If the µC input port is of the latched type, the switch can be interfaced to it directly. Otherwise the latch has to be provided externally.

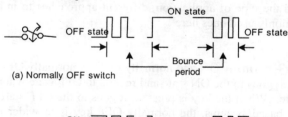

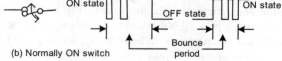

Figure 10.25 Switches and their bounce characteristics

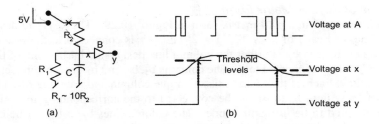

Figure 10.26 Hardware debounce scheme: (a) circuit (b) associated waveforms

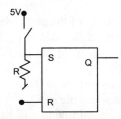

Figure 10.27 Use of RS latch to do debounce

10.5.4.3 Interface of Isolated Keys

If the application concerned has only a few keys, they can be interfaced to the µC directly. The interface can be through an external interrupt line. µC can identify the source and carry out the service necessary . The same procedure can be adopted for all the interrupts. Routines for source classification and branching to relevant service routine are required only if more than one key is present.

10.5.4.4 Interface of 8 Keys

If the number of keys present is 8 or less, they can be interfaced through a port dedicated to the purpose. A possible scheme is shown in Figure 10.28; bits $b_7 - b_0$ of a port are connected to eight keys. All the bits are to be configured in input mode. All are grounded through resistors. If a key is pressed the corresponding bit goes high; otherwise it remains low. The µC can read the byte at regular intervals by polling. Change in status of any bit can be identified by comparison with the last read value of the byte; Specific keys pressed can be identified by checking the bits separately. One can identify multiple key closures also.

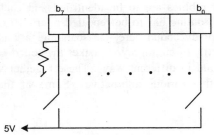

Figure 10.28 A scheme of interfacing 8 keys to a µC

10.5.4.5 Interface of Keys in Matrix Form

A set of 16 keys is arranged in a 4X4 matrix form in Figure 10.29. The keys are designated '0', '1', '2', - - and 'f'. One lead each of keys '0', '1', '2' and '3' is connected to the row designated b_0 – similarly with rows b_1, b_2, and b_3. Further the column line designated b4 has one of the second set of leads of '0', '4', '8' and 'c' connected to it. The columns b_5, b_6, and b_7 too are connected to the second set of leads of each key. Each of the 16 keys has a unique column and a unique row identified with it. In the µC interface, row lines b_0 to b_3 can be connected to the corresponding four bits of a port and these port bits configured to be in output mode. The column lines b_4 to b_7 can be connected to the corresponding bits of the port and these bits configured to be in input mode. Further all the four input lines are grounded through resistors. Key closures are identified by the µC through a repeated 'output-read' sequence as follows:

- Output the nibble 1000 on lines b_3 b_2 b_1 b_0 of the port – designed as W_0.
- Read the nibble b_7 b_6 b_5 b_4 of the port – designated as R_0.
- Output the nibble 0100 on lines b_3 b_2 b_1 b_0 of the port – designated as W_1. Read corresponding nibble b_7 b_6 b_5 b_4 as R_1.
- Use 0010 for output nibble W_2 and read nibble as R_2.
- Use 0001 for output nibble W_3 and read nibble as R_3.

If a key is pressed, the corresponding returned nibble will have its column position identified by a 1 bit; all other bits of the read nibble will be 0. One can go through a 'closed key identification' routine to identify any key closure. Two observations are in order here:

- Multiple key closures can be identified by the scheme with some restrictions
- Debouncing demands that the 'output – scan' cycle has to be repeated after 10 ms and a decision regarding valid key closure taken after that

An alternate scheme of interfacing the matrix key board uses keys with 2 contacts each. The scheme is shown in Figure 10.30. Each key has a row contact and a column contact. One lead each of these is grounded. The other set of leads are brought out and connected as shown. Thus contact leads R_0, R_1, R_2 and R_3 are connected together and brought out as b_0. Bit lines b_1, b_2, and b_3 are connected to the remaining rows in the same fashion. The column leads C_0, C_4, C_8, and C_c are connected together and brought out as b_4. The remaining three column lines are formed in the same manner and brought out as b_5, b_6, and b_7. One can see that the contact pair – [R_0- C_0] – represent the digit '0'; the contact pair – [R_0- C_1] – represent the digit '1' and so on; the contact pair – [R_4- C_4] – represent the hex digit 'f'

All the eight lines b_7 to b_0 are to be pulled up to the high logic level (Use of internal pull-up – if available – eliminates the need for additional hardware). The lines b_7 to b_0 are to be connected to the corresponding bit lines of a µC port. The port is to be configured to be in input mode. The key closure identification requires two steps to be carried out. Firstly, the byte has to be read at the port; secondly the 0s is the upper and lower nibbles are to be identified and each zero pair used to extract the corresponding key value. The routine has to be repeated after ~10 ms to do debounce. Unlike the previous case, the keys are already coded here; the scanning task is eliminated. Provision of twin contacts and associated complexity of circuit layout makes the coded keyboard more costly. The ports in PIC® series can be configured in different ways. These together with the possibilities of keypad interfacing discussed above offer various alternatives. Some of these are illustrated through the applications in Chapter 12.

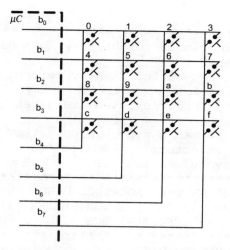

Figure 10.29 Scheme of interfacing 16 keys to the μC in a 4X4 matrix form

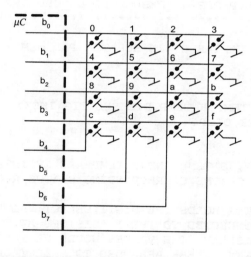

Figure 10.30 Scheme of interfacing 16 coded keys to the μC in a 4X4 matrix form

Program 10.1

```
;******************************************************************
;       Notes: This program is for displaying a data (23.45) in a
;       four digit display
;       Timer 0 has been used for display switching, and each digit
;        will be enabled for  approximately 5ms. Details of other port
;       assignments and variables used are given below.
;
;       Display data (segments a,b,..,g,dp) - RB7 to RB0
;       Display digit selectors    - RC4 to RC7
;       Frequncy of crystal oscillator   - 10MHz
;******************************************************************
LIST    P=16F877   ;DEFINE THE PROCESSOR USED
#INCLUDE <P16F877.INC>    ; INCLUDE THE HEADER FILE HAVING THE SFR
                          ;DECLARATIONS
;******************************************************************
; SETTING THE CONFIGURATION BITS

    __CONFIG _CP_OFF & _WDT_OFF & _BODEN_ON & _PWRTE_ON & _HS_OSC &
_WRT_ENABLE_OFF & _LVP_OFF & _DEBUG_OFF & _CPD_OFF
;******************************************************************

;DECLARE A BLOCK OF VARIABLES IN SCRATCH PAD MEMORY, WHOSE STARTING
; ADDRESS IS 20H

 CBLOCK 0X20
    W_TEMP,P_TEMP,S_TEMP,F_TEMP,DIG_SEL,DIG_INDX,
    DISP1,DISP2,DISP3,DISP4
 ENDC

; THE W_TEMP, P_TEMP, S_TEMP AND F_TEMP ARE REGISTERS WHICH HOLD
; THE PUSHED VALUES OF THE CORRESPONDING STATUS REGISTERS DURING
; INTERRUPTS
; THE DIG_SEL REGISTER HOLDS THE DIGIT SELECT VALUE, WHICH WILL BE
; MOVED TO PORTC, CONNECTED TO THE TRASISTORS AT THE COMMON CATHODE
; TERMINAL OF THE DISPLAY. THE VALUES WOULD BE 1000_XXXX, 0100_XXXX
; 0010_XXXX, 0001_XXXX. THESE ARE USED TO SELECT ONE PARTICULAR
; DIGIT DISPLAY AT A TIME IN SEQUENCE
; THE DIG_INDX REG HOLDS THE NUMBER CORRESPONDING TO THE POSITION
; OF THE CURRENTLY ENABLED DIGIT DISPLAY. (0 FOR FIRST DIGIT, 1 FOR
; SECOND DIGIT, 2 FOR THIRD DIGIT AND 3 FOR FOURTH DIGIT
; DISP1, DISP2, DISP3 AND DISP4 HOLDS THE DATA TO BE DISPLAYED ON
; THE CORRESPONDING DIGITS. THIS VALUE WILL BE MOVED TO PORTB WHICH
; WILL BE CONNECTED TO THE SEVEN SEGMENTS (a,b,..g,dp) OF THE
; DISPLAY DIGITS.
;******************************************************************
;USER DEFINED CONSTANTS
DP  EQU 0A ; A CONSTANT SPECIFYING DECIMAL POINT
```

```
; THIS CONSTANT VALUE INDICATES THE LOCATION OF THE SEVEN SEGMENT
; CODE EQUIVALENT OF A DECIMAL POINT IN THE LOOK UP TABLE
;*********************************************************************
; SET THE ORIGIN OF THE PROGRAM
 ORG 0X0000
    CLRF    PCLATH
    GOTO    START   ; BRANCH TO THE STARTING OF THE MAIN PROGRAM


;*********************************************************************
; PUSH AND POP MACROS USED TO SAVE AND RETRIEVE IMPORTANT REGISTERS
; DURING AN INTERRUPT

PUSH    MACRO
    MOVWF   W_TEMP      ; Save W register
    MOVFW   STATUS
    CLRF    STATUS
    MOVWF   S_TEMP      ; Save STATUS register
    MOVFW   PCLATH
    MOVWF   P_TEMP      ; Save PCLATH
    CLRF    PCLATH
    MOVFW   FSR
    MOVWF   F_TEMP      ; Save FSR
    ENDM


POP MACRO
    MOVFW   F_TEMP      ; POP FSR
    MOVWF   FSR
    MOVFW   P_TEMP      ; POP PCLATH
    MOVWF   PCLATH
    MOVFW   S_TEMP      ; POP STATUS
    MOVWF   STATUS
    MOVFW   W_TEMP      ; POP W register
    ENDM
;*********************************************************************
; INTERRUPT SERVICE ROUTINE LOCATED AT 04H

 ORG 0X04
ISR             ; INTERRUPT SERVICE ROUTINE
    BTFSS   INTCON,T0IF    ; CHECK FOR TIMER0 INTERRUPT
                ; TIMER 0 INTERRUPT OCCURS EVERY 3.2768mS
    GOTO    RET_ISR

TMR0_ISR
    PUSH
    BCF INTCON,T0IF    ; CLEAR THE INTERRUPT FLAG
    MOVLW   DISP1      ; GET THE STARTING ADDRESS OF THE DISPLAY ;
        DATA
    MOVWF   FSR
    MOVFW   DIG_INDX
```

```
    ADDWF   FSR,F        ; ADD THE DIGIT INDEX AND MOVE IT TO FSR
    MOVFW   INDF
    MOVWF   PORTB        ; GET THE CORRESPONDING DATA AND MOVE
;                 IT TO PORTB
    MOVFW   DIG_SEL
    MOVWF   PORTC        ; GET THE CORRESPONDING DIGIT ENABLE
;                 VALUE AND MOVE IT TO PORTC
    INCF    DIG_INDX,F ; INCREMENT DIGIT INDEX TO POINT NEXT
;                 DIGIT
    RRF DIG_SEL,F  ; ROTATE RIGHT DIGIT SELECT TO SELECT
;                 NEXT DIGIT
    BTFSS   DIG_SEL,3  ; IF LAST DIGIT IS REACHED..
    GOTO    RET_ISR

    CLRF    DIG_INDX     ; CLEAR DIGIT INDEX TO POINT FIRST DIGIT
    MOVLW   0X80
    MOVWF   DIG_SEL      ; SET '1000_0000' IN DIG_SEL TO SELECT
;                 THE FIRST DIGIT


RET_ISR
    POP
    RETFIE           ; Return from interrupt

;*****************************************************************
;       SEVEN SEGMENT DISPLAY LOOK UP TABLE
;*****************************************************************
TABLE
    ADDWF   PCL,F
 DT B'11111100', B'01100000', B'11011010', B'11110010', B'01100110',
B'10110110', B'10111110', B'11100000', B'11111110', B'11110110',
B'00000001'


;*****************************************************************
;       MAIN PROGRAM --   INITIALISATION
;*****************************************************************
; CONFIGURING PORT PINS
START
    CLRF    PORTA
    CLRF    PORTB
    CLRF    PORTC
    CLRF    PORTD
    CLRF    PORTE
    BANKSEL TRISB    ; PORT B AND PORT C CONFIGURED AS OUTPUTS
    CLRF    TRISB
    CLRF    TRISC

; CONFIGURING TIMER0

    MOVLW   0XD4     ; INTERNAL CLOCK, PRESCALER OF 1:32
```

```
        MOVWF   OPTION_REG
        BANKSEL TMR0
        CLRF    TMR0

; RAM INITIALISATION
        MOVLW   0X20    ; CLEAR ALL RAM LOCATIONS FROM 20H TO 7FH
        MOVWF   FSR
CLR_LOOP
        CLRF    INDF
        INCF    FSR,F
        BTFSS   FSR,7
        GOTO    CLR_LOOP

        CLRF    DIG_INDX
        MOVLW   0X80
        MOVWF   DIG_SEL

; CONFIGURING INTERRUPTS

        MOVLW   0X20    ; TIMER 0 INTERRUPT IS ENABLED
        MOVWF   INTCON
        CLRF    STATUS
        BSF INTCON,GIE ; GLOBAL INTERRUPT ENABLED
;********************************************************************
;               MAIN PROGRAM
;********************************************************************
; PUT CORRESPONDING SEVEN SEGMENT CODES FOR DIGITS 23.45

        MOVLW   2   ; MOV VALUE TO BE DISPLAYED INTO 'W'
        CALL    TABLE ; GET THE CORRESPONDING 7 SEG EQUIVALENT
        MOVWF   DISP1 ; MOVE IT TO FIRST DIGIT DISPLAY REGISTER

        MOVLW   3
        CALL    TABLE
        MOVWF   DISP2 ; SECOND DIGIT DISPLAY REGISTER
        MOVLW   DP
        CALL    TABLE
        IORWF   DISP2,F; ADD DECIMAL POINT

        MOVLW   4
        CALL    TABLE
        MOVWF   DISP3   ; THIRD DIGIT DISPLAY REGISTER

        MOVLW   5
        CALL    TABLE
        MOVWF   DISP4   ; FOURTH DIGIT DISPLAY REGISTER

        GOTO    $   ; HALT (ENDLESS LOOP)
END
```

CHAPTER 11

SERIAL COMMUNICATION

11.1 AN OVERVIEW OF SERIAL COMMUNICATION

Bytes can be exchanged between two processors through parallel ports and one or two supervisory signals. The communication here is fast, but it requires around 10 lines of connection between the two processors; further both processors have to be kept close to each other to ensure reliable transmission and also to maintain the high speed of transmission (with all these, it is more often preferable to assign all activities to one processor than to do such splitting!) When two processors are to communicate, more often the communication is organized in a bit serial fashion. A physical link of two or three wires is established between them; the bits are transported over the link bit by bit in serial fashion. Different types of applications exist. A printer may be interfaced to a controller based system. Once the printer is ON and active, the system can request for a print and send down the file to be printed as a sequential file of bytes. The communication is of a blind one way type – called 'simplex transmission'. Similar interface to a PC monitor is another example for simplex transmission.

There are situation where one processor is involved in a supervisory activity and sheds the drudgery of 'lower level' activities to another. The supervisor may give 'guidelines' at intervals and ask for regular reports from the other processor – may be called a 'slave processor'. The scheme can be implemented in a 'half duplex' mode. In half duplex mode when one processor is in transmit (talk) mode, the other is in receive (listen) mode. The latter can transmit (talk) only when the former has stopped transmission (talking). At a time only one transmits (talks). A properly executed telephone link can be serviced by a half duplex scheme.

A multi-processor scheme can be more demanding than what a half –duplex scheme can offer. Here each processor will be involved in its own task; each may have to communicate to other processors on a continuing basis. Processor A may be transmitting data and processor B receiving. Simultaneously (in a parallel line) processor B may be transmitting data and processor A receiving it. Such a scheme is called a full duplex scheme of transmission.

11.1.1 Asynchronous Communication

Many microcontrollers in use have in-built circuits to do serial transmission. At the transmitter side the circuit will convert a byte to a serial bit sequence and transmit it. At the receiver side, it will receive the serial bit sequence representing a byte and convert it into a byte. Minimal support for both conversions will be available. The full transmission implementation requires substantial additional tasks to be implemented. Normally these are carried out by dedicated serial communication support routines at 'higher layers' within the processors. Consider serial transmission between two processors A and B as shown in Figure 11.1. The processors are connected through a pair of wires – one being a ground (GND) line and the other a logic signal line. When no transmission takes place, the signal line is held at state '1' (5V). When A wants to transmit (and B is to receive) a byte, it will output a '0' on the signal line for a definite period – called the 'bit period'; A transmits a '0' bit on the line signifying start of transmission. This '0' at the beginning is called a 'start' bit. Note that the start bit will always be a '0' bit. At the receiving end B senses the change in the logic signal line from '1' to '0' state and interprets it as the start bit. Immediately following the start bit, the eight bits of a byte are transmitted as a serial sequence on the logic signal line. Once the full byte is transmitted, a 'stop' bit is also transmitted. The stop bit is a '1' bit. Following this the logic signal line goes high and remains high.

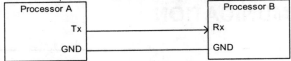

Figure 11.1 Connection for simplex serial transmission between two µCs

If A wants to transmit a sequence of bytes to B, a start bit and a stop bit are to be inserted on either side of every byte; then the set of 10 bits is to be transmitted. The two processors should communicate bit streams in the above manner at a prearranged bit rate – we call this the 'baud rate'; for our purposes baud rate represents the number of bits per second at which the transmission and reception between the two processors take place. Commonly used baud rtes have specific nominal values; 0.3, 0.6, 1.2, 2.4, 9.6, 19.2, 28.8, 38.4, 57.6, 76.8, 96, 115.2, 250, 625, 1250 K-baud are in common use. With the improvements in processor speeds and their clock stability, the lower values are no longer in vogue. Consider the implementation of a scheme at 9.6 K baud. It implies transmission at 9600 bits per sec; each bit takes 104.25µs for transmission. In the scheme of Figure 11.1 both the µCs will have a clock of 104.25µs time period; it is called the 'baud rate generator'; often the baud rate generator output will be derived from the main processor clock by division using a suitable counter.

Figure 11.2 shows a hypothetical situation of transmission at 9.6 Kbaud. Transmitter A has a baud rate generator with a pulse cycle period of 104.25µs. The narrow periodic pulses signify the start of transmission of each bit starting with the start bit. The receiver at B senses the 0 to 1 transition in the signal line at point P and interprets it as the start bit. It synchronizes its own baud rate generator to this transition and outputs pulses at intervals of 104.25µs starting from point-Q. Such a synchronization of the baud rate generators is essential to ensure that the receiver captures the bit transitions in the transmission line faithfully.

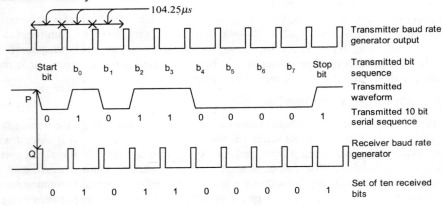

Figure 11.2 Asynchronous transmission of a byte at 9.6 kbaud

Normally the processor clocks at A and B will be at least 8 times faster than the baud rate generator; within the 104.23 μs interval between points Q and R the processor B will go through at least 8 periods of its clock. Processor B has to use the received waveform and with the help of its own baud rate generator and main clock extract the received bit sequence (010110001 including start and stop bits) with some level of reliability. Processors do this in different ways; one method is illustrated here. To facilitate discussion Figure 11.2 is reproduced in Figure 11.3. The segment related to the first bit received at B is shown in an expanded form. Assume its own clock rate to be eight times that of baud rate. B samples the received waveform at 3 consecutive periods in the middle of the interval Q R; the sampling instants are shown as S, T and U in the figure. If two (majority) of the sample values are at 0 level, the bit received is declared as zero; else it is declared erroneous. With every subsequent bit received, the bit value is decided as '0' if at least two of the samples are at '0'; else it is decided as '1' since two of the sample values are at '1'. In the scheme discussed above the transmitter baud rate generator and the receiver baud rate generator function independently; P – the starting instant of the start bit is the reference point for the receiver to synchronize its baud rate generator to that of the transmitter. The scheme is referred to as 'asynchronous' mode of transmission. Since transmission takes place from A to B in a unilateral manner, it is of the 'simplex' type.

Figure 11.4 is a modification of that in Figure 11.1 considered earlier. Both processors – A and B have been modified to provide transmission and reception. Each has a transmitter terminal and a receiver terminal. The receiver is of the open collector type. At each end the Rx and Tx terminals are joined together and a resistor (Ra and Rb) connected to supply. The two junctions are connected together through a logic signal line; the two ground terminals are also connected together. The 2-wire scheme here is set to carry out asynchronous half duplex serial communication. When processor A is in 'transmit' mode, its Rx terminal is disconnected internally; simultaneously processor B is set to receive mode by disconnecting its Tx terminal internally. With this connection, processor A can carry transmit bytes to processor B as explained earlier. When processor A has completed transmission, it can set itself to receive mode; simultaneously processor B can set itself to transmit mode. With the change on either side, processor B can transmit bytes to processor A.

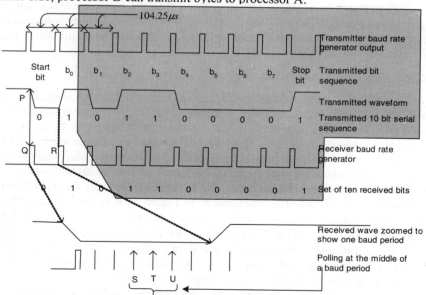

Figure 11.3 Region of one bit of the received bit stream to show sampling & polling to decide the value of the bit

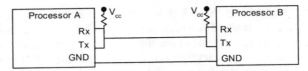

Figure 11.4 Connection for half duplex serial transmission between two μCs

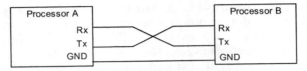

Figure 11.5 Connection for full duplex serial transmission between two μCs

The scheme has been further modified and shown in Figure 11.5. The Tx terminal of processor A is directly connected to the Rx terminal of processor B; similarly, the Tx terminal of processor B is directly connected to Rx terminal of processor A. If the receive terminal Rx is of the open collector type, necessary pull-up has to be provided. We have a 3-wire scheme of connection between the processors. Both processors can transmit and receive simultaneously. The scheme carries out asynchronous full duplex serial communication.

Processor A can transmit serial data to processor B at its Tx terminal; simultaneously it can receive at its Rx terminal serial data transmitted from processor B. Because both activities are to be carried out simultaneously, the necessary overhead tasks will be conspicuously higher and the same have to be handled by the processor. In a similar manner processor B too transmits and receives simultaneously; and it has to be geared up for the enhanced level of handling the overheads. There is the need for the two processors to synchronize – at least for their independently functioning baud rate generators to fall in step. Hence the communication scheme requires start and stop bits to be inserted for every byte– as discussed earlier with the simplex scheme. This puts a constraint on the maximum attainable speed of transmission. The attainable baud rate is one order lower than the processor clock rate; it is 10 to 20 times less than the processor clock rate. However at such high rates of communication the processor will be unduly strained to handle the communication overheads and still keep pace with the communication channel.

11.1.2 Synchronous Communication

The scheme in Figure 11.6 is a slightly modified version of that in figure 11.4. An additional clock line has been added here. The baud rate generator of processor A supplies the baud rate clock to processor B. With this addition, 3 wires are used for the serial communication. Since the baud rate clock used by A to generate the bit sequence waveform is known and available to B, the same need not be generated locally. Further, the bit to bit transition points in the received waveform are definitely known to B. The received bit stream and the clock are already in synchronism with a much higher level of certainty than with the asynchronous scheme considered so far. The scheme is referred as the 'synchronous' mode of serial communication. Specifically the arrangement in Figure 11.6 is for half duplex synchronous serial communication. On either side one lead is committed to the clock; the Tx and Rx leads are bunched together into one. The μC will assign Tx or Rx function to the pin as required. At start processor A will be in transmit mode and its lead will be configured for Tx; processor B will be in receive mode and its lead will be configured for Rx. As communication progresses, the assignments will be swapped depending on the needs and according to the protocol.

With synchronous communication one can reduce the bit period and check and poll for the bit level at the receiver with a better level of certainty than with the corresponding asynchronous case; it results in communication speed being improved 4 to 5 times. Once a start bit is received, a much larger number of bits than that with a byte can be sent by A and received by B, before the stop bit is inserted. It essentially eliminates the need for an additional pair of bits for every 8 bits. The gain in transmission rate is about 20% (slightly lower than 2/8 = 25%). Considering both the factors, the overall improvement in communication speed is 5 to 6 times. In fact under the best of conditions the baud rate can be the same as the μC clock rate. The clock can be directly used to shift the received bit stream serially into the receiver buffer; the start bit is followed by a large serial bit file and a stop bit or other structured bit sequence at the end. The net data rate is very close to the clock rate itself.

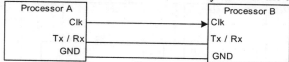

Figure 11.6 Connection for synchronous half duplex transmission between two μCs

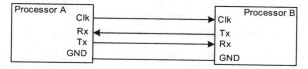

Figure 11.7 Connection for synchronous full duplex transmission between two μCs

The full duplex scheme for synchronous' mode of serial communication is shown in Figure 11.7. It needs 4 wires for communication. Note that processor A supplies the master baud rate clock for half duplex as well as full duplex schemes – even when processor B is in transmission mode.

11.1.3 Communication amongst Multiple Processors

When more than two processors are interconnected for serial communication, the communication structure has to be clearly defined. Priority, source identification, proper addressing etc., are some of the issues involved. Books on computer networking discuss these in greater detail. We introduce one scheme here to emphasize the viability and bring out the need for protocol.

The scheme in Figure 11.8 has four processors connected to a two wire scheme. Each can communicate with another in asynchronous half duplex mode. Each is assigned a one byte address. For example one possible assignment is as shown in Table 11.1. One viable protocol can be as follows:

- Normally processor A is in transmit mode; its Rx pin is disabled and Tx pin enabled; all other processors have the Tx pins disabled and Rx pins enabled; all of them are in receive mode
- A common baud rate is agreed and set in all the processors.
- Processor A transmits a byte stream to Processor B; the first byte is the address of processor B. The second byte is the total number of bytes sent in the current session. The rest of the bytes follow. At the end of the session, Processor A goes to receive mode by deactivating its Tx terminal and activating the Rx terminal
- All other Processors – B,C and D receive the above byte stream addressed to B from A. All of them 'study' the first byte namely the address byte: Processors C and D check the address in software and revert to idle mode or other tasks on hand.
- Processor B receives the next byte signifying the number of bytes; all the following bytes are also received by B. At the end of the session B goes to transmit mode by deactivating its own Rx terminal and activating the Tx terminal.
- Processor B sends address of processor A as first byte and the number of bytes as second byte; additional bytes follow as desired. At the end of the session Processor B reverts to receive mode.

Table 11.1 Address assignments to different μCs in Figure 11.8

Processor	Address
A	01h
B	02h
C	03h
D	04h

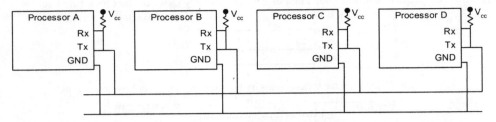

Figure 11.8 Four μCs connected for communication amongst them in half duplex asynchronous mode

- All other processors – A, C and D receive the byte stream. Processors C and D check the address byte and realize they are not involved. They revert to idle mode or other active tasks of their own.
- Processor A receives the byte stream. At the end of the session, it reverts to transmit mode.
- Processor A can address processor C or processor D and send a byte stream as described above. The same has to be followed by a return byte stream being sent by processor C or processor D as the case may be.

Some observations on the scheme are in order here:

- Up to 256 processors can be accommodated in the network, provided the Tx leads have necessary drive capacity.
- Processor A has the role of a 'master' here. It decides whom to talk to and who can be allowed to talk.
- As the number of processors in the network increases, the communication task of processor A becomes more time consuming; eventually processor A may have to be assigned the role of a 'communication processor' with no other assigned task.
- If processor B has to communicate to processor C, the same has to be routed through processor A; two pairs of sessions are involved – one from processor B to processor A and the other from processor A to processor C.
- As mentioned earlier, all overheads of communication are to be handled in software separately by each processor.
- The scheme functions satisfactorily if processors B, C, D etc., have similar roles and the role of processor A is essentially supervisory in nature.

11.2 USART AND OVERHEADS

Most microcontrollers support serial communication. The support is by way of a built in module which carries out the serial communication. Such a module is widely known as a 'Universal Synchronous Asynchronous Receiver Transmitter (USART)'. The simplest UART will have the following features.

- A baud rate generator: It provides a baud rate clock output. The baud rate can be programmed in software.

- Parallel to Serial Converter: A shift generator which can be loaded with the byte to be transmitted. The loading is done in parallel as a whole byte. The loaded byte is serially shifted out through the 'Transmit Pin'.

- Insertion of additional bits: If necessary the Start Bit and Stop Bit will be inserted at the beginning and end of each byte before transmission. Some schemes will have the option to insert a parity bit after the eighth bit and before the stop bit. The parity bit is the 9th bit. Alternatively the ninth bit can indicate the byte to be an address or data; if it is an address, the bit can be used to latch the address into a separate address latch.

- Data recovery at Receiver: A scheme of synchronizing the receiver baud rate generator to the transmitter baud rate generator will be provided for asynchronous transmission. A logic to sense each bit pulse at predetermined instants and deciding whether it represents '0' or a '1', will be provided.

- Programmability: Selection of synchronous or asynchronous scheme of communication will be programmable. Tx and Rx pin enabling and disabling will be done in software. The baud rate will be programmable. Option to use parity bit will be programmable.

- Bit checks: Identification of Start Bit and Stop Bit will be done by the UART. Parity checking too will be carried out.

- Serial to parallel converter: The received bit stream is to be loaded into a shift register in serial form, information byte extracted and readied for parallel loading; these are within the scope of this unit.

- Buffers: An additional buffer may be provided at transmitter end; it may store the 'byte-in-waiting' to be transmitted when one byte is under transmission. A similar buffer will be provided at the receiver end. To accommodate the lack of predictability of the instant of receipt of the 'byte-to-be-received-next', the receiver buffer may be one or two bytes deeper than the transmitter buffer. The receiver buffer stores the bytes received until the same are read by the processor.

- Interrupts: The UART may generate an interrupt signal at transmitter side when the transmitter buffer is empty. It is a signal to the processor that 'I am ready for next byte and waiting'. The receiver side may generate an interrupt when its buffer is full. It is a signal to the processor that 'I am full; off-load me'. The processor can sense these interrupts and do necessary service.

- Error Indication: Errors through parity check, improper receipt of Start / Stop bits etc., may be indicated through respective error flags and interrupts.

- Different schemes of representing the zero and one bit values are in vogue. The simplest represents them directly at the two logic levels; it is called the 'Non-Return Zero (NRZ)' scheme. All USARTs will support at least the NRZ scheme.

11.2.1 Other Overheads

The features mentioned above are the bare minimum to transmit bytes of data. In computer networking parlance all the features discussed so far together, is often referred to as the 'physical layer'. For the communication to be effective some amount of additional support is necessary; these can be in terms of the number of bytes transmitted, signaling the end of transmission, incorporating error checks, carrying out error checks and so on. Request for specific data, specification of data format, request for retransmission are additional features often desired. Specific characters in the ASCII character set [see Appendix A] can be directly used for some of these. Other features have to be built in through additional software. Depending on the scope of communication, such software can be provided at different levels called 'layers'. Well known internationally accepted standards specify the scope and functions of each such layer.

11.3 USART in PIC®16 SERIES OF PROCESORS

The PIC®16 series have an USART built in with a transmitter block and a receiver block within. The baud rate generator is common for both. Both blocks have necessary programming facilities.

11.3.1 Baud Rate Generator

The baud rate is derived from the processor clock by dividing it by a selected byte – X – in the 0 to 255 range. The number X is to be loaded into SPBRG register (at SFR address 0x99h). It is done at initialization time. In synchronous mode the baud rate is given by

$$baudrate = \frac{F_{osc}}{4(X+1)}$$

and in asynchronous mode

$$baudrate = \frac{F_{osc}}{16(X+1)}$$

Or

$$baudrate = \frac{F_{osc}}{64(X+1)}$$

The bit TXSTA<b2> (at SFR address 0x98h) designated BRGH decides the choice between the above two rates in asynchronous mode; BRGH = 0 selects the lower baud rate and BRGH = 1 selects the higher baud rate. Depending on the system clock frequency, one can select an X which gives a baud rate closest to the nominal value. Note that in asynchronous mode, BRGH = 1 leads to less deviation from nominal value; hence BRGH=1 may be preferred wherever possible.

The baud rate is generated in every µC communicating in asynchronous mode; but in synchronous mode, the baud rate generator is used only when the processor is in master mode; TXEN (See Figures 11.9 & 11.10) gates the baud rate generator output to the TSR register. In slave mode the local baud generator is ignored; baud rate clock is provided by the master.

11.3.2 Transmitter

The USART transmitter is shown in block diagram form in figure 11.9. Data byte to be transmitted is loaded into TXREG register (at SFR address 0x19h). The processor will automatically transfer it to the TSR register in parallel mode. If transmission is enabled by setting bit TXEN, the byte is serially shifted out through the Tx pin. The Start and Stop bits are inserted at TSR. When TXREG content is shifted to TSR register, the transmit interrupt flag TXIF is set. It is an indication to the processor that the transmitter unit is ready to receive the next byte for transmission. If the transmission is to be continued, the processor can load TXREG with the next byte to the transmitted. It can be done in an interrupt service routine. The bits– TX9D, TX9, TMRT, and SPEN are in the TXSTA and RCSTA registers (at SFR addresses 0x98h and 0x18h respectively). Either of these can be accessed for writing (to change the transmitter configuration) and for reading (to know the status).

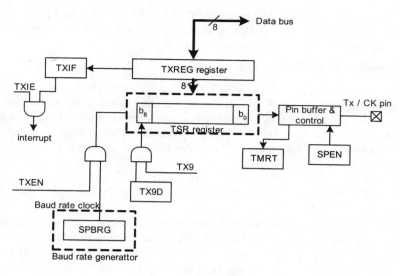

Figure 11.9 USART transmitter in PIC®16series in block diagram form

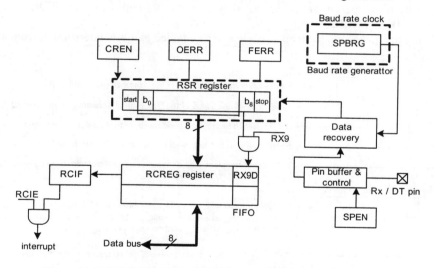

Figure 11.10 USART receiver in PIC®16series in block diagram form

11.3.3 Receiver

The receiver is shown in block diagram form in Figure 11.10. The received serial signal is input at Rx. The baud rate generator – local or externally supplied as the case may be – is used to recover the data from it. Data byte is extracted from it and transferred to RSR register; the data byte – and if necessary the 9th bit – are loaded into RCREG register. Loading of RCREG register causes the interrupt flag RCIF to be set. RCREG can store two successive received bytes while the third is being loaded into RSR Register. In fact RCREG is a pair of registers of the FIFO type, both with the same address (at SFR address 0x1Ah). The RCIF flag getting set is an indication to the processor that a data byte is received and it may be read by the processor. These bytes can be read from RCREG if necessary through an interrupt service routine. The bits– RX9D, RX9, CREN, OERR, FERR, and SPEN are in the TXSTA and RCSTA registers. Either of these can be accessed for writing (to change the transmitter configuration) and for reading (to know the status).

11.3.4 Asynchronous Communication

The asynchronous transmission is in full duplex mode. The registers TXSTA and RCSTA have bits to configure the transmission scheme. They also have bits representing the status and result of the transmission. The bits concerned are to be defined as part of initialization before embarking on the transmission. Six bits are to be assigned values. Figure 11.11 shows the assignments to set up asynchronous transmission for 8-bit scheme. The connection between two processors to implement the full duplex scheme is shown in Figure 11.12.

Apart from the assignments as in Figure 11.11 TXIE, RCIE, PEIE and GIE are to be enabled. TXIE and RCIE are in the PIE register; their specific positions in PIE are device dependent. PGPIE (b6) and GIE (b7) are in INTCON register.

- TXIE enables TXIF; TXIF is set when the transmitter is ready to accept next data byte to be transmitted; the processor can load TXREG with the same through an interrupt service routine.
- RCIE enables RCIF; RCIF is set when the receiver is ready to supply the next byte of received data to the processor. The processor can read this byte from its RCREG in an interrupt service routine.
- TXIE and RCIE are peripheral interrupts; they are gated through PEIE and PEIE is to be enabled to enable TXIE and RCIE
- GIE – being the global interrupt enable – has to be enabled to allow interruption from any source

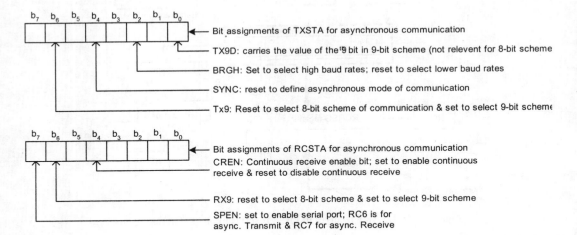

Figure 11.11 Assignments to the bits of the control registers – TXSTA and RCSTA – to define 8-bit asynchronous communication

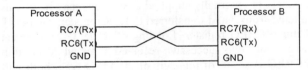

Figure 11.12 Connection of two processors for full duplex asynchronous scheme of communication

The processor can effect transmission of a byte by setting TXEN [TXSTA<b5>] and loading TXREG with the byte to be transmitted. Along with TXEN (in TXSTA) SREN (RCSTA<b5) in RCSTA is to be set to enable reception of the byte at the receiving end. In fact SREN should be set before embarking on the transmission; this makes the receiver ready to receive the byte before the same is transmitted from the transmitter end.   Figure 11.13 shows the transmitter side and receiver side waveforms for transmission of a single byte. The transmission and reception sequences are as follows:

11.3.4.1 Transmission Sequence

- At   instant A, SREN is set; receiver is ready to receive a byte on the Rx pin
- At instant B, TXEN is set; the transmitter is ready to transmit a byte through Tx pin
- At instant C the processor writes a word into TXREG register. The transmitter loads TSR register with the byte.
- From instant D (beginning of the next period of the BRG shift clock) the byte is transmitted serially through Tx pin. The start bit and stop bit are inserted at the beginning and end of the bit sequence representing the byte.
- Soon after the byte is shifted to TSR register from TXREG register, TXIF is set. Processor can sense this and supply the next byte to be transmitted; it can be done at any time before the stop bit is shifted out of TSR register.
- Soon after TSR register gets the byte to be transmitted, TRMT bit (TXSTP<b1>) is reset; it indicates that TSR register is having data and is busy with serial shifting out (i.e. transmission).
- When the byte transmission is complete, the TRMT bit is set.

11.3.4.2 Reception Sequence

- At the receiver, firstly the Rx line goes low signifying reception of Start Bit; this happens at instant   F (Instant F will follow instant D above with a few µs delay – decided by the communication link)
- Soon after, the data recovery circuit goes into action. At the middle of received bit it samples the bit three times on three successive clock periods (See Figure 11.3). Through polling the received bit is identified as a '0'. The same is loaded into RSR register.
- The bit reception and identification procedure is repeated for all the 10 bits; if the stop bit is detected as at level one (as it should be with error free reception), the bit sequence b7 – bo is transferred in parallel mode to the RCREG register.
- As soon as a data byte is available in RCREG register, RCIF goes high; it indicates that RCREG has data and the processor can read it – may be through an Interrupt Service Routine.
- RCREG is two bytes deep. The processor can start reading RCREG even after b7 of the following byte has been received and the same led into RCREG.
- If the last bit received is not a '1' bit, the framing error bit [RCSTA<FERR>] is set; processor can read RCSTA, identify the error and take any corrective action necessary.
- With RCREG being full with 2 bytes, if they are not read and RCREG flushed out, the following byte being received in RSR Register cannot be accommodated. The overflow error bit [RCSTA<OERR>] is set. Processor can read RCSTA contents and identifying the presence of an overflow error. It can also carry out any remedial action necessary.

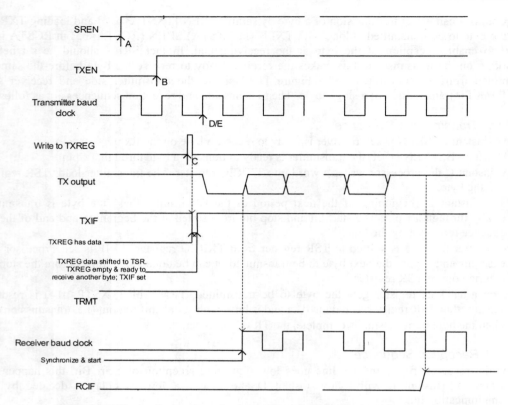

Figure 11.13 Transmitter and receiver waveforms for transmission of a single byte.

11.3.4.3 Nine-Bit Transmission

One can opt for 9 bit transmission instead of the 8 bit scheme described above. Tx9 [TXSTA<b6>] and RX9 [RCSTA<b6>] are to be set to select the 9 bit scheme. The bit assignments of Figure 11.11 are to be modified correspondingly. This constitutes the initialization for 9 bit transmission. The 9th bit can be used at least for two purposes. One possibility is to incorporate parity checking in the scheme; b8 can be the parity bit. For every byte to be transmitted parity checking has to be done in software and the value of b8 decided. It has to be loaded into TX9D [TXSTA<bo>] prior to the byte being loaded into TXREG; it will be inserted automatically into the serial bit stream at b8 position, during transmission. At the receiver, b8 will be received and stored into RX9D [RCSTA<bo>]. The bit value can be read and parity check carried by the processor at the receiving end in software. Discarding the byte if the parity check shows error and (if necessary) asking for retransmission etc., has to be done by software at a higher layer. Some PIC® series processors have an added facility with b8. More than two processors can be connected to the serial line; one can play the role of a master and all others can be slave processors. The receiving modules in all slave processors can be assigned addresses. Each unit can store its own address. Communication can be controlled by the master. When the master sends out bytes, the first one can be the address. Bit – b8 – can be set indicating that the associated byte is an address. All slaves sense b8; if b8 = 1, they check for an address match; each can accept further bytes only if it finds an address match. The flag ADDEN [RCSTA<b3>] is a replica of b8 and helps in gating data bytes in case of an address match.

11.3.5 Synchronous Communication

For synchronous communication, one processor is to be designated as the master and the other as the slave. The master supplies the baud rate clock; it is supplied to the slave. The slave uses the baud rate clock supplied by the master for its own reception as well as transmission activity. The bit assignments required to set up 8 bit synchronous transmission are as shown in Figure 11.14. Three bits in TXSTA (b7, b6, and b4) and four bits in RCSTA (b7, b6, b5, and b4) are the ones involved. The baud rate has to be set to the same value for the master as well as the slave. Apart from these TX1E, RCIE, PEIE, and GIE are to be enabled as explained above with synchronous transmission. The connection between the master and the slave is as shown in Figure 11.15. If processor A is assigned the master status and processor B the slave status, A generates baud rate clock and supplies it to B at its CK output terminal – RC6. B receives the same at its clock input terminal – RC6. RC7 – designated as DT – is common for transmission and reception. When data is being transmitted from processor A, reception at the pin is disabled. Similarly when processor A is receiving data, transmission at RC7 is disabled. Assignment of values to TXEN and RCEN is to be coordinated between both ends to ensure flawless operation.

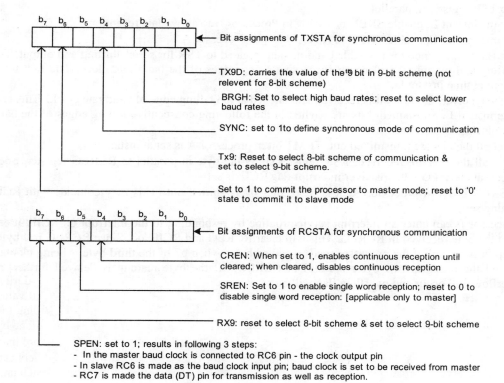

Figure 11.14 Bit assignments to set up 8 bit synchronous transmission

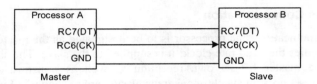

Figure 11.15 Connection between master and slave for synchronous

11.3.5.1 Transmission and Reception Sequences

Set SPEN – both in master as well as slave – to commit RC7 and RC6 for serial transmission. TX9 [TXSTA<b6>] and RX9 [RCSTA<b6>] are reset in master as well as in slave to commit both for 8 bit transmission. SYNC [TXSTA<b4>] bit is set in master as well as slave to opt for synchronous transmission. SCRC [TXSTA<b7>] is set in master (Processor A) to enable and output band rate clock. SCRC (TXSTA<b7>]in slave is reset (Processor B) to disable local baud rate and use baud rate clock supplied by master through pin CK [RC6]. All these are part of initialization. Subsequent signal sequence for master to transmit and slave to receive is as follows (See Figure 11.16):

- At time instant A load TXREG of Processor A with the byte to be transmitted. The same is loaded into TSR register in parallel.
- At time instant B, enable SREN or CREN in Processor B to allow data reception.
- At time instant C, enable TXEN [TXSTA<b5>] of Processor A
- The Baud rate generator is enabled and its output gated to TSR for serial shifting and output. The shifting and output take place on successive rising edges of the baud rate clock. The first bit is output at time instant D.
- At the receiver, the first bit is received and latched at the falling edge of baud rate clock. This is at time instant E. Subsequent bits are latched in the following consecutive falling edges of the baud rate clock.
- After all the bits are transmitted out, TRMT bit in processor A is set at instant F
- After all the bits are received, RSR register shifts the byte in parallel to RCREG register. Soon after – at instant G – the receiver interrupt flag RCIF is set
- If SREN is set in processor A, it is reset after b7 is shifted out (This may coincide with RCIF getting set).
- Processor B can enter an interrupt service routine by responding to the interrupt on RCIF; it can read the byte received in RCREG. This will clear RCREG and RCIF too. Since RCREG is 2 bytes deep, it is enough if the processor reads received bytes when b7 of the third byte is being clocked in and the previous two bytes are still in RCREG. If the byte reading is delayed further, an overflow will occur and overflow flag will be set.

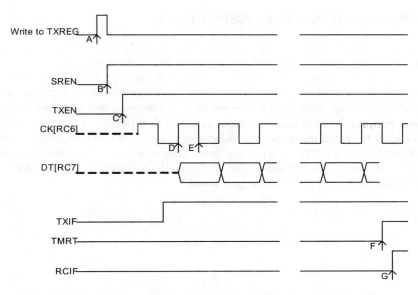

Figure 11.16 Synchronous communication – transmitter & receiver waveforms

11.3.5.2 Nine-Bit Transmission

Nine bit transmission scheme can be implemented with synchronous transmission scheme also. TX9 [TXSTA<b6>] and RX9 [RCSTA<b6>] are to be set to select the 9 bit scheme. The bits assignments for initialization shown in Figure 11.14 are to be modified accordingly. The 9th bit can be used for parity checking as explained earlier.

Observations:

- In a synchronous transmission scheme, serial shifting is done at the rising edge of baud rate clock; receiver latches the received data in the immediately following falling edge of the baud rate clock. Sampling in successive periods and polling are not called for, since clock is the same for transmitter as well as receiver.

- The simplicity of receiver circuit and the much higher level of certainty in reception allow higher data rates compared to the asynchronous scheme. In fact one can use the system clock directly for synchronous transmission; it allows a baud rate equivalent to the instruction cycle of the processor itself (fosc/4). At such high data rates reading the received data in time from RCREG on a continuing basis and processing it on-line, can be challenging tasks.

- The baud rate clock becoming active is an indication of start of transmission. End of receipt of b7 or b8 as the case may be, can be directly used for identifying end of the word. Insertion of Start and Stop bits is not called for.

- Start of transmission and end of transmission from master are to be properly coordinated with those from the slave. It has to be done in firmware at a higher layer. A preferred scheme is for the master to ask for response / data from slave and slave to respond to it; here the initiative is with the master.

- Since the slave receives the baud rate clock input from the master, the slave can be made to receive serial communication when in sleep mode. It may wake up through the interrupt service routine.

11.4 SYNCHRONOUS SERIAL PORT (SSP) MODULE

Different schemes of implementing serial transmission are available. Each has associated protocols. As the number of wires increases beyond two or three, the additional wires are used for peripheral tasks like start communication, stop communication, indication of errors, indication of need for retransmission etc. Relegation of such tasks to other wires contributes to increase in data rate but makes the interface correspondingly cumbersome both in hardware and software. The PIC® series have a separate peripheral called the 'Synchronous Serial Port (SSP)'. By proper configuration the module can be used for serial transmission conforming to either of two protocols; one of these is the 'Serial Peripheral Interface (SPI)' – originally from Motorola – and the other the 'Inter-Integrated Circuit (12C)' – originally from Philips: The SSP module is different from USART. Details are available in the data sheets and the Mid-Range Manual.

CHAPTER 12

OTHER FACILITIES AND APPLICATIONS

12.1 ADCs

Many applications of the μC require analog signals to be directly accepted by it. The PIC®16 series have built-in AD converter modules to facilitate direct interface of analog inputs; necessary control and configuration are done by programming. The μC can accommodate a maximum of eight analog inputs. Since the port pins have multiple assignment possibilities, they are to be made analog inputs as part of configuration; the pins to be so assigned and the SFRs for the same are dependent on the selected device. The ADC functions in successive approximation mode; the module does conversion in 11 clock cycles and – being a 10-bit ADC – returns a 10-bit digital output. Two SFRs – ADCON0 and ADCON1 – are provided to do the configuration and control. The converted outputs are stored in two SFRs – ADRESH and ADRESL.

The ADC module in the PIC®16 series is shown in block diagram form in Figure 12.1. The choice of reference for conversion, number of lines to be committed to analog input, and the mode of loading the converted number into the result register are decided by the bits of the ADCON1 register. The conversion rate can be set in terms of the internal clock or an RC oscillator available within the ADC module. The bits b7 and b6 of ADCON0 – designated as ADCS1 and ADCS0 – decide the clock source. The ADC operation is carried out through the other bits of the ADCON0 register.

12.1.1 ADCON1 Register

The assignments of ADCON1 register are as shown in Figure 12.2. The reference for conversion can be the internal supply voltage or it can be supplied externally. With the internal selection the converted digit is a relative figure. It suffices to compare samples of different analog signals or samples of the same signal at different instants of time. The internal selection makes two more lines available for IO. If the reference is supplied externally, the converter output can be used as an absolute value [of course relative to the external reference]; the conversion accuracy is traceable to the reference voltage source. With single sided reference one can have a maximum of seven analog channels. For the converter output to be fully meaningful, the precision, stability, and traceability of the reference have to be compatible with the accuracy of the converter – 0.05%. Two-sided balanced reference provides best accuracy for the 10-bit conversion used. The bits <b3:b0> of ADCON1 register – designated as <PCF3:PCF0> – decide the assignments to the reference as well as the demarcation between analog inputs and digital IOs; details are available in the data sheets. ADCON<b7> is designated as ADFM. It decides the format of loading the 10-bit number into the result register. ADCON1 is to be loaded with the configuration data as part of the initialization procedure.

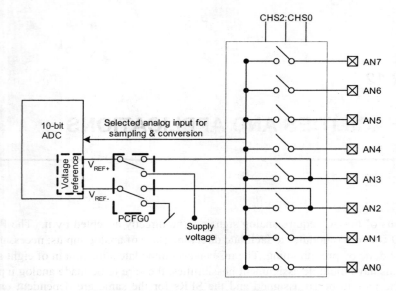

Figure 12.1 ADC module in block diagram form

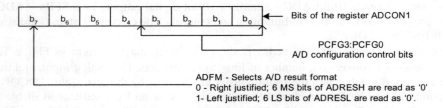

Figure 12.2 Bit assignments of ADCON1 register

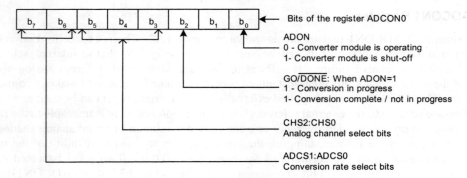

Figure 12.3 Bit assignments of ADCON0 register

12.1.2 ADCON0 Register

The bit assignments of ADCON0 register are shown in Figure 12.3. As mentioned earlier, the bits b7 and b6 decide the clock source (See Table 12.1). They are to be decided and loaded as part of initialization. The bits b5-b3 – designated <CH2:CH0)> – select the analog channel for sampling and conversion (See Table 12.2). Conversion is initiated by setting bit b2 – designated 'GO / $\overline{DONE}$. When the conversion is complete, the module resets b2 to 0. The bit b0 – designated ADON – disables the module when not in use. It reduces the power consumption of the device. Enabling and disabling of the device is to be done in software.

Table 12.1 A/D conversion rate selection

		Clock rate
ADCS1:ADCS0 values	00	Fosc/2
	01	Fosc/8
	10	Fosc/32
	11	FRC

Table 12.2 Channel selection for sampling & conversion

		Channel selected
CHS2:CHS0 bit values	000	Ch. 0-(RA0/AN0)
	001	Ch. 1-(RA1/AN1)
	010	Ch. 2-(RA2/AN2)
	011	Ch. 3-(RA3/AN3)
	100	Ch. 4-(RA5/AN4)
	101	Ch. 5-(RE0/AN5)
	110	Ch. 6-(RE1/AN6)
	111	Ch. 7-(RE2/AN7)

12.1.3 AD Conversion Operation

ADC operation has two Parts to it – configuration and conversion. The configuration can be done as part of the initialization of the µC. It involves the following steps:
- Decide on the number of channels to be used in analog mode and select the pins for the same. The pins are to be configured as input pins by setting the bits in the concerned TRIS registers.
- Do port configuration by assigning suitable values to the bits – <PCFG3:PCFG0> – in ADCON1 register.
- Load ADFM bit in the ADCON1 register with the desired value along with the above. The 10-bit result of AD conversion will be loaded into <ADRESH:ADRESL> register pair , the format being decided by the value of the ADFM bit (See Figure 12.4 for details).
- Select AD conversion rate by assigning values to <ADCS1:ADCS0> of ADCON0 register. Table 12.1 gives the details.
- Set the ADON bit in the ADCON0 register to '1' state to activate the ADC module. This can be carried out along with the last step.
- The above steps constitute the configuration. If the interrupt mode of reading the converted value is to be used, the following additional steps are needed.
- Clear the ADIF bit.
- Set the ADIE and GIE bits to enable the interrupts concerned.

Regular operation involves the following:
- Select the AD channel to be sampled and converted by assigning values to <CHS2:CHS0> in ADCON0 register. If the application uses only one channel, the channel selection can be done as part of initialization. If more than one analog channel is in use, the channel selection is done preferably prior to start of conversion as explained here.
- Set the GO bit to start conversion.

- Read the <ADRESH:ADRESL> register pair to get the 10-bit output value. The reading can be done in interrupt mode or by polling the GO bit.

Timing details of ADC operation are given in Figure 12.5. The input signal is connected to the holding capacitor after channel selection. One has to wait for enough time for the charging to be complete before embarking on conversion (by setting the GO bit). Once the GO bit is set, the holding capacitor is disconnected from the input side with a delay of ~100ns. AD conversion starts at the following conversion clock period. At the end of conversion the <ADRESH:ADRESL> register pair is loaded with the result in the desired format. The GO bit is cleared and ADIF is set; these signify that the conversion is complete. The holding capacitor is connected back to the selected channel and it starts charging again.

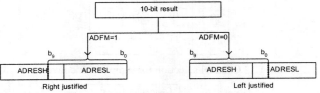

Figure 12.4 Steering of AD conversion result into the ADRES registers decided by ADFM bit

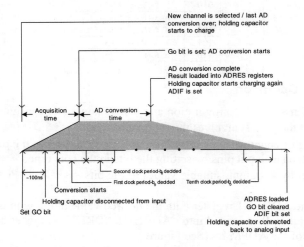

Figure 12.5 Timing details and sequence of AD conversion for one sample

12.2 EEPROM BASED DATA MEMORY

Some of the PIC®16CXX devices have 'Electrically Erasable and Programmable Memory (EEPROM)'. A definite but limited number of bytes can be stored there. Data written in is retained even when power supply is switched off to the device. Constants specific to an application, sample values collected at a specific time, location, application environment etc., to be retained for later use, can be stored there. Reading and writing are done in an indirect manner. Four SFRs are involved in this, namely:

- EEDATA register: EE DATA is the data source for writing; it is the automatic data destination for data read operation.

- EEADR register: The EE ADR register is to hold the address of the location at which data is to be written as well as the address of the location from which data is to be read.
- EECON1 register: The EE CON1 register has five bits which control read and write operations; the other bits (b5, b6, and b7) are not used.
- EECON2 register: EE CON2 register is used during EE PROM data memory write cycle. It has no other function. It is not accessible for user for any other purpose.

The bit assignments of EECON1 REGISTER are as follows:

- b0(RD): It is set to read the EEPROM data. Once Read operation is completed, the processor resets the bit.
- b1(WR): It is set to write data into EEPROM data memory location. When the write operation is complete, the bit is reset by the processor.
- b2(WREN): The bit is set to allow writing into the EE PROM data memory; reset on successful completion of write operation. If WREN is reset, writing to EE PROM data memory is inhibited. It is desirable to keep it reset and set it before any specific writing.
- b3(WRERR): EEPROM error flag: Set if write operation is prematurely terminated by a RESET due to any reason.
- b4(EEIF): The interrupt flag is set when the write operation is completed. It must be cleared in software. The write operation extends over a few instruction cycles and its completion is indicated by the interrupt flag going high. In fact the interrupt service routine can test WRERR flag and initiate another write cycle in case of an erroneous writing

12.2.1 Read Operation

Reading a location in EE PROM DATA memory area involves three steps;

- Load the desired source address into EEADR register; since EEADR is in memory Bank0, the same should be selected in advance.
- Read the data in the selected location; the read data is automatically loaded into EEDATA register; RD bit in EECON1 has to be set for the read operation; EECON1 being in Bank1, one has to switch to Bank1 to access EECON1<RD>
- Transfer EEDATA to the desired location. EEDATA being in Bank1, the same has to be selected to access EE DATA.

A minimum of seven instructions has to be executed to do a read operation. If the active bank is BANK0 prior to the Read Operation, one need to execute only 6 instructions for the Reading .

12.2.2 Write Operation

Write sequence involves five distinct steps.

- Disable all interrupts by clearing GIE. Bank 1 should be selected for this to access GIE bit in INTCOM register
- Enable access for 'write' operation by setting EECON<WREN>.
- Carry out the following write sequence

    $0X55h \rightarrow$(EECON2)

    $0XAAh \rightarrow$ (EECON2)

The sequence is specified by the manufacturer.

- Carry out write operation by setting EE CON1 <WR>
- Enable interrupts by setting GIE.
- Disable access for 'write' operation to EEPROM by resetting EECON<WREN>.

The write operation calls for the execution of at least nine instructions including bank1 selection, disabling interrupts prior to the write operation, and enabling them after the write operation. The data to be written and the destination address are assumed to have been loaded in EEDATA and EEADR respectively, prior to execution of the sequence. A dedicated timer within the processor controls the Write Operation; further during the EEPROM memory write, processor clock is halted. The clock becomes active and the processor resumes operation only when Write operation is completed. This explains the need to keep all interrupts disabled during a write operation. Some devices in the series have EEPROM for program memory. They also have the facility to access and rewrite program memory segments. Necessary additional SFRs are provided in such devices.

12.3 WATCH DOG TIMER

A μC based system can have an assorted set of components with different levels of reliability. A likely erroneous operation of one of them can result in the malfunction of the system. The concept of a 'watch dog' is invoked to monitor the system continuously and detect such malfunctions. A corrective action may be taken or at least an emergency action initiated at the instance of the watch dog. The PIC®16 series has a 'Watch Dog Timer (WDT)' dedicated for the purpose. It is a simple 'one shot' type timer. Figure 12.6 shows the operation of the WDT in a typical μC application.

InPIC16CXXX, the Watch Dog Timer (WDT) is a separate dedicated on-chip oscillator – different from the μC clock – and a counter working together, to generate a pulse with a time delay. The pulse can be used to reset the processor. If the processor is healthy and functioning properly, it resets the watchdog timer at regular intervals, well before the WDT timer times out. The interval is chosen to be shorter than the WDT delay. WDT starts afresh after such a reset – if necessary the watch dog timer pulse can be used to generate an alarm also.

In most processor applications one sweep of the main program will be completed in one ms or so (with an 8 MHz clock, it allows 2000 instructions to be executed). One can include one 'CLR WDT' instruction in the main program; it will be executed once in every sweep of the main program and the WDT reset. Figure 12.6(a) shows such a regular function of the μC and the WDT working together. The μC resets the WDT at point P once in every cycle of its operation. WDT starts afresh from that instant.

The WDT time period (in PIC®16 series) is normally 18 ms (If necessary the time delay can be increased by tying the prescaler to the WDT). In case of a malfunction, the μC may stop functioning altogether. It may not continue with the main routine. The WDT is not reset any longer. The situation is shown in Figure 12.6(b). WDT timer continues functioning until it times out on its own. It resets the μC and the μC starts functioning afresh at point S. The initialization routine following a reset can be configured in different ways. It can give out a warning, carry on a health check, go into a safe mode of operation, enter manual mode and so on.

12.4 REFERENCE VOLTAGE MODULE

The module provides a reference voltage output. It is derived from the power supply to the μC through a potential divider. The module is shown in block diagram form in Figure 12.7. The SFR CVRCON decides and controls the operation. The module can be turned off if not used by the application program. The bit CVRCON<b7> – designated CVROE – is used to switch on (CVROE = 1) the module or to keep it turned off (CVROE = 0). By turning it off, the power drain to the module can be eliminated. The module provides output in two ranges decided by CVRR. In each range one can have 16 equally spaced outputs by switching it to a selected tap on the potential divider chain. The tap selection is decided by the four bits CVRCON<b3:b0> – designated CVR3, CVR2, CVR1, and CVR0. The status of bit – CVRCON<b5> – designated CVROE – is used to turn on the reference to the output pin. The reference can also be used as an internal reference to the comparator module.

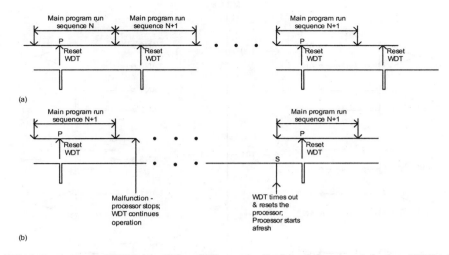

Figure 12.6 Timing sequence of WDT (a) µC functioning normally (b) µC function defective; WDT takes over & resets the µC

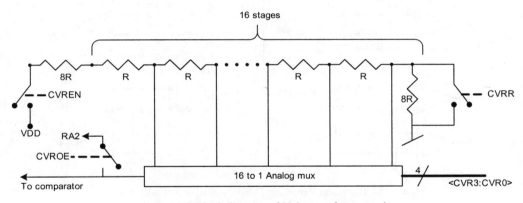

Figure 12.7 Block diagram of Voltage reference unit

12.5 COMPARATOR MODULE

A comparator compares two signals represented by respective analog voltages. If x(t) and y(t) are the analog voltages, the comparator provides a logic 1 output if x(t) > y(t); else – that is if x(t) < y(t) – it gives a logic 0 output. The module has two comparators within. Their input and output assignments are done in software. The CMCON register (SFR) is dedicated to the comparator module. CMCON<b6> is assigned to the output of comparator 1 and CMCON<b7> assigned to the output of comparator 2. The bits CMCON<b3:b0> decide the outputs of the two comparators (See Figure 12.8). When external inputs are compared with an internal reference, the output of the reference voltage module is used as the reference. As mentioned earlier, the internal reference can be selected from a number of possible reference values. An internal interrupt flag is dedicated to the comparator module. The interrupt is set whenever the interrupt output status changes. The interrupt flag is cleared whenever the CMCON register is read.

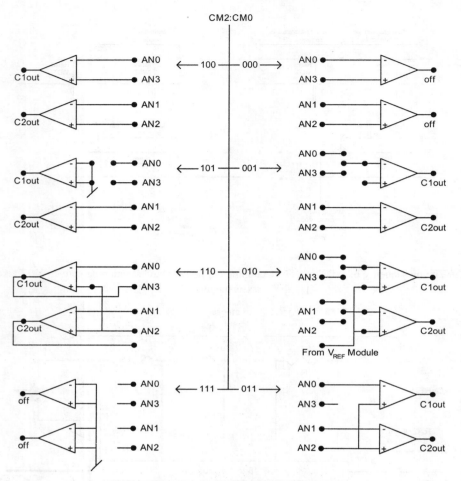

Figure 12.8 Comparator module: Different assignment possibilities

12.6 CCP MODULE

The Capture, Compare, and PWM (CCP) module is a multi-function unit built around a pair of SFRs. The µC has two CCPs which function essentially identically. The SFRs are designated as CCPR1H and CCPR1L for one module and CCPR2H and CCPR2L for the other module. The CCPRs can be configured in different ways; Operating along with Timer1 and Timer2, they provide three different modes of operation. Each CCPR has its own control register – SFRs designated as CCP1CON and CCP2CON. The bits of CCPCON decide the mode of operation as well as other operational details.

12.6.1 Capture Mode

The Capture mode is intended to capture the time of occurrence of events extraneous to the µC. The module configured for operation in capture mode is shown in block diagram form in Figure12.9. The module has a prescaler which can be set to select one of three scales. – 1:1, 1:4, and 1:16. A rising edge on the input line is interpreted as an event. When in the 1:1 mode, at the occurrence of the event – at the rising edge of the input – the content of TMR1 is captured and loaded into the register pair –

<CCPRxH:CCPRxL>. Simultaneously CCPRxIF – the CCPRx Interrupt flag – is set. µC can read the content of <CCPRxH:CCPRxL> in an ISR and use the same suitably. The interrupt flag is to be reset in the ISR to facilitate the next capture operation.

When the pre-scale value is set to 1:4, the capture is done at every 4th rising edge of the input. Similarly with a pre-scaler value of 1:16 the capture is at every 16th rising edge on the input. For the specific case of a prescale setting of 1:1 one has the option of selecting between the rising edge and the falling edge of the input as the event for capture operation. The CCPRx pin has to be configured in input mode for the module to accept the input signal and operate.

12.6.2 Compare Mode

In a way the COMPARE mode is the dual of the CAPTURE mode. Here an event is initiated at a set time. The operational circuit for the mode is shown in Figure 12.10 in block diagram form.

A 16-bit number is loaded into the register pair – <CCPRxH:CCPRxL> – and Timer1 is set to operate. When the content of Timer1 matches that of the <CCPRxH:CCPRxL> pair, the interrupt flag – CCPRxIF – is set. The µC can carry out the designated activity in an ISR. The interrupt flag is to be reset within the ISR itself.

Depending on the values assigned to the mode select bits, three additional activities can be carried out in COMPARE mode.

- TMR1 register – the 16-bit period register –is reset.
- The AD conversion can be started (along with the resetting of TMR1). The facility is available only with one of the CCPR modules.
- A hardware trigger signal is provided at the assigned pin – designated CCPx. The pin has to be configured to be in output mode as a prerequisite.

12.6.3 PWM Mode

A Pulse width modulated (PWM) signal is a train of periodic pulses, the pulse width being proportional to a modulating signal. As the modulating signal magnitude varies, the pulse width also varies accordingly. Figure 12.11 shows a typical example of a test signal and the corresponding PWM output.

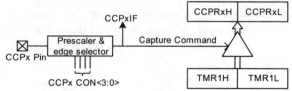

Figure 12.9 A CCP unit configured to function in Capture mode

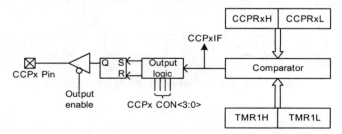

Figure 12.10 A CCP unit configured to function in Compare mode

The rising edge of the PWM pulse train is to occur at regular fixed intervals, the interval being the period of the carrier frequency. The falling edge of a pulse is decided by the magnitude of the signal during the interval. As the signal value changes, the location of the falling edge – within the pulse period – changes accordingly.

The CCPR module can function as a PWM modulator using Timer2 as the basic clock. A module configured to function in this manner is shown in block diagram form in Figure 12.12. The PWM output pulse train is shown in Figure 12.13. The quantities deciding the transitions and the associated events are also indicated in the figure.

Content of PR2 decides the pulse repetition frequency. Whenever the content of Timer2 matches that of PR2, flip flop Fm is set. The prescaler is tagged to Timer2; it also provides limited scaling levels for the modulator. Fm is reset whenever the content of TMR2 matches that of CCPRxH. The matching is for a 10-bit width – It is between the content of Timer2 and two bits of the prescaler on the one hand and the content of CCPRxH and two additional latches within, on the other hand. Thus in each period the pulse width is decided by the content of CCPRxH and the two additional bits.

CCPRxL and two bits of CCPCON register – CCPCON<b6:b5> – together form a 10-bit register. It functions as a buffer for CCPRxH. CCPRxH and the 2 latches are updated with the content of <CCPRxL, CCPCON<b6:b5>> soon after Fm is set. This ensures that CCPRxH cannot be changed directly and prevents any glitches in the PWM output.

12.7 APPLICATIONS

A set of typical application examples is considered here as examples. They illustrate the use of different peripherals – configuration as well as application. A practical application requires formation of modules of the type discussed here and their use in combinations. A brief description of each example follows. The program listings are given subsequently.

12.7.1 AD Converter

Program_2 converts a signal of bandwidth of 100 Hz (approx.) into digital form; twenty successive samples are taken at 1ms intervals, converted and saved in a set of sequential register pairs in the scratchpad memory area. Timer0 provides the 1 ms timing for sampling. The interrupt from it is used to initiate AD conversion. The main program polls the ADIF flag; when the flag is set the converted data is read and saved. After acquiring 20 such sample values, the program halts. Program is organized as follows:

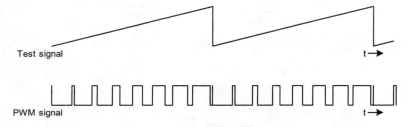

Figure 12.11 A test signal and its PWM representation

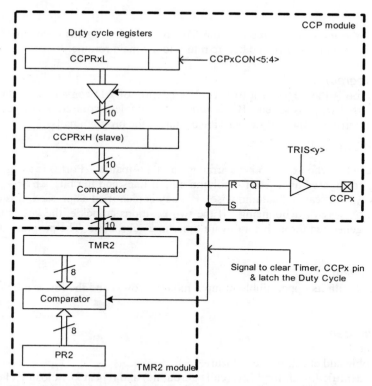

Figure 12.12 A CCP module configured to function as a PWM modulator

12.7.1.1 Initialization

Port A is configured as input port. RA0 is the analog input channel. The internal supply voltage is taken as the reference for conversion. The conversion rate is set as (fosc/32). The converted result is right justified. Timer0 uses the system clock with a prescale factor of 4; it is set to interrupt the µC at 1ms intervals.

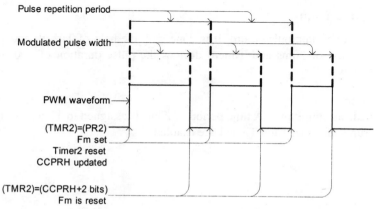

Figure 12.13 A PWM pulse train showing related events in the CCP module

12.7.1.2 ISR
The ISR identifies the interrupt source as from Timer0: the key registers are saved and conversion started. The saved registers are retrieved before return to the main routine.

12.7.1.3 Main Program
The main program polls the ADIF flag to check completion of AD conversion. Once completion is detected, the converted value is saved. If the number of samples acquired is not yet 20, the program waits for the next interrupt. Once 20 samples are acquired, the program halts.

12.7.2 Keyboard Scan

Program_3 is to scan a keyboard of 16 keys arranged as a 4×4matrix. Port B is assigned the task. The bits – b3-b0 – are configured as a set of 4 column output lines and the bits – b7-b4 – as a set of row input lines. The keys are scanned column wise. If a key is identified as pressed, it is checked again in a scan 15ms later, by way of debouncing. If the two positions coincide, the key is taken as pressed. The 15ms delay is generated through a software timer in a subroutine. The program organization is as follows:

12.7.2.1 1Initialisation
Port B is configured with its upper nibble in input mode (4 rows) and the lower nibble in output mode (4 columns).

12.7.2.2 Main Program
Start scan at column 1.
Read the return nibble and check for key closure.
If a key closure is identified, wait for 15ms (call subroutine) and repeat key closure check
If the two key closures coincide, identify key position and save the same.
If there is no key closure, or if two successive key closures do not coincide, proceed with the next column.
Repeat the scan for all 4 columns cyclically.

12.7.2.3 Subroutine
Delay is created by a set of 2 nested loops decrementing to 0.

12.7.3 Periodic Pulse Train

Program_4 generates a periodic pulse train. It is done with the help of Timer0 and Timer1. Timer1 provides the pulse period of 20ms and Timer0 decides the pulse duration of 100 µs. The program organization is as follows:

12.7.3.1 Initialisation
Pin7 of Port B is made as output pin. A time period of 20ms is assigned to Timer1 and a period of 100 µs to Timer0. Interrupts of Timer0 and Timer1 are enabled.

12.7.3.2 ISR
Identify the interrupt source as Timer0 or Timer1. If it is Timer1, set output to 1. If it is Timer0, reset output to 0.

12.7.3.3 Main Program
Wait for an interrupt

12.7.4 Time Delay

Many situations demand a digital output to be set to '1' state, after a digital input goes to '1' state. The transition in the output is to be done with a definite time delay. Program_5 achieves this. The bit b0 of port B is treated an interrupt input. Timer0 provides the delay. As soon as the interrupt input goes high, the timer starts operation. On its overflow the output is set to '1' state. Starting the timer and setting the output – both are done in interrupt mode. Program organization is as follows:

12.7.4.1 Initialization
Port B –b7 – is configured as output pin.
Port B – b0 – is configured as external interrupt input pin.
Internal clock is connected to timer). Prescaler is not used.

12.7.4.2 ISR
If the interrupt is from the external source, enable interrupt from timer0. Clear the timer and return.
If the interrupt is from timer0, set output bit high and return.

12.7.4.3 Main Program
Wait for interrupt
 Program_6 is alternative version of the delay task. The routine uses Timer1 to realize the time delay; in all other respects the program is similar to Program_5.

12.7.5 Serial Transmission

Illustration of serial transmission is done through a simple sample program – Program_7.

12.7.5.1 Initialization
Set baud rate to 9.6Kbaud.
 Set asynchronous transmission mode with 9 bit word.

12.7.5.2 Subroutines
Generation of parity bit
Loading the transmit buffer with next data.

12.7.5.3 Main Program
Transmit 20 bytes of data successively and halt.

```
;**********************************************************************
;    PROGRAM_2                                *
;    Notes: This program reads an analog signal on RA0 every 1mS.   *
;    The SOC is asserted every 1mS by timer0. The converted digital *
;    values are stored in an array of length 20. The timer is
;    stopped and conversion is stopped after acquiring 20 samples.  *
;           OSCILLATOR FREQUENCY = 4.096 MHz          *
;**********************************************************************

 list  p=16f877
#include <p16f877.inc>


 __CONFIG _CP_OFF & _WDT_OFF & _BODEN_ON & _PWRTE_ON & _HS_OSC &
_WRT_ENABLE_OFF & _LVP_OFF & _DEBUG_OFF & _CPD_OFF

 CBLOCK 0X20    ;BLOCK OF 20 WORDS (40 BYTES) OF MEMORY FOR ADC
                ;SAMPLES STARTING FROM 20H
     SAMPLES
 ENDC

 CBLOCK 0X50
     W_TEMP, S_TEMP, P_TEMP  ; FOR CONTEXT SWITCHING DURING
                             ; INTERRUPTS
     COUNT ; FOR COUNTING SAMPLES
 ENDC

#DEFINE CHANNEL0  0X81        ; VALUE TO BE LOADED IN ADCON0
#DEFINE SAMPLE_LENGTH D'20'   ; NO OF SAMPLES TO BE TAKEN IN WORDS


PUSH    MACRO
    MOVWF   W_TEMP
    MOVF    STATUS,W
    CLRF    STATUS
    MOVWF   S_TEMP
    MOVF    PCLATH,W
    MOVWF   P_TEMP
    CLRF    PCLATH
    ENDM

POP MACRO
    MOVF    P_TEMP,W
    MOVWF   PCLATH
    MOVF    S_TEMP,W
    MOVWF   STATUS
    MOVF    W_TEMP,W
    ENDM

;   DECLARE PROGRAM ORIGIN
```

```
 ORG 0X00
    CLRF    PCLATH
    GOTO    START

;   INTERUPT SERVICE ROUTINE FOR TIMER0

 ORG 0X04
ISR
    PUSH
    BTFSS   INTCON,T0IF         ; CHECK FOR TIMER0 INTERRUPT
    GOTO    RET_ISR

    BCF INTCON,T0IF             ; Clear timer 0 interrupt flag
    BSF ADCON0,GO_DONE          ; ASSERT START OF CONVERSION
RET_ISR
    POP
    RETFIE

;   MAIN PROGRAM

 ORG 0X30
START
;   INITIALIZATION

;   CONFIGURING PORT PINS

;   RC2/CCP1   - USED FOR PWM OUTPUT

    CLRF    PORTA
    CLRF    PORTB
    CLRF    PORTC
    BANKSEL TRISC
    MOVLW   0X0F
    MOVWF   TRISA               ; CONFIGURE PORTA AS INPUT PORT

;   CONFIGURING ADC

;   AD CLOCK    - FOSC/32
;   AD INPUT    - CHANNEL 0
;   AD FORMAT   - RIGHT JUSTIFIED
;   AD REFERENCE   - VDD-VSS

    BANKSEL ADCON1
    MOVLW   0X8E
    MOVWF   ADCON1
    CLRF    ADRESL
    MOVLW   0XD1            ; TIMER0 IN INTERNAL CLOCK AND PRESCALER 1:4
    MOVWF   OPTION_REG
```

```
        BANKSEL ADRESH
        CLRF    ADRESH
        MOVLW   SAMPLE_LENGTH ; INITIALIZE SAMPLE LENGTH TO COUNT
        MOVWF   COUNT
        MOVLW   SAMPLES ; PUT STARTING ADDRESS OF SAMPLES ARRAY IN FSR
        MOVWF   FSR
        CLRF    TMR0

;    CONFIGURE INTERRUPTS

        MOVLW   0X20    ; ENABLE TIMER0 INTERRUPT
        MOVWF   INTCON
        BSF INTCON,GIE

;    MAIN PROGRAM
MAIN
        MOVLW   CHANNEL0    ; SELECT CHANNEL 0 FOR CONVERSION
        MOVWF   ADCON0
CONVERT
      . BTFSS   PIR1,ADIF  ; CHECK FOR AD COMPLETE (POLL AD
                    ; INTERRUPT FLAG)
        GOTO    $-1

        BCF PIR1,ADIF  ; CLEAR AD INTERRUPT FLAG
        BANKSEL ADRESL
        MOVFW   ADRESL
        BANKSEL ADRESH      ; SAVE LOWER BYTE FIRST AND HIGHER BYTE NEXT
        MOVWF   INDF        ; IN THE SAMPLES ARRAY, IN THE LOCATION
        MOVFW   ADRESH      ; POINTED BY FSR AND FSR+1 RESPECTIVELY.
        INCF    FSR,F
        MOVWF   INDF
        INCF    FSR,F
        DECFSZ  COUNT,F     ; CHECK WHETHER SAMPLE COUNT IS REACHED
        GOTO    CONVERT     ; CONTINUE OTHERWISE
        CLRF    INTCON      ; IF YES, DISABLE ALL INTERRUPTS AND
        GOTO    $           ; HALT
    END
```

```
;*********************************************************************
;     Program_3
;        Notes: This program scans a matrix keyboard (4x4) of 16 keys.
;        portb bits 7:4 is used as scan lines(rows) and bits 3:0 is
;        used for drive lines (columns). The columns are scanned
;        sequentially from col1(RB0) to col4(RB3) and the rows
;        are read in parallel(RB7:RB4) and key numbers are
;        assigned according to their position.
;        If a key is pressed and
;        held, the next column is not scanned until the key is   *
;        released. If multiple keys are pressed together, the    *
;        priority encoding gives higher priority to the key      *
;        which has the highest position number.           *
;        frequency of crystal oscillator : 10Mhz          *
;*********************************************************************
 LIST   P=16F877             ; DEFINE THE PROCESSOR USED
#INCLUDE <P16F877.INC>       ; INCLUDE THE HEADER FILE HAVING THE SFR
                             ; DECLARATIONS

;*********************************************************************
;    SETTING THE CONFIGURATION BITS
 __CONFIG _CP_OFF & _WDT_OFF & _BODEN_ON & _PWRTE_ON & _HS_OSC &
_WRT_ENABLE_OFF & _LVP_OFF & _DEBUG_OFF & _CPD_OFF
;*********************************************************************
;    DECLARE A BLOCK OF VARIABLES IN SCRATCH PAD MEMORY, WHOSE
;    STARTING ADDRESS IS 20H
 CBLOCK 0X30
    TEMP, KEY_DAT, KEY_TMP, SCAN, DEL1, DEL2
 ENDC
;    TEMP IS A REGISTER USED TO STORE TEMPORARY DATA KEY_DAT IS
;    A REGISTER WHICH WILL HOLD THE KEY NUMBER ;CORRESPONDING TO
;    ITS POSITION IN THE MATRIX AFTER THE KEY PRESS HAS BEEN
;    SUCCESSFULLY CONFIRMED.
;    KEY_TMP IS A REGISTER USED TO STORE THE SCANNED DATA FROM PORTB,
;    AS SOON AS A KEY PRESS IS RECOGNIZED. ITS VALUE IS CHECKED
;    ONCE AGAIN
;    WITH A FRESH SCANNED VALUE AFTER A DEBOUNCE
;    DELAY OF 15mS TO CONFIRM A KEY PRESS.
;    SCAN REGISTER HOLDS THE VALUE TO BE PUT ON PORTB TO SCAN
;    THE COLUMNS SEQUENTIALLY i.e.
;    XXX_1000 FOR COL1, XXXX_0100 FOR COL2, XXXX_0010 FOR COL3 &
;    XXXX_0001 FOR COL4
;    DEL1 AND DEL2 ARE USED FOR REGIGTER DELAY
;**********************************
;    SET THE ORIGIN OF THE PROGRAM
 ORG    0X00
    CLRF   PCLATH
    GOTO   START   ; BRANCH TO THE STARTING OF THE MAIN PROGRAM
;    STARTING OF THE MAIN PROGRAM
```

```
 ORG    0X300
START
;*****************************************************************
;       PORTS INITIALIZATION
;*****************************************************************
;   CONFIGURING PORT PINS

    CLRF    PORTA
    CLRF    PORTB
    BANKSEL TRISB
    MOVLW   0XF0    ;MAKE UPPER NIBBLE OF PORTB AS INPUTS AND LOWER
                    ; NIBBLE AS OUTPUTS
    MOVWF   TRISB
    BANKSEL PORTA
    MOVLW   B'00001000'    ;INITIALIZE SCAN REG TO COL1 VALUE
    MOVWF   SCAN


;*****************************************************************
;       KEY BOARD SERVICE ROUTINE
;*****************************************************************
KEY_SERV
    MOVF    SCAN,W
    MOVWF   PORTB   ; MOVE SCAN REG TO PORTB
    NOP
    MOVF    PORTB,W ; READ PORTB AND SAVE IT IN TEMP
    MOVWF   TEMP
    ANDLW   0XF0
    BTFSC   STATUS,Z   ; CHECK WHETHER ANY KEY IS PRESSED (UPPER

                       ; NIBBLE IS NONZERO)
    GOTO    SCAN_NEXT  ; IF NO GOTO SCAN NEXT COLUMN

    MOVF    TEMP,W     ; STORE TEMP IN KEY_TMP
    MOVWF   KEY_TMP
    CALL    DELAY_15MS ;WAIT FOR DEBOUNCE DELAY 15MS
    MOVF    PORTB,W    ; READ PORTB AGAIN
    XORWF   KEY_TMP,W  ; COMPARE THE READ VALUE WITH KEY_TMP
    BTFSS   STATUS,Z
    GOTO    KEY_SERV   ;IF NOT EQUAL, CONTINUE WITH NEXT SCAN

;----KEY PRESS IS CONFIRMED
    BTFSC   KEY_TMP,7  ; IF EQUAL, FIND THE CORRESPONDING KEY
                  ; POSITION
    MOVLW   D'01'      ; NUMBER USING THE BITS SET IN THE
                  ; KEY_TMP AND
    BTFSC   KEY_TMP,6  ; AND SAVE IT IN KEY_DAT
    MOVLW   D'05'
    BTFSC   KEY_TMP,5
```

```
        MOVLW   D'09'
        BTFSC   KEY_TMP,4
        MOVLW   D'13'
        BTFSC   KEY_TMP,2
        ADDLW   0X01
        BTFSC   KEY_TMP,1
        ADDLW   0X02
        BTFSC   KEY_TMP,0
        ADDLW   0X03
        MOVWF   KEY_DAT      ;THE NUMBER OF THE KEY PRESSED IS
                             ;STORED IN KEY_DAT

; --------ACTION TO BE TAKEN CORRESPONDING TO THE KEY PRESSED

        GOTO    KEY_SERV     ;CONTINUE WITH NEXT SCAN
SCAN_NEXT
        RRF SCAN,F           ;ROTATE SCAN REG TO RIGHT ONCE TO
                             ;SELECT NEXT COLUMN
        BTFSS   STATUS,Z     ;CHECK WHETHER LAST COLUMN HAS BEEN
                             ;SCANNED (SCAN == 0)
        GOTO    KEY_SERV
        MOVLW   B'00001000'  ;IF YES, RESET SCAN REG TO FIRST
                             ;COLUMN
        MOVWF   SCAN
        GOTO    KEY_SERV     ; CONTINUE WITH KEY SERVICE

;*****************************************************************
;   DELAY FOR 15mS USED FOR KEY DEBOUNCING
;       FOSC    10MHz
;   FCY =   ---- = ----- = 2.5MHz, WHICH GIVES TCY = 0.4µS
;   44 NO OF INSTRUCTION CYCLES TO BE EXECUTED TO GET
;   15mS    = (15mS /0.4µS) = 37500
;           = 0x927C = 0x96 * 0xFA

DELAY_15MS
        MOVLW   0X96      ; load 96h in del1
        MOVWF   DEL1
LOOP
        MOVLW   (0XFA / 3)-1  ; load 0xFA/3 in del2
        MOVWF   DEL2
        DECFSZ  DEL2,F ; inner loop - decrement del2 until it becomes
                       ; zero
        GOTO    $-1
        DECFSZ  DEL1,F ; outer loop - decrement del1 until it becomes
                       ; zero
        GOTO    LOOP   ; ie. repeat inner loop del1 times.
        RETURN         ; return
    END
```

```
;*****************************************************************
;   Program_4
;       Notes: This program generates a pulse train on the pin
;       portb,7. The period of the pulse is decided by the
;       value loaded into timer1 and the pulse width is decided
;       by the value loaded into timer 0.
;       Frequency of oscillation        - 10 MHz
;*****************************************************************

 LIST  P=16F877   ;DEFINE THE PROCESSOR USED
#INCLUDE <P16F877.INC>    ; INCLUDE THE HEADER FILE HAVING THE SFR
DECLARATIONS

;*****************************************************************
; SETTING THE CONFIGURATION BITS

 __CONFIG _CP_OFF & _WDT_OFF & _BODEN_ON & _PWRTE_ON & _HS_OSC &
_WRT_ENABLE_OFF & _LVP_OFF & _DEBUG_OFF & _CPD_OFF
;*****************************************************************

#DEFINE OUTPUT         PORTB,7
#DEFINE PERIOD         0XC350 ; VALUE CORRESPONDING TO 20mS
#DEFINE PULSE_WIDTH  0XFA   ; VALUE CORRESPONDING TO 100µS

; SET THE ORIGIN OF THE PROGRAM
 ORG 0X00
   CLRF    PCLATH
   GOTO    START

; INTERRUPT SERVICE ROUTINES
 ORG 0X04
ISR                        ; INTERRUPT SERVICE ROUTINE
   BTFSC  PIR1,TMR1IF     ; CHECK FOR TIMER1 INTERRUPT
   GOTO   TMR1_ISR
   BTFSS  INTCON,T0IF
   RETFIE

TMR0_ISR                   ; timer overflow time = 102.4µS
   BCF INTCON,T0IF         ; Clear timer 0 interrupt flag
   BCF OUTPUT
   BCF INTCON,T0IE
   RETFIE

TMR1_ISR                   ; timer overflow time = 26.2144mS
   BCF PIR1,TMR1IF         ; Clear timer 1 interrupt flag
   BSF OUTPUT
   MOVLW  LOW(~PERIOD)
   MOVWF  TMR1L
   MOVLW  HIGH(~PERIOD)
```

```
        MOVWF    TMR1H
        MOVLW    ~PULSE_WIDTH
        MOVWF    TMR0
        BCF INTCON,T0IF
        BSF INTCON,T0IE
        RETFIE                  ; Return from interrupt

 ORG 0X30
START
;               INITIALISATION

; CONFIGURING PORT PINS

    CLRF    PORTA           ; Clear port pins
    CLRF    PORTB
    BANKSEL TRISA
    CLRF    TRISB
    CLRF    TRISA

; CONFIGURING TIMER

    MOVLW   0X40            ;PERIPHERAL INTERRUPT ENABLED
    MOVWF   INTCON
    MOVLW   0XDF            ;timer0 internal clock , no prescaler
    MOVWF   OPTION_REG
    BSF PIE1,TMR1IE         ;TIMER 1 INTERRUPT ENABLED
    BANKSEL T1CON
    CLRF    TMR1L
    CLRF    TMR1H
    MOVLW   0X01
    MOVWF   T1CON
    BSF INTCON,GIE      ; Enable global interrut
    GOTO    $           ; Stop
END
```

```
;*********************************************************************
;    Program_5
;    Notes: This program detects the rising edge of a digital
;    input and sets a digital output high after a fixed time
;    delay. The time delay is achieved by one overflow time of the
;        timer0 without prescaler.
;        The digital input is taken as the external interrupt
;        INT.
;        OUTPUT = PORTB,7
;        Frequcy of oscillation         - 10 MHz
;*********************************************************************

 LIST   P=16F877              ;DEFINE THE PROCESSOR USED
#INCLUDE <P16F877.INC>    ; INCLUDE THE HEADER FILE HAVING THE SFR
;DECLARATIONS


;*******************************************************************
; SETTING THE CONFIGURATION BITS

__CONFIG _CP_OFF & _WDT_OFF & _BODEN_ON & _PWRTE_ON & _HS_OSC &
_WRT_ENABLE_OFF & _LVP_OFF & _DEBUG_OFF & _CPD_OFF
;*******************************************************************


 #DEFINE OUTPUT PORTB,7

; SET THE ORIGIN OF THE PROGRAM
 ORG 0X00
    CLRF    PCLATH
    GOTO    START

; INTERRUPT SERVICE ROUTINES
 ORG 0X04
ISR                          ; INTERRUPT SERVICE ROUTINE
    BTFSC   INTCON,INTF
    GOTO    INT_ISR
    BTFSS   INTCON,T0IF     ; CHECK FOR TIMER0 INTERRUPT
    RETFIE

TMR0_ISR                     ; timer overflow time = 102.4µS
    BCF INTCON,T0IF          ; Clear timer 0 interrupt flag
    BSF OUTPUT
    BCF INTCON,T0IE
    RETFIE                   ; Return from interrupt

INT_ISR
    BCF INTCON,INTF
    CLRF    TMR0
    BCF INTCON,T0IF
    BSF INTCON,T0IE
```

```
       RETFIE

 ORG 0X30
START
;    INITIALISATION

;    CONFIGURING PORT PINS

       CLRF    PORTA          ;Clear port pins
       CLRF    PORTB
       BANKSEL TRISA
       MOVLW   0X01
       MOVWF   TRISB          ;RB0 - INPUT
       CLRF    TRISA

; CONFIGURING TIMER

       MOVLW   0XDF           ;timer0 internal clock , no prescaler
       MOVWF   OPTION_REG
       MOVLW   0X10
       MOVWF   INTCON         ;ENABLE EXTERNAL INTERRUPT INT
       BANKSEL TMR0
       CLRF    TMR0
       BSF INTCON,GIE         ;Enable global interrut
       GOTO    $              ;Stop
END
```

```
;****************************************************************
;    Program_6
;         Notes: This program detects the rising edge of a digital
;         input and  sets a digital output high after a fixed time
;         delay. The time delay is achieved by one overflow time of the
;         timer1 without prescaler.
;         the digital input is taken as the external interrupt
;          INT.
;         OUTPUT = PORTB,7
;         Frequncy of oscillation        - 10 MHz
;****************************************************************

 LIST   P=16F877   ;DEFINE THE PROCESSOR USED
#INCLUDE <P16F877.INC>    ; INCLUDE THE HEADER FILE HAVING THE SFR
;DECLARATIONS


;****************************************************************
; SETTING THE CONFIGURATION BITS

 __CONFIG _CP_OFF & _WDT_OFF & _BODEN_ON & _PWRTE_ON & _HS_OSC &
_WRT_ENABLE_OFF & _LVP_OFF & _DEBUG_OFF & _CPD_OFF
;****************************************************************


 #DEFINE OUTPUT PORTB,7

;   SET THE ORIGIN OF THE PROGRAM
 ORG 0X00
    CLRF    PCLATH
    GOTO    START


;    INTERRUPT SERVICE ROUTINES
 ORG 0X04
ISR              ; INTERRUPT SERVICE ROUTINE
    BTFSC   INTCON,INTF
    GOTO    INT_ISR
    BTFSS   PIR1,TMR1IF    ; CHECK FOR TIMER1 INTERRUPT
    RETFIE

TMR1_ISR                    ; timer overflow time = 26.2144mS
    BCF PIR1,TMR1IF         ; Clear timer 1 interrupt flag
    BSF OUTPUT
    BCF T1CON,TMR1ON        ; STOP TIMER
    RETFIE                  ; Return from interrupt


INT_ISR
    BCF INTCON,INTF
    CLRF    TMR1L
    CLRF    TMR1H
    BSF T1CON,TMR1ON
```

```
      RETFIE

  ORG 0X30
START
;                INITIALISATION

; CONFIGURING PORT PINS

     CLRF    PORTA  ; Clear port pins
     CLRF    PORTB
     BANKSEL TRISA
     MOVLW   0X01
     MOVWF   TRISB  ; RB0 - INPUT
     CLRF    TRISA

; CONFIGURING TIMER

     MOVLW   0X50    ;EXTERNAL INTERRUPT ENABLED
     MOVWF   INTCON
     BSF PIE1,TMR1IE
     BANKSEL T1CON
     CLRF    T1CON
     BSF INTCON,GIE ; Enable global interrupt
     GOTO    $       ; Stop
END
```

```
;******************************************************************
;   Program_7
;        Notes: This program transmits a set of 20 values (WORDS)
;        stored in  the scratch pad memory through USART serial port
;        continuously.
;        9 bit (along with parity) asynchronous transmission is used.
;        Transmission baudrate = 9.6Kbaud
;   OSCILLATOR FREQUENCY = 10 MHz
;******************************************************************

 LIST   P=16F877    ;DEFINE THE PROCESSOR USED
#INCLUDE <P16F877.INC>    ; INCLUDE THE HEADER FILE HAVING THE SFR
;DECLARATIONS


;****************************************************************
; SETTING THE CONFIGURATION BITS

 __CONFIG _CP_OFF & _WDT_OFF & _BODEN_ON & _PWRTE_ON & _HS_OSC &
_WRT_ENABLE_OFF & _LVP_OFF & _DEBUG_OFF & _CPD_OFF
;****************************************************************
 CBLOCK 0X20
    TX_DATA      ;BLOCK OF 20 WORDS (40 BYTES) STARTING AT 20H
 ENDC

TEMP1   EQU 0X50
TEMP2   EQU 0X51
COUNT   EQU 0X52

 #DEFINE DATA_LENGTH D'20'*2 ; LENGTH OF DATA IN BYTES

 ORG 0X00
    CLRF    PCLATH
    GOTO    START


 ORG 0X20
START
;            INITIALISATION

; CONFIGURING PORT PINS
;   RA6/TX - USED FOR SERIAL TRANSMISSION
;   ALL OTHER PORT PINS ARE DIGITAL OUTPUTS

    CLRF    PORTA   ; Clear port pins
    CLRF    PORTB
    CLRF    PORTC
    BANKSEL TRISC
    CLRF    TRISC
    CLRF    TRISB   ; All are output pins
```

```
;    CONFIGURING SERIAL PORT
;
;    9 BIT TRANAMISSION USED
;    ASYNCHRONOUS MODE
;    HIGH SPEED BAUD RATE (BRGH) = FOSC/(16(SPBRG+1)) = 9.6 KBAUD

    BANKSEL SPBRG
    MOVLW   D'64'
    MOVWF   SPBRG
    MOVLW   B'01100110'
    MOVWF   TXSTA

    BANKSEL RCSTA       ; ENABLE SERIAL PORT
    MOVLW   B'10000000'
    MOVWF   RCSTA

; MAIN PROGRAM
    MOVLW   TX_DATA     ; LOAD STARTING ADDRESS OF DATA TO FSR
    MOVWF   FSR
    MOVLW   DATA_LENGTH    ; LOAD LENGTH OF DATA TO COUNT
    MOVWF   COUNT
;               UPDTE
;   CHECK PARITY AND TRANSMIT
    MOVFW   INDF
    MOVWF   TEMP1       ; FETCH THE DATA INTO TEMP1
    CALL    PARITY      ; CALL PARITY GENERATION SUBROUTINE
    MOVFW   INDF
    MOVWF   TEMP1       ; GET DATA IN TEMP1
    CALL    TRANS       ; CALL TRANSMIT SUBROUTINE
    INCF    FSR,F       ; INCREMENT FSR
    DECFSZ COUNT,F      ; CHECK WHETHER ALL DATA ARE
                ; TRANSMITTED
    GOTO    UPDTE
    GOTO    $          ; STOP

; SUBROUTINE FOR PARITY GENERATION

PARITY
    MOVLW   0X08     ;LOAD A COUNT 8(NO OF BITS) IN TEMP2
    MOVWF   TEMP2
    CLRW             ; CLEAR WORK AND CARRY
    CLRC
L1
    RRF TEMP1,F ; ROTATE DATA AND CHECK FOR CARRY
    BNC L2      ; CARRY WILL CONTAIN THE LSB OF DATA
    XORLW   0X01
    GOTO    L3 ; IF CARRY IS '1' XOR '1' WITH WORK L2
    XORLW   0X00    ; IF CARRY IS '0' XOR '0' WITH WORK
L3
```

```
        DECFSZ TEMP2,F ; CONTINUE TILL ALL 8 BITS ARE OVER
        GOTO   L1
        MOVWF  TEMP2
        BTFSS  TEMP2,0 ; CHECK THE LSB OF WORK AND MOVE THE PARITY BIT
        GOTO   L4      ; TO 9TH BIT FOR TRANSMISSION IN TXSTA
        BANKSEL TXSTA
        BSF TXSTA,TX9D
        GOTO   L5
L4
        BANKSEL TXSTA
        BCF TXSTA,TX9D
L5
        BANKSEL TEMP1
        RETURN     ; RETURN

; SUBROUTINE FOR TRANSMITTING THE DATA

TRANS
        BANKSEL TXSTA
        BTFSS  TXSTA,TRMT ; CHECK FOR TRANSMIT BUFFER EMPTY
        GOTO   $-1
        BANKSEL TEMP1      ; IF EMPTY, LOAD DATA TO BE TRANSMITTED
                           ; INTO TXREG
        MOVF   TEMP1,W
        MOVWF  TXREG
        RETURN          ; RETURN
    END
```

CHAPTER 13

8051 SERIES OF CONTROLLERS

13.1 INTRODUCTION

Different companies offer μCs which are enhanced versions of a basic μC –8031. All have the same processor core and the same instruction set [6]. They differ in details of number of internal registers, types and number of peripherals built in, size of internal program memory and possibilities of extended hardware layout.

The PIC® 16 series discussed in the last few chapters too have similar possibilities. The differences are essentially in details. The PIC® series is organized as a RISC Processor. It has a compact instruction set. It enables the designer or programmer to commit all instructions to memory easily and recall each with equal ease. Further the instruction set is of a 'minimal' nature. An instruction used in a context cannot be easily replaced by another without increasing the size or extent of one or more of the resources used. Such increase can be in terms of increase in number of instructions, execution time, number of registers used, or their combinations. Hence the instruction set is also called an 'orthogonal instruction set'. With each step being of such a minimal nature, it is reasonable to expect the overall firmware too to be optimal (This is not necessarily true; recall that steepest descent is not necessarily the shortest or optimal descent!).

The 8051 family and many other processors are of the CISC category. A CISC processor has a comprehensive instruction set. Some of the instructions – like those used for multiplication or division of bytes are compact subroutines. If the instructions for a task are selected with enough care at the time of coding, the firmware realized becomes compact. It requires the designer or programmer to be fully familiar with the instruction set; he should be able to recall each with equal ease. A programmer's selection of instructions for a task often proceeds as a reflex action. Hence, his selection of an instruction or a group of instructions to execute a task need not necessarily be optimal. In fact, a RISC implementation is justified on the premise that a programmer uses only 25% to 35% of the instructions of the comprehensive instruction set of a CISC processor in the firmware designs.

13.2 ARCHITECTURE OF 8051 PROCESSOR

The processor in the 8051 series is shown in Figure 13.1 in block diagram form [6]. The PC is directly tied to the program memory through a Program Address Register (PAR). An 8-bit data bus is the common bus for data transfer amongst different registers and ALU. The timing and control unit provides the control signal for all the functional units. The temporary register tied to the accumulator and the program memory register tied to the program memory are not accessible for reading or writing. The program memory is accessible only for reading; it is in write mode only when the program is being written into it. Other units are accessible for reading and writing.

13.2.1 Program Memory

Every location in program memory is one byte wide. The simplest version of the Controller (8031) does not have program memory within it. All other versions have it – it is 4K long in the 8x51 series. In others, it can extend up to 16 K bytes. Further, all the versions have the facility to extend the program memory to a segment outside. Simplest of the instructions are one byte long. Others are of two or three bytes length. Depending on the instruction length, the PC gets incremented automatically for each instruction.

13.2.2 Data Memory

The data memory can be only internal or extended to an externally provided RAM. Many applications can be realized without the need for external RAM. The internal RAM has different segments each with its own features; a low end µC has 128 bytes of internal RAM.

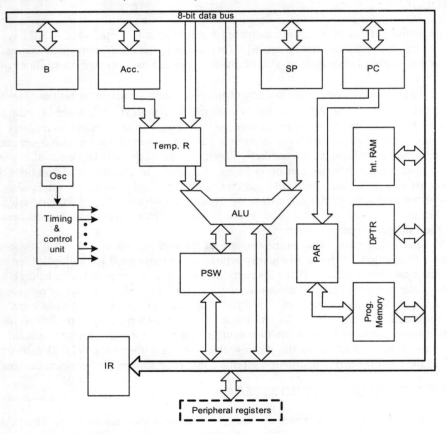

Figure 13.1 Block diagram of the 8051 processor core

13.2.2.1 Internal RAM

The internal RAM is of 128 bytes; any of them can be specified and selected by specifying the absolute address – 7 bits for 128 byte RAM. The RAM area is segmented as shown in Figure13.2. The segmentation is based on modes of accessing for instruction execution. The area from 000h to 01Fh is divided into four banks – each of 8 bytes. The bank address is specified by 2 bits of a Program Status Word (PSW); PSW is discussed later. When a specific register within a bank is to be addressed, a 3-bit address – which can span the 8 locations – is used for the same. Thus the accessing is done in two parts. Normally the active bank address is kept loaded and ready and the registers of interest are (if possible) within that. If one has to access a register in a 'dormant' bank, its bank address is brought into the PSW before the register concerned is addressed. The RAM area – 020h to 02Fh – is bit addressable. A set of bits here can be earmarked as flags, bits to represent status, give commands, etc. Each of them – 128 in total – can be accessed separately for instruction execution.

13.2.3 Special Function Registers (SFR)

The registers A, B, PSW, DPTR, and SP are assigned specific addresses in the RAM area. Some other registers with specific functions related to the peripherals are also assigned such addresses. All these registers are called SFRs. All SFR addresses are in the 080h to 0FFh range in 8051. Details of the registers and their respective address assignments are given in Table 13.1. Note that all addresses are not assigned here. Some devices in the family which have enhanced features have additional registers with address assignment in the SFR area. Some of the SFRs require direct access to specific bits in them. They are grouped together separately and provided with direct bit address; these are not to be confused with the bit addressable region in the scratchpad area.

13.2.4 Addressing Modes

The addressing mode is implicit in some instructions like those involving the accumulator, (in some cases) B register, Data Pointer, etc. Apart from these four addressing modes are possible for different instructions.

13.2.4.1 Direct Addressing

The operand address is available in the instruction itself. In case the operand is within the active bank of 8 registers in the internal RAM, the address can be specified by three bits of the instruction itself. As an example consider the instruction.

```
MOV A, Rn
```

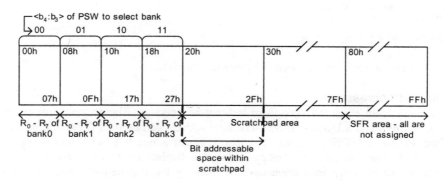

Figure 13.2 Segmentation of internal RAM: Details shown are for 8051

It signifies the operation

$$(R_n) \rightarrow (A)$$

that is, the content of register R_n is moved into A register.

The instruction is encoded as the byte

1110 1 fff

The 3 bits 'fff' specify the source register within the active bank. The active bank is specified separately by the two bits in PSW assigned for it.

The internal RAM and SFR area can be accessed directly with one byte long address. For example consider the instruction

MOV   A, RR

where RR is a byte

The instruction is coded as

1110 0101 RR

The operation carried out is

$$(RR) \rightarrow (A)$$

Content of internal RAM location or SFR as the case may be is brought and loaded into the accumulator. The instruction is 2 bytes long.

Table 13.1 Addresses of SFRs – all addresses are hex values

SFR	Acc.	B	PSW	IP	IE	P0	P1	P2	P3
Address	E0	F0	D0	B8	A8	80	90	A0	B0
SFR	SCON	SBUF	TCON	TMOD	TL0	TL1	TH0	TH1	SP
Address	98	99	88	89	8A	8B	8C	8D	81
SFR	DPL	DPH	PCON						
Address	82	83	87						

13.2.4.2 Indirect Addressing

In the indirect mode of addressing the address of data concerned is specified indirectly. One of the SFRs specifies the address. Different possibilities exist. Details are discussed later

13.2.4.3 Immediate Addressing

The data to be used is specified as part of the instruction itself. The data byte follows the instruction opcode. Consider the instruction

MOV A1 # 064h

It is coded with two successive bytes as

0111 0100 0110 0100

The operation carried out is

$$064h \rightarrow (A)$$

The instruction is two bytes long, the second byte being the data.

13.2.4.4 Indexed Addressing

Indexed addressing is an extended version of indirect addressing. The address concerned is in two parts – a Base Address and a Displacement. The base address can be the PC content or the content of the 16-bit Data Pointer (DTPR) Register. The displacement is one byte wide and is specified within the instruction itself. The address is formed by adding the displacement to the base address. The byte-wide displacement follows the instruction opcode, making the instruction a two byte one. But the address formed is 2 bytes wide since the base address itself is two bytes wide.

13.2.5 Registers for Program Control

The SFRs mentioned earlier are of three categories:

- the accumulator (referred as A register) and the B register are part of the Scratch Pad Register available within the μC (See Table 13.1). The accumulator is in close link with the ALU; its privileged status makes it the most versatile of the registers. For many of the instructions one of the operands is in A; it is implied in the opcode itself and need not be separately specified. The B register is used in a similar manner for Multiply and Divide instructions.
- A set of registers dedicated to the peripherals are of the second category. Some of them decide the assignment details of the peripherals and the others reflect the peripherals' status on a continuing basis. These registers are discussed along with the peripherals concerned.
- The third category aid program flow in different ways – The Program Status Word (PSW), the Data Pointer (DPTR), and the Stack Pointer are the three registers of this category.

13.2.5.1 PSW

The bit assignments of PSW are shown in Figure13.3. The bits b_0, b_2, b_6 and b_7 reflect the result of instruction execution.

- b_0 is the parity bit. With every instruction executed, the parity bit is refreshed. It is set to '1' state if the total number of 1-bits in the accumulator is odd; else it is reset to '0' state. It can be used for all parity checks directly.
- b_2 is the overflow flag. An overflow as a result of an arithmetic operation decides the flag status.
- The bit pair – <b4 : b3> – decide the active bank. The value sets 11, 10, 01 and 00 represent addresses of bank B3, B2, B1 and B0 respectively. The active bank can be switched from one to another by reloading these bit values.
- b_6 is the auxiliary carry (AC) bit; arithmetic operations that cause a carry bit generation from b_3 to b_4 change the auxiliary carry bit status. The AC flag is used in BCD algebra.
- b_7 is the carry flag. Arithmetic operations that result in a carry bit generated from MSB position or those which cause a bit to be borrowed to the MSB position change the carry bit status. Carry bit is used for different types of tasks including multi-byte algebra.
- The bits b_1 and b_5 have no specific assignment. Their function can be defined by the user and used accordingly; they are like the other addressable bits in the internal RAM area.

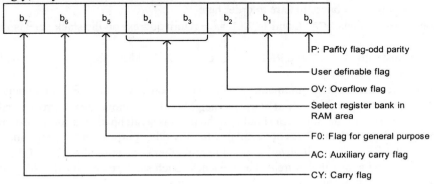

Figure 13.3 Bit assignments of PSW register

13.2.5.2 DPTR

A pair of registers – designated DPH and DPL – together forms the data pointer. The two registers can be independently addressed and altered as any other scratchpad register. In addition the register pair taken together – designed DPTR – can be used to store a 16 bit address. It is used by some instructions in indirect mode and others in indexed mode.

13.2.5.3 Stack Pointer

A selected segment of the internal RAM is kept apart for stack. The µC has an SFR dedicated to the stack called the 'Stack Pointer (SP)'. The SP stores the current address of the top of the stack. At the time of Power-ON and at Reset, the µC initializes the SP to 0x07h of the internal RAM. Left to itself, the stack functions from that location. µC uses the stack to save return address as well as to retrieve it. At every subroutine call, the SP is incremented to point to the next available location on the stack and the PC content (2 bytes) stored in 2 successive locations in the stack. Again the SP points to the top of the stack. Similarly when returning from the subroutine, the stored address is retrieved from the top of the stack and SP decremented to point to the top of the stack. For example consider execution of the instruction.

ACALL   ABD

The instruction is two bytes long. It is encoded as

a10	a9	a8	1	0	0	0	1	a7	a6	a5	a4	a3	a2	a1	a0

Here the bits <a10:a0> represent the 11 LS bits of the subroutine starting address. The MS bits are those of the PC itself. Execution of the instruction result in the following sequence of operations

(PC)	(PC) + 2
(SP)	(SP) +1
((SP))	(PC<b0:b7>)
(SP)	(SP) +1
((SP))	(PC<b8:b15)
(PC10 : 0)	new address specified <a10: a0 >

Note that the return address of 2 bytes has been saved on the stack and SP made to point to the new top of the stack. Execution of a RET(urn) instruction retrieves the saved address from the stack and loads PC with it. SP is decremented twice to point to the new value of top of the stack. The following observations regarding stack operation are in order here.

- Execution of instructions ACALL and LCALL save return address on the stack. SP is incremented twice.

- Execution of instructions RET and RETI retrieve the saved address from stack and load it into the PC: SP is decremented twice.

- Interrupt service starts with an automatic LCALL instruction execution. SP is incremented twice.

- Since SP is initialized to 0x007h at start, the neighbourhood above this address is taken as the stack. To avoid any mix-up, this area should not be used as scratchpad area for the program. It also means that (at least) the bank R0 is not available for any other purpose in the program.

- If necessary, the SP can be initialized to point to any other convenient location of the RAM area. The neighbourhood above that is treated as the stack. Such stack relocation should be done in the program as part of initialization.

13.2.5.4 System Clock and Machine Cycles

The system clock can be generated internally through a crystal oscillator or supplied from an external source. Each instruction is fetched and executed in one or more machine cycles. Duration of each such machine cycle is 12 clock periods. The processor operation takes place in six phases – designated φ1, φ2, φ3, φ4, φ5, and φ6. Each such phase lasts for two successive clock periods (See Figure 13.4). There are two instructions – MUL and DIV of- of four machine cycle duration each. All other instructions are fetched and executed in one or two consecutive machine cycles. These are again of 3 categories.

- One byte one cycle instruction
- One byte two cycle instruction
- Two bytes one cycle instruction

The fetch-execute sequence for all the three categories of instructions are shown in Figure 13.4; in each case specific instructions are taken for illustrative purpose. Operation is characterized by the following:

- In each case the first byte of the instruction (B1) is fetched during phase φ1.
- In the following φ4, the next byte (B2) in program memory is also fetched routinely.
- If the instruction is of the single byte single cycle category – like INC A – the second byte B2 (opcode of the following instruction) – though – fetched, is discarded. Instruction execution is continued and completed within φ6. Byte fetching from program memory and instruction execution after decoding are carried out by different hardware segments of the μC. They can function in parallel. Hence fetching byte B2 and discarding it takes place concurrently with instruction execution.
- If the instruction is of one byte two cycle category – like INC DPTR – B2 fetched during φ4 is discarded. Further, the same byte will be fetched in the next machine cycle – during the φ1 phase and φ4 phase as well. Since execution of instruction extends over the whole of this second machine cycle, B2 will be discarded in both the cases. In total the same byte may be fetched four times – three to be fetched and discarded and the fourth time to be fetched and executed. As mentioned earlier the parallel hardware streams for program memory fetch and execution facilitate this without the need for extending the time for execution.
- The third category of instructions – two bytes one cycle like ADD 0x34h – use the byte B2 fetched during φ4, for execution. Since the second byte is also of the same instruction, the opcode for the following instruction is fetched in the next machine cycle, for execution. It requires the PC to be incremented once again before the instruction is fetched. Amongst the categories considered, this is the only case where the second byte fetched is used and not discarded.
- In applications where the μC accesses external memory, the execution sequence can be different. It is not dealt with here.

13.3 INTERRUPTS

An interrupt to the μC can be from five sources; two of these are generated internally from the two timer peripherals. The third one is from the serial communication peripheral – it can be from the transmitter side or receiver side. Apart from these three internally generated interrupts, the μC has the provision to accept two external interrupts. All the three internal interrupts are generated by respective flags going high. In the case of the timers, the interrupt flag is reset internally and automatically when the μC embarks on the related service routine. The interrupt flag from the serial communication unit is not reset automatically. It has to be cleared in software.

The external interrupts – INT0 and INT1– have an added level of flexibility. Either can be made sensitive to level or to change. When INT0 is made level sensitive, its transition to '0' state signifies an interrupt. The μC may sense it and go through the service routine. But the interrupting source has to clear the interrupt request; the request will remain active until that time. If the interrupt is made edge sensitive; a 1 to 0 transition is latched by the μC as an interrupt flag. The same is automatically reset when the µC embarks on the service routine. The second interrupt –INT1– too can be activated in two ways; it can be made edge sensitive or level sensitive. Two bits – designated IT0 and IT1- in the TCON register (an SFR) decide the assignments to the two external interrupts.

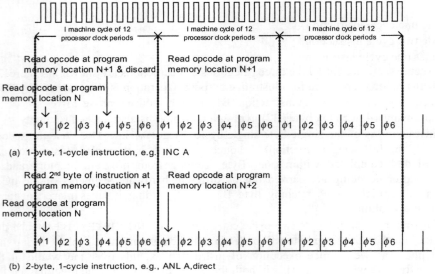

(a) 1-byte, 1-cycle instruction, e.g. INC A

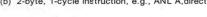

(b) 2-byte, 1-cycle instruction, e.g., ANL A,direct

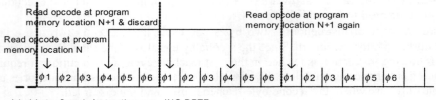

(c) 1-byte, 2-cycle instruction, e.g., INC DPTR

Figure 13.4 Instruction execution sequences for different types of instructions

13.3.1 Interrupt Masks

The µC has an Interrupt Enable Register (SFR); five of its bits form enable bits of the five interrupts; a sixth bit is a global enable bit. Each interrupt can be selected separately and masked or enabled; similarly, the whole set of interrupts can be masked by resetting the global enable bit.

13.3.2 Interrupt Prioritizing

The facility to selectively mask the interrupts allows prioritizing. Apart from this the interrupt response structure allows two tiers of prioritizing. The interrupt source can be selectively assigned to one of two layers – a high priority layer and low priority layer (See Figure 13.5).

- On receipt of an interrupt request (if global enable bit is set) the µC scans the high priority layer first, for valid interrupts. All the valid requests there are responded to first. Only after completion of such responses, the µC responds to the valid requests in the low priority layer. Responding selectively in the two layered manner is one tier of prioritizing. The second tier is in the polling for service. The valid interrupts are polled in a predetermined sequence for response. The polling sequence is shown in Figure 13.5.

- Masking, assignment to one of the two layers, and prioritizing these two together can be used judiciously to prioritize response in different ways. A few possibilities are given here by way of illustration.

- All the five interrupts can be assigned high priority (or all assigned low priority). The serial port interrupt will have the highest priority and external interrupt 0 will have the lowest priority – all decided by polling

- External interrupt 0 can be assigned high priority level. All others can be assigned low priority. μC will give first priority to external interrupt 0; the second priority will be to serial port interrupt and so on.

- Serial port interrupt can be assigned to low priority level and all others to high priority level. With such an assignment, the serial port interrupt will be given lowest priority.

- All except external interrupt 0 may be masked by resetting respective enable bits. μC will respond only to the external interrupt 0. Once the service routine is completed, other interrupts can be enabled selectively.

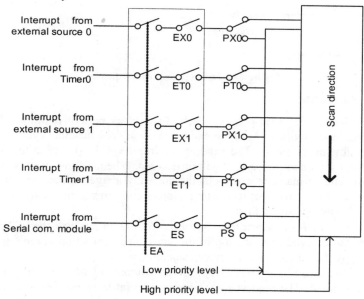

Figure 13.5 Interrupt enabling & prioritizing: EX0, ET0, EX1, ET1, and ES are the respective interrupt enable bits. EA is the global enable bit. PX0, PT0, PX1, PT1, and PS assign the interrupts to high (if the bit value is 1) or low (if the bit value is 1) level.

13.3.3 Interrupt Vectoring

Each interrupt is assigned a starting address in the program memory as shown in Table 13.2. When the μC responds to an interrupt request, it saves the return address and loads the PC with the starting address of the service routine. The priority scheme discussed above decides the interrupt source to be attended to. Depending on the source, μC loads the PC with the starting address of the service routine. The starting addresses are spaced eight bytes apart. If the service routine is short enough, it can be accommodated locally; else through a jump instruction the µC has to transfer control to the service routine positioned suitably in the program memory area.

The power-on reset initializes the PC to the first location in program memory – namely 0 00h. Since the interrupts vector to locations 0003h to 0023h, it is essential in most applications to locate the main routine above this area. Often it may start at 0030h. The first instruction executed on 'power-on' can be a 'jump to 0030h', located at 000h, to ensure proper start.

Table 13.2 Vector addresses for different interrupt sources

Interrupt source	External-0	Timer-0	External-1	Timer-1	Serial comm.
Vector address	003h	00Bh	013h	01Bh	023h

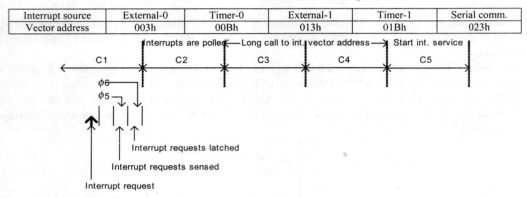

Figure 13.6 Interrupt response sequence

13.3.4 Interrupt Response

Each machine cycle lasts through six phases – each of two clock periods duration. The µC senses the interrupt request flag (See Figure 13.6) at the middle of φ5 of each machine cycle. The interrupt will be latched in the following φ6 phase. The pending interrupts will be polled in the following machine cycle and the interrupt source to be serviced will be decided. The machine cycle is designated as C2 in the figure. If at the end of machine cycle C2, the currently active instruction is fully executed, µC executes a jump to the service routine; its starting address conforms to the values given in Table 13.2. However there are two exceptions to this:

- *The instruction executed is a return from interrupt (RET1)*: in this case the PC will be loaded with the return address retrieved from the top of the stack. The instruction at the returned address will be executed. Only then the interrupt service can commence.
- *The currently active program is an interrupt service routine of higher priority*: in this case the new request will be ignored. The request source has to wait (at least) until the service routine is fully executed.

The interrupt service sequence is shown in the form of a flowchart in Figure 13.7. As a rule the µC completes the currently active instruction before embarking on an interrupt service. Hence the response time ('interrupt latency') can vary depending on the currently active instruction. The following observations are in order here:

- If the instruction under execution during the machine cycle C2 is of one machine cycle duration, the interrupt service will start at machine cycle C5 (after the 'jump to subroutine is executed during C3 and C4). C2 may be the last machine cycle of a currently active instruction: in that case too, interrupt service will commence to machine cycle C5 itself.
- The instruction under execution during machine cycle C2 may extend to C3 or beyond. In such a case interrupt service will be correspondingly delayed. The worst case can happen with the

multiply and divide instructions. If such an instruction had been fetched for execution during the C1 phase, interrupt service will start only during the C7 phase. In general the interrupt response time will be 3 to 9 machine cycles,.

- Once an interrupt is recognized during φ5 of a machine cycle, μC does the polling every time the polling is done anew.
- The interrupt service routine starts with the μC loading its starting address into the PC. During this time, the concerned interrupt flag will be automatically cleared by the μC with some sources. With others, the flag clearing has to be done separately.
- The last instruction to be executed in an interrupt service routine can be RET or RET I. Execution of RET returns control to main program. Execution of RET I returns control to main program if no other interrupt of lower (or same) priority has its service pending or incomplete.
- The μC senses the interrupt flag during φ5 of each machine cycle. Since all internal interrupts are of latched type, they are sure to be sensed. An external interrupt − if programmed to be level sensitive − has to be high for at least one machine cycle to be sensed it.
- If the external interrupt is of the negative edge sensing type, it has to be high for at least one machine cycle and low for at least one machine cycle, to be sensed by the μC.
- With level sensitive external interrupts, if the conditions are not favourable, μC may not respond to the interrupt in the immediately following machine cycles. In such cases the interrupt source has to remain active until μC responds to it; else an interrupt service response may not be generated at all !
- Any status saving for a running routine (prior to embarking on interrupt service) has to be done in software − if necessary at the beginning of the interrupt service routine itself. Often PSW may have to be saved. If necessary, such saving can be hastened by resorting to bank switching within the internal RAM.

13.4 INSTRUCTION SET

8051 and its enhancements have a common instruction set. It is reproduced in Appendix B in summarized form. The detailed operation of each instruction is available in the user manual. Salient features of the instructions are summarized here.

13.4.1 Arithmetic Instructions

All the arithmetic instructions are listed in a compact form in Table 13.3. The operations, addressing modes, and number of machine cycles of execution are also given in the table.
Observations:

- For ADD and SUBtract instructions, one of the operands is A. the other operand is decided by the addressing mode. The result of operation is stored back in A. The instruction is fetched and executed within one machine cycle.
- The INCrement and DECrement instructions can have A as the operand or one of the registers. If a register is the operand, the accumulator is not involved in the execution of the instruction. The instruction is fetched and executed in one machine cycle.
- The INC DPTR instruction increments the 16 bit Data Pointer. Execution lasts for two machine cycles. The instruction is useful for external memory access.
- The MULtiply and DIVide instructions use A and B registers as operands. The results are also stored back in A and B registers. These are the only instructions which take four machine cycles for execution.

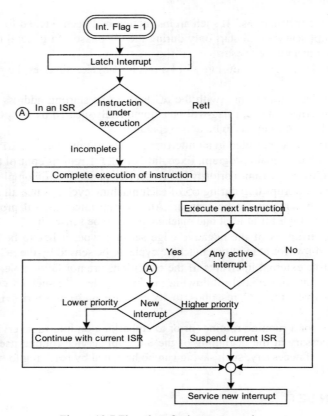

Figure 13.7 Flowchart for interrupt service

Table 13.3 Arithmetic instructions of 8051 Series ('*' signifies availability of the instruction in the mode. The execution time is given in terms of the number of machine cycles)

Mnemonic	Operation	Addressing modes				Execution time
		Dir.	Ind.	Reg	Imm	
Add A, <byte>	(A)+<byte>→(A)	*	*	*	*	1
ADDC A, <byte>	(A)+<byte>+C→(A)	*	*	*	*	1
SUBB A, <byte>	(A)–<byte>–C→(A)	*	*	*	*	1
INC A	(A) + 1 → (A)	Accumulator only				1
INC <byte>	<byte> + 1 →<byte>	*	*	*		1
INC DPTR	(DPTR) + 1 → (DPTR)	Data pointer only				1
DEC A	(A) – 1 → (A)	Accumulator only				1
DEC <byte>	<byte> – 1 →<byte>	*	*	*		1
MUL AB	(B) × (A) → (B):(A)	ACC & B only				4
DIV AB	Int[(A)/(B)] → (A) Mod[(A)/(B)] → (B)	ACC & B only				4
DA A	Decimal adjust	Accumulator only				1

13.4.2 Logical Instructions

The instructions are presented in a compact form in Table 13.4; for each category of instruction, mnemonic, operation, addressing modes, and execution time are given in the table.
Observations:

- Instructions to clear the accumulator, complement accumulator content, rotate accumulator content left or right by one bit position, and the instruction to swap the nibbles are all of one machine cycle duration each. They affect only the accumulator content.
- The instructions for AND, OR, and XOR operations are carried out on a bit by bit basis. There are two categories:
- The first category involves the accumulator and a second operand in one of the modes of addressing. Instruction fetching and execution are completed within one machine cycle.
- The second category involves a data in a register and a byte in immediate mode. Accumulator is left undisturbed. The instructions are 3 bytes long and execution extends to two machine cycles.

Table 13.4 Logical instructions of 8051 Series ('*' signifies availability of the instruction in the mode. The execution time is given in terms of the number of machine cycles)

Mnemonic	Operation	Addressing modes				Execution time
		Dir.	Ind.	Reg.	Imm.	
ANL A, <byte>	(A) AND <byte> → (A)	*	*	*	*	1
ANL <byte>, A	(A) AND <byte> → <byte>	*				1
ANL <byte>, #data	<byte> AND #data → <byte>	*			.	2
ORL A, <byte>	(A) OR <byte> → (A)	*	*	*	*	1
ORL <byte>, A	(A) OR <byte> → <byte>	*				1
ORL <byte>, #data	<byte> OR #data → <byte>	*				2
XRL A, <byte>	(A) XOR <byte> → (A)	*	*	*	*	1
XRL <byte>, A	(A) XOR <byte> → <byte>	*				1
XRL <byte>, #data	<byte> XOR #data → <byte>	*				2
CRL A	00h → (A)	Accumulator only				1
CPL A	NOT (A) → (A)	Accumulator only				1
RL A	Rotate ACC left by 1 bit	Accumulator only				1
RLC A	Rotate ACC left by 1 bit thro' carry	Accumulator only				1
RR A	Rotate ACC right by 1 bit	Accumulator only				1
RRC A	Rotate ACC right by 1 bit thro' carry	Accumulator only				1
SWAP A	Swap nibbles in A	Accumulator only				1

13.4.3 Data Transfer Instructions

All the instructions for data transfer are shown summarized in Table 13.5.
Observations:

- Three of the instructions are to move a specified byte to accumulator or from the accumulator or to exchange data with accumulator. All these are executed within one machine cycle.
- One instruction to move a byte specifies the source and destination addresses directly. It is a three byte instruction executed in two machine cycles. The accumulator is left undisturbed.
- The POP instruction pops byte from stack to the address specified directly. The stack pointer is decremented once correspondingly. The PUSH instruction pushes a byte into the stack from the

specified address directly. SP is incremented subsequently. Both are executed in two machine cycles.

- MOV DPTR instruction loads a 16 bit number (often address) into the DPTR. The number is specified in immediate mode. Execution takes two machine cycles.
- One instruction exchanges the low order nibble of accumulator with the corresponding nibble of R0 or R1 as specified.

Table 13.5 Data transfer instructions of 8051 Series ('*' signifies availability of the instruction in the mode. The execution time is given in terms of the number of machine cycles)

Mnemonic	Operation	Addressing modes				Execution time
		Dir.	Ind.	Reg.	Imm.	
MOV A, <src>	<src> → (A)	*	*	*	*	1
MOV <dest>, A	<dest> → (A)	*	*	*		1
MOV <dest>, src	<dest> → <src>	*	*	*		2
MOV DPTR, #data16	16-bit imm. Constant → (DPTR)				*	2
PUSH <src>	INC SP, ((SP)) → <src>	*				2
POP <src>	<src> →((SP)), DEC SP	*				2
XCH A, <byte>	(A) ↔ <byte>	*	*	*		1
XCHD A, @Ri	(A) ↔ @Ri (only low nibbles)		*			1

Table 13.6 Boolean instructions of 8051 Series ('*' signifies availability of the instruction in the mode. The execution time is given in terms of the number of machine cycles)

Mnemonic	Operation	Execution time
ANL C, bit	(C) AND bit → (C)	1
ANL C, /bit	(C) AND .NOT bit → (C)	1
ORL C, bit	(C) OR bit → (C)	1
ORL C, /bit	(C) OR .NOT bit → (C)	1
MOV C, bit	bit → (C)	1
MOV bit, C	(C) → bit	1
CLR C	0 → (C)	1
CLR bit <byte>	0 → bit	1
SETB C	1 → (C)	1
SETB bit	1 → bit	1
CPL C	NOT C → (C)	1

13.4.4 Boolean Instructions

A set of instructions available to do Boolean algebra on a selected bit, are given in Table 13.6. The carry bit C has the role analogous to that of the accumulator. The bit address is specified in the instruction. It refers to a specified bit in the addressable bit area of internal RAM. Wherever the operation specified involves only the carry bit, execution time is limited to one machine cycle. In all other cases it extends to two machine cycles.

13.4.5 Instructions for Branching

The instructions for program branching change the PC content; execution continues from the new location pointed by PC. Such instructions are of two categories – those for conditional branching and those for unconditional branching; conditional branching instructions are given in Table 13.7 and the unconditional branching instructions in Table 13.8.

Observations:

- SJMP, LJMP and AJMP execute an unconditional jump to the new location; Execution extends to 2 machine cycles. SJMP is 2 bytes long: The jump is relative to the present PC content. The first byte of the instruction is its opcode and the second one the displacement. The displacement byte is treated as a signed 7-bit number. The jump can be to any location within the span of (PC) - 128 to (PC) + 127.

- The LJMP instruction is 3 bytes long; the latter two bytes specify the 16-bit absolute address to which the jump has to take place.

- The AJMP instruction is 2 bytes long; The 11-LS bits of the location are specified in it. Rest of the MS bits remains unaltered. The jump span is 2 k bytes.

- The JMP @A+DPTR executes a jump to a location whose absolute address is computed at execution time. It is a single byte instruction executed in two machine cycles. The instruction is ideally suited to realize 'case jump' through a destination address table.

- CALL and RET are for subroutine call and return from subroutine actions. LCALL specifies a 16 bit full address. The ACALL specifies a 11-bit LS address – similar to AJMP.

- RET is used to return from subroutines and RET1 to return from interrupt services.

- Some instructions in Table 13.7 test the specified bit and execute a relative jump depending on the status of the concerned bit. The destination address span is (PC) - 128 to PC) + 127.

- Other instructions in Table 13.7 check a condition and execute a jump on the condition being satisfied. The jump is relative with a span of (PC)-128 to (PC) + 127

Table 13.7 Conditional branch Instructions in of 8051 Series ('*' signifies availability of the instruction in the mode. The execution time is given in terms of the number of machine cycles)

Mnemonic	Operation	Addressing modes				Execution time
		Dir.	Ind.	Reg.	Imm.	
JZ rel	Jump if (A) = 0	Accumulator only				2
JZ rel	Jump if (A) 0	Accumulator only				2
DJNZ <byte>, rel	Decrement & jump if not zero	*		*		2
CJNE A, <byte>, rel	Jump if (A) <byte>	*			*	2
CJNE <byte>, #data, rel	Jump if <byte> #data		*	*		2
JC rel	Jump if (C) = 1	All these instructions are bit based				2
JNC rel	Jump if (C) = 0					2
JB rel	Jump if bit = 1					2
JNB rel	Jump if bit = 0					2
JBC rel	Jump if bit = 0; clear bit					2

Table 13.8 Instructions for unconditional branching in of 8051 Series (The execution time is given in terms of the number of machine cycles)

Mnemonic	Operation	Execution time
JMP addr	Jump to addr	2
JMP @A+DPTR	Jump to(A)+DPTR	2
CALL addr	Call subroutine at addr	2
RET	Return from subroutine	2
RETI	Return from interrupt	2

13.4.6 NOP Instruction

The NOP instruction is not listed in any table above. It is a single byte instruction. It can serve two purposes: One is to cause an intentional and specific delay in a routine. The second purpose is to fill the gaps in an existing program which has been modified – for example at debugging stage.

13.5 PORTS

The µC has four 8-bit wide ports. All are bi-directional ports. The ports are designated as P0, P1, P2 and P3. Each port has an associated SFR. Writing to the SFR of P0 is equivalent to outputting a byte to port 0. Reading P0 is equivalent to data being input from P0. Similarly, with the other three ports too. Figure 13.8 shows the functional block diagram of the circuit associated with one bit of Port 1 (The circuits of other ports are only marginally different).

When the control signal 'Write to latch' is exercised, the data bit in the data bus line is latched and is available as output. Similarly, the control signal 'Read pin' is exercised to read data input at the port pin. When in output mode, a port pin may be loaded by the device connected to it and its voltage level may droop. Reading it can be erroneous. Hence whenever the port bit is to be read for modification, the port latch output (and not the pin status) is to be read. For all such instructions the µC automatically activates the 'Read Latch' and not the 'Read pin' signal. If necessary the ports P0 and P2 can be used for external memory access – to supply address and to output or input data byte. Two bits of port1 and all the bits of port 3 have alternate assignments. The hardware details of these are more enhanced than those in the figure here; it is to accommodate such alternate functions. Details are available in the user manual.

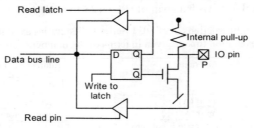

Figure 13.8 Functional block diagram of one bit of Port 1

13.6 TIMERS

The µC has two dedicated registers; each can be made to function as a timer or as a counter. Each has four operating modes – designated as mode 0, mode 1, mode 2, and mode 3. The mode selection and assignment to counter or timer function are programmable. Two SFRs – TMOD and TCON – are assigned for this. The function in each of the modes is explained here.

13.6.1 Modes 0 and 1

Consider the timer 0. The operation conforms to the scheme shown in Figure 13.9. TH1 (8 bits) and the MS five bits of TL1 together form a 13-bit counter register. With every pulse input at point P, the count in the register increases by one. As the register overflows from the all 1's state (01FFFh) to all 0's state, the interrupt flag is set. The other 3 bits of TL1 are not used. The TL1 bits form the 5LS bits of the 13-bit counter; they play the role of a divide by 32 pre-scaler. The b6 bit in TMOD register decides the function as timer or counter. If set as timer, counting takes place once every machine cycle. If set as counter, T1 pin is the input. A 1 to 0 transition on the pin is counted. For a transition to be sensed, the input should be high for at least one machine cycle. After the transition, it should be low for at least one machine cycle. The timer/counter pulses are gated to the counter register, if the TRI control bit is high and GATE = 0 or Int1 = 1. The unit can be used with different combinations of control signals. For example with GATE = 0, one can load a number N into TH1 and set the unit to timer mode. The timer will become active as soon as TR1 is set. TF1 goes high with a delay given by

$$\text{Delay} = T_{mc} \times 32 \times N \text{ seconds}$$

Where T_{mc} is the machine cycle period (= 12 Oscillator periods). With a 12 MHz clock one gets a minimum delay of 32 µs and a maximum delay of 32 ms. Operation in mode 0 is identical for Timer 0 and Timer 1. Mode 1 uses all the 16 bits of TL and TH registers cascaded. In all other respects it is identical to Mode 0.

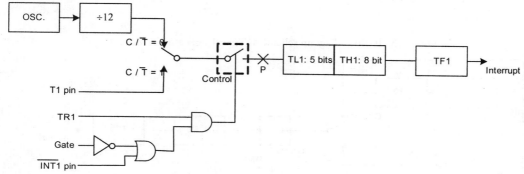

Figure 13.9 Block diagram scheme of Timer1 operating in Mode 0

13.6.2 Mode 2

The timer/counter functional block diagram operating in Mode 2 is shown in Figure 13.10. The control is identical to Mode 0. TL1 (TL0) has the role of the counter register. On overflow of TL1 (TL0) content of TH1 (TH0) is automatically transferred to TL1 (TL0). The count-up continues from this preset number. TH1 (TH0) may be changed as desired by software. If a number N is loaded into TH1 (TH0), an interrupt is generated for every (255-N)10 counts.

13.6.3 Mode 3

TH0 and TL0 function as two independent 8 bits units (TH1 and TL1 are not involved here). Figure 13.11 shows the configuration. TH0 acts as an 8-bit timer. The clock is internal and TR1 is the gating input to it. TL0 is configured as another timer / counter unit. Operation is similar to that in mode 0 without the prescaler.

13.7 SERIAL COMMUNICATION INTERFACE

The port bits P3.0 and P3.1 have alternate function assignment for serial communication. P3.0 is for reception and P3.1 for transmission. They are designated as RXD and TXD respectively. Two SFRs – SBUF and SCON – are for serial communication. SCON is to decide the communication details. SBUF is a buffer pair which holds the data for transmission and reception. The byte to be transmitted is written to SBU, it is transmitted serially LSB being the first. The serial data received is shifted into a buffer. It can be read at the address SBUF; Thus through the address for writing and reading is the same, the physical locations are different. Transmission and reception can take place simultaneously. The interface provides full duplex facility. Four different modes of operation are possible.

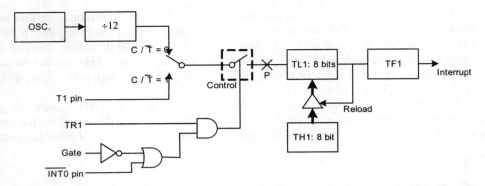

Figure 13.10 Block diagram scheme of Timer1 operating in Mode 2

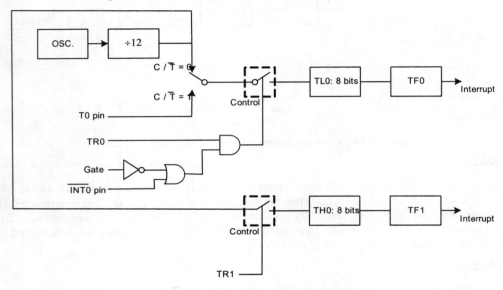

Figure 13.11 Block diagram scheme of Timer1 operating in Mode 3

13.7.1 Mode 0

The mode uses RXD pin for transmission as well as reception. TXD is the clock source. Transmission and reception occur at the fixed band rate of $1/12^{th}$ of the µC Oscillator frequency – one bit is transferred every machine cycle. Transmission is in synchronous mode at the highest possible band rate.

13.7.2 Mode 1

For every byte to be transmitted µC inserts the start bit (0) and the stop bit (1). At the receiver the stop bit is loaded into RB8 in the SFR SCON. Timer 1 overflow provides the baud rate clock for the system. The mode is intended for asynchronous transmission. The receiver functions with a basic clock rate 16 times the band rate. Once the baud rate is set the receiver waits for a 1-to-0 transmission in the data line; it signifies a start bit. As soon as the start bit transition is detected, the receiver clock is reset and started afresh. The bit is sampled at the 7th, 8th, and 9th counter states of the clock. The bit value is decided based on the majority – if at least 2 of the 3 samples coincide that value is taken as the bit value.

13.7.3 Mode 2

The baud rate is fixed at $1/32^{nd}$ or $1/64^{th}$ of the Oscillator frequency. The SMOD bit (in PCON) decides the band rate. Band rate is $1/64^{th}$ of oscillator frequency if SMOD is set. It is $1/32^{nd}$ of Oscillator frequency if SMOD is reset. Operation is suited for asynchronous transmission. For every byte 11 bits are transmitted. It includes the start bit (0) and the stop bit (1). The additional bit (9th bit) can be decided by user. It can be the parity bit; in that case P (in PSW) can be moved into TB8 (in SCON) position for every byte transmitted. At the receiver the 9th bit is loaded into RB8 (in SCON) position. Parity checking has to be done separately in software. The transmit side is similar to that in mode 1; need to accommodate the 9th bit is additional. The receive side is identical to that in mode 1 with the basic clock being 16 times the band rate. Synchronization and bit identification procedure remains the same.

13.7.4 Mode 3

Mode 3 is similar to mode 2. It has the additional flexibility of a programmable baud rate. As with mode 1 timer 1 overflow decides the baud rate. The scheme uses 11 bits for every byte transmitted – the start bit (0), the stop bit (1) and the 9th bit being the additional ones. Insertion and detection procedure are similar to those in mode 2.

CHAPTER 14

PROJECTS

14.1 LOOK-UP TABLES

RETLW instruction can be used to handle look up tables in an effective manner [12]. Angle a is available in degrees in the range 0° to 90° at intervals of 1°. Return with sin a

14.1.1   Prepare a table of $a \sim \sin a$. Execute the program in Figure 14.1. Register Ra has a. Move a to W and get sin a. Here it is assumed that the decimal point is at the left of the number returned.

$$\text{Sin } 90° \quad = 0\text{XFFh} = 0.\text{FFh}$$
$$[0.\text{FFh} \quad = 1 - 1.526 \times 10^{-5}]$$

14.1.2   Obtaining intermediate values – For small values of b

$$\sin(a+b) \quad = \sin a \cos b + \cos a \sin b$$
$$= \sin a + b \cos a.$$

14.1.3   The above formula can be used for b up to 1°. A second table of cos a can be prepared and used to compute $b \cos a$.

14.1.4   The overall program is as follows:-

14.1.5   Get the angle in degrees. Split it into parts a and b [The angle is expressed as '$a.b°$']. Use the look-up table for sin a and return the value of sin a. Use the look-up table for cos a. Return the value of cos a and compute $b \cos a$. Add sin a and $b \cos a$ to get the value of sine $(a+b)$. For the example here, cos a table need not be stored separately. Table for sin a can be used with the relationship

$$\cos a \quad = \sin(90 - a).$$

Similar look-up tables can be created and used to obtain logarithms, powers, square root, cube root and so on.

```
         MOVF Ra, 0;
GETSN    CALL SINE;
                  .

                  .
                  .
SINE     ADDWF PC;
                  RETLW K0;
                  RETLW K1;
         .  .  .  .  .  .  .
                  RETLW K90;
```

Figure 14.1 Program to get the sine of an angle in degrees

14.2 DA CONVERTER

The circuit in Figure 14.2 is a commonly used DA Converter [15]. If $b_7 = 1$, the switch is connected to the input side of the opamp. If $b_7 = 0$, the switch is connected to ground. Similar connection is made for all the other bits also. Show that current I is proportional to the binary number represented by the 8 bits used to do switching. Show that V_o is proportional to the number converted in this manner. Obtain the relation between V_o and V_R. Implement the DAC using an output port of the µC.

14.3 WAVEFORM GENERATION (WAVEFORM SYNTHESIZER)

14.3.1   Configure a port of the µC in output mode. Use the output bits and set up a DAC as in Figure 14.2.

14.3.2   Clear a scratchpad register. Load its contents to the output port. Increment the register content and output to the port. Do this continuously at a basic clock interval. V_o will be a staircase waveform as shown in Figure 14.3. When the register content overflows from 0FFh to 000h, the waveform output starts afresh from zero.

14.3.3   Add a suitable filter capacitor across the feedback resistor – 3R. V_o becomes (almost) a saw tooth waveform.

14.3.4   Modify step 3: When the register content becomes 05Ah (90°), clear it and start afresh. After incrementing the register, use its value along with the look-up table in Project 14.2 above for obtaining sin a. Output sin a to the port.

14.3.5   Modify step 4: Continue incrementing the register from 00h to 05Ah. As soon as it attains the value 05Ah, decrement the register content at every time step. When it becomes 00h, resume incrementing. Vo becomes a triangular wave

14.3.6   Modify step 5: Use the look-up table and for every value of a, get sin a and output the same. We get the plot of a sine wave.

Different waveforms can be formed and output in a similar manner. Alternately the waveform can be generated using two timers. One timer (Timer2) can output a pulse at the basic interval and the other timer (Timer1) can do incrementing and counting.

14.3.7

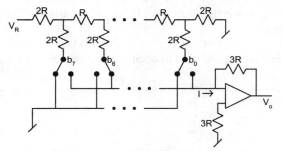

Figure 14.2 A commonly used DA Converter circuit

Figure 14.3 Staircase waveform generated by the DA Converter

14.4 RANDOM NUMBER GENERATOR

14.4.1   Consider the 15 bit number of Hex value 000BAh. Shift the bits left by one position. Make
$b_0 = b_{14} \oplus b_{15}$.

14.4.2   Repeat the shifting and XOR operation at regular intervals. We get a maximum length random binary sequence; the 15 bit sequence will go through all the possible 2 15 −1 states in a random manner and then starts afresh. The unit is called a Pseudo Random Binary Sequence (PRBS) generator.

14.4.3   Use two successive scratch pad registers to form a 15 bit sequence and implement the PRBS.

14.4.4   After every shift, output the 15 bit number on a pair of ports. We get a 15 bit pseudo random number sequence

14.4.5   Make a 15 bit DAC by modifying the scheme of Figure 14.2. Use the 15 bit PRBS output sequence as input to it and covert the number sequence into equivalent analog waveform.

14.4.6   Put a capacitor C across the feedback resistor 3R to filter out the 'step' type error in the output. The capacitance value should be such that 3RC should be 3 to 5 times the bit shift period.

14.4.7   Random signals have various applications. One application is in testing systems. Take a system of bandwidth $2\pi f_n$ rad/s. To test the system a random signal can be used as input. It should meet the following two criteria.

14.4.7.1   The bit shift rate used to produce the random sequence should be greater than $10f_n$.

14.4.7.2   The time duration of the random sequenced should be greater than $10/f_n$ s.

14.4.8   Confirm that the above sequence satisfies the second criterion, if the shift rate is selected to satisfy the first one. Verify that an 8-bit PRBS will suffice for the purpose here.

The bit shift rate can be set to any desired value by adding a timer to the sequence generator unit. It can be a hardware timer (Timer 0) or a software timer.

14.5 DISPLAY OF TRAIN DETAILS

Major railway stations have a dot matrix LED based display system in operation in the station platforms. The system is composed of a set of identical display units – each corresponding to one coach of the train. There can be 25 of these units. Each unit is kept aligned to the position of one coach of the train when the train stops at the platform. The system will be activated before the arrival of a train on the platform. The display unit will show the train number and the coach details – like 'AS1', 'S10', 'AC2' and so on – alternatively. Design and layout of the scheme can be on the following lines:

14.5.1   General Arrangement: Each display unit is to have a 3-character dot matrix based LED display. In addition two lights – one to display 'Train No.' and the other 'COACH' – are to be provided; these two can be of the back lit type. All the 25 display units are in communication with a PC (Figure 14.4). The train No. and coach details are entered in the PC in a structured format. Then a responsible functionary can download them into the individual display units (Access to the PC software can be through a password).

14.5.2   Display Unit: The unit can function around a µC. The dot matrix LED display can be interfaced through dedicated ports as in Section10.5.2. The unit can have 3 modes:

14.5.2.1   Mode S: The mode is to set a serial number to the unit, the number can have any value, from 1 to 25. One port pin can be committed for this and set to '1' or '0' state manually. Default state is '0' – 0V. When the switch is put to '1' state, 5V is connected to the pin. To set the serial number, the unit can be put in serial communication with a PC. The set bit can be set to high state. The serial number can be downloaded through the link from the PC. The number can be stored in the EEPROM area by the unit. After this, the 'S mode' can be deactivated by making the set bit low. Whenever the unit is in set mode, the byte received through the serial communication link is taken as the newly assigned serial number.

14.5.2.2   Mode R: This is to be the default mode. The unit will display the train no. and coach details alternatively for 15 sec. These are to be fetched from allotted scratchpad memory locations and displayed.

14.5.2.3   Mode P: It is the program mode. New data from the PC (master) has two parts – train number and coach details. The unit receives them. Soon after, the unit enters the program mode. It turns off the display, updates scratchpad area with new train number and coach details and exits program mode. When it exits program mode, default mode – mode R– is automatically activated.

14.5.3   Communication: The PC is to be set in asynchronous communication mode at 9.6 Kbaud with 9-bit mode of transmission. PC can address each Display Unit and send the data. Each unit can receive the assigned information, do parity check and if data is OK accept the same; else reject it. Acceptance or rejection is informed to PC by sending 055h or 0BBh respectively.

Implement the whole scheme including the communication protocol between the PC and the display units.

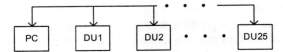

Figure 14.4 Scheme of connection of the display units to the PC

14.6 POWER LINE METER

A power line meter can be constructed around a μC. It will display frequency, current, voltage, power, power factor and cumulative energy consumed. It will have a 4 digit LED display and a function selector unit. A substantial amount of analog signal processing is required. The same is not discussed here; books on analog electronics and instrumentation may be referred for the purpose. The discussion here is limited to the outline of μC activities. The analog interface requirements are shown in block diagram form in Figure 14.5. Their scope is outlined.

14.6.1   Analog Interface: A step down transformer T has one secondary winding, which is the source of power. The voltage is rectified and a power supply unit –PS–made to function from it. PS provides all power necessary for analog circuits and the μC.

14.6.1.1   Another secondary winding provides the voltage signal. It has a peak voltage of 2.5 V at the maximum value of the power supply voltage. Analog unit B adds a bias of 2.5V DC to it. The supply voltage is presented to the μC as an analog input; its value can vary from 0V to 5V and back to 0V in each cycle of supply.

14.6.1.2   Analog unit B also provides a square wave on line F of 0v and 5V levels. It is generated from the AC input to unit B itself. The transition instants are co-incident with the zero-crossings of the supply.

14.6.1.3   Analog unit A has the line current as input. It can be the voltage drop across a shunt or a Hall Effect based sensor. Unit A processes the current signal and outputs a voltage on line I; its value varies from 0V to 5V and back to 0V in each cycle of supply – for the full rated current. The signal I is 2.5 V for zero current. As the current increases I swings more and more around its steady value of 2.5V. The signal is in phase with the line current – the 2.5 V levels coinciding with the zero- crossings of the current signal. .

14.6.2 Frequency Measurement: Signal F can be given as digital input to a port bit. The number of pulses on it can be counted for 10s. The same can be used to display frequency with 0.1 Hz precision. At 50 Hz, the displayed frequency value is 50.0.

14.6.3 Sampling Period Generation: Time period of signal F can be measured with the help of a timer. The timer can be configured to use the μC clock. All pulses for the period when F is in high state can be gated to the timer – only for one pulse period on F. With 10 MHz clock for the system, the timer (when active) increments every 400 ns. At 50Hz of supply frequency the timer counts for 10 ms (half period). The count value attained is 25,000. The number can be divided by 4 (2 right shifts) and the result used as sampling period; it results in 8 samples of current and voltage per cycle of supply.

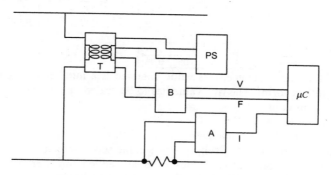

Figure 14.5 Analog interface for the power line meter in block diagram form

14.6.4 RMS Value Calculation: Take 8 consecutive samples – all of 10-bit width. Square each and get the sum. The sum may be stored as a 24-bit number. Ignore the LS 12 bits.

14.6.5 From the Hex representation of the MS 12 bits, obtain the RMS value of the signal with the procedure given below. The sums value is given in volts. It corresponds to the signal of amplitude 2.5 being taken as the maximum one with rms value of 1.58V. Confirm the same by your own analysis.

If the number N > 0F00h, rms value = 1.53+0.01×integer part of (N-0 F01Ah)/033h

If 0F00h > N > 0E00h, rms value = 1.48+0.01×integer part of (N-0 E0Ah)/031h

If 0E00h > N > 0D00h, rms value = 1.43+0.01×integer part of (N-0 D1Bh)/030h

If 0D00h > N > 0C00h, rms value = 1.37+0.01×integer part of (N-0 C08h)/02Eh

If 0C00h > N > 0B00h, rms value = 1.31+0.01×integer part of (N-0 B00h)/020h

If 0B00h > N > 0A00h, rms value = 1.25+0.01×integer part of (N-0 A04h)/02Ah

If 0A00h > N > 0900h, rms value = 1.19+0.01×integer part of (N-0 913h)/028h

If 0900h > N > 0800h, rms value = 1.12+0.01×integer part of (N-0 80Ah)/026h

If 0800h > N > 0700h, rms value = 1.05+0.01×integer part of (N-0 711h)/023h

If 0700h > N > 0600h, rms value = 0.97+0.01×integer part of (N-0 608h)/021h

If 0600h > N > 0500h, rms value = 0.89+0.01×integer part of (N-0 514h)/01Eh

If 0500h > N > 0400h, rms value = 0.79+0.01×integer part of (N-0 400h)/01Bh

If 0400h > N > 0300h, rms value = 0.69+0.01×integer part of (N-0 300h)/019h

If 0300h > N > 0280h, rms value = 0.62+0.01×integer part of (N-0 277h)/015h

If 0280h > N > 0200h, rms value = 0.56+0.01×integer part of (N-0 203h)/013h

If 0200h > N > 01A0h, rms value = 0.50+0.01×integer part of (N-0 19Ah)/011h

If $01A0h > N > 0150h$, rms value = $0.450 + 0.01 \times$ integer part of $(N-0\ 14Ch)/010h$

If $0150h > N > 0120h$, rms value = $0.42 + 0.01 \times$ integer part of $(N-0\ 121h)/00Fh$

If $0120h > N > 00F0h$, rms value = $0.38 + 0.01 \times$ integer part of $(N-0\ 0ECh)/00Fh$

If $00F0h > N > 00C0h$, rms value = $0.34 + 0.01 \times$ integer part of $(N-0\ 0BEh)/0Ch$

000h to 0C0h use look-up table – Table 14.1.

Table 14.1 Look-up table for getting the rms values in Project 14.5

N value	B3	A8	9E	94	8A	81	78	6F	67	5E	57	4F
Rms value	0.33	0.32	0.31	0.30	0.29	0.28	0.27	0.26	0.25	0.24	0.23	0.22

N value	48	42	3B	35	2F	2A	25	20	1C	18	14	10
Rms value	0.21	0.20	0.19	0.18	0.17	0.16	0.15	0.14	0.13	0.12	0.11	0.10

N value	D	B	8	6	4	3
Rms value	0.09	0.08	0.07	0.06	0.05	0.04

14.6.6   In the above look-up sequence, rms value obtained is a number in decimal form. All other numbers are in hex form. The rms value has to be scaled up suitably – to current in amperes and line voltage in volts.

14.6.7   KVA Calculation: $KVA = I\,V/1000$

14.6.8   Power Calculation: The instantaneous values of current and voltage samples are multiplied. The products over 10 periods of supply – 80 sample sets – are summed – designated as S. For 50 Hz supply, five consecutive values of S are added together; it represents the power consumed in 50 cycles or 1s (For other frequencies, the summing has to be suitably modified). The 12 MS bits of the power value can be taken and the rest ignored. This number has to be suitably scaled and converted to a decimal number.

14.6.9   Power Factor: (Power / VI) gives the power factor.

14.6.10   Implementation: A 4 digit display and a selector switch – to select the function to be selected are to be provided in addition to the analog interface described above.

14.6.11   One of the internal timers can be used along with a software timer to provide the command to carry out the measurement sequence – once a minute or so.

14.6.12   Every measurement sequence involves measurement of frequency, measurement of time period, obtaining sampling period, sampling, ADC and above mentioned computations.

14.7 BAR GRAPH DISPLAY

Consider a register of 7 flip-flops. The register can store a number N within the range 0 to 127. Develop a logic circuit to have 127 outputs designated $O_0, O_1, O_2, .. O_{126}$ and O_{127}. For any value of N all outputs from O_0 to O_N are to be 1 and all others zeroes. Provide an additional output set $P_0, P_1, P_2,$ .. P_{126} and P_{127} such that only P_N is 1 and all other Ps are 0.

A set of 128 LEDs can be arranged in a line with O_0 driving the first LED, O_1 driving the 2nd LED and so on up to O_{127} driving 127th LED. The arrangement represents a 'lit-bar'; its length represents the value of N. Similarly P's can be used to drive a set of 128 LEDs in a line. The single 'lit LED' represents a moving point, its position corresponding to the value of N.

14.8 SOFT SELECTOR – FUNCTIONS / KEYS

Applications often require the provision of a selector switch. In the simplest of cases, it involves selection of range for a display or selection of a quantity from a given set, for display. A more involved case can have two selector switches – one to select a quantity from an available set and the other to decide its range. Other instances requiring more than two selectors are also common. We consider here the provision of one set of selector switches and their realization in software.

14.8.1   Simple Selector: A meter can have one display. One from a given set of quantities is to be selected and displayed.

Typical examples:

- A line power meter has to display, frequency, current voltage, power or power factor.

- A process instrument has to display, temperature, pressure, ph value, flow rate, and purity of fluid in a pipeline.

- There are other applications too – not necessarily of display type; for example, open one selected valve from a given set; when a quantity like pressure is measured and displayed, the displayed figure can be in one of different possible units – inches of Hg, mm of Hg, bars, pascal, psi, p absolute, kg/mw, or Nw/m2. The unit can be selected manually.

A 'soft selector' type of selector switch can be implemented in a μC application with the port assignments shown in Figure 14.6.

14.8.1.1   Consider a port of a μC – port B: bit B $< b0>$ is connected to a push button switch S. The other seven bits are connected to seven LEDs designated $D_1, D_2, \ldots D_6$, and D_7.

14.8.1.2   Configure b_0 as input and all other bits as outputs. b_0 is to be treated as an external interrupt input.

14.8.1.3   Configure $b_1, b_2, \ldots b_6$, and b_7 as outputs. Make $b_1=1$ and all other bits as zero. LED D_1 is ON; all other LEDs are off.

14.8.1.4   When b_0 is pressed, μC enters an ISR. b_1 is made 0 and b_2 made equal to 1. All other bits are at 0. LED D_2 connected to b_2 becomes ON. All other LEDs are OFF.

14.8.1.5   Repeat the above type of ISR for every key closure at b_0. With every key closure, the bit selected for display advances by one. When the bit selected is b_7, next time it advances to b_1 (and not b_0).

14.8.1.6   The advancement can be carried out using 'Rotate Left' instruction. Port-B register can be read, and the high bit identified; based on this, the μC can vector to a selected routine.

Implement the Soft Selector. In effect the Soft Selector selects and carries out one of seven pre-defined activities.

14.8.2   Some applications require a soft selector to be implemented in a 2-tier fashion. As an example a PID controller has 3 constants associated with it.

- KP – proportionality constant

- KI – Integration action constant

- KD – Derivative action constant

Each of the three constants has to be separately selected and set. The selection and adjustment are normally done in a 'Set' mode. Once the set operation is complete, the controller can resume operation in a regular 'operate' mode.

   Consider a more versatile scheme. The scheme requires a set of six parameters to be used. Their values are to be adjusted in set mode. For this the μC is to enter a set mode, select the parameters desired and increase or decrease it through prompting. The parameter values are to be stored in a set of 6 registers. A set of six pointer bits is to be used to point to these registers. A register can be selected for alteration by making its pointer bit high. Two other bits can be used as indices to indicate whether the value (of selected parameter) is to be increased or decreased. Thus one can enter a 'set' mode, 'select' a parameter, and alter its value. These are done sequentially by activating a set of switches dedicated for the respective purposes. Implement a 2-tier soft selector as outlined below (See Figure 14.6):

14.8.2.1   Configure b_0, b_1, b_2 and b_3 bits of Port B as input bits.

14.8.2.2   Configure b_4, b_5, b_6, and b_7 as output bits. Connect one LED to each of these bits.

14.8.2.3   Configure all 6 bits of Port A as outputs. LEDs D_0, D_1, .. D_4, and D_5 are connected to these.

14.8.2.4   A push button switch – designated 'Set / Operate' – is connected to b_0. Whenever activated, external interrupt input flag is set. Every time the Set/Operate switch is pressed, bit b_3 should change state. b_3 and its complement are used to turn ON LEDs designated 'Set' and 'Operate'.

14.8.2.5   Normally only one of the bits of Port A is high; all others are low. In turn only one of the 6 LEDs out of D_0, D_1, … and D_5 is ON. All others are in off state. A push button switch is connected to b_1 – designated – 'Select Parameter'. Whenever the 'Select Parameter' switch is pressed and the µC is in 'Set' mode, the ON bit of Port A advances by one bit position to the left. When bit b_1 is high, LED connected to output bit b_6 turns ON, indicating 'Select Parameter' is ON.

14.8.2.6   Have 6 scratchpad registers – R_0, R_1, R_2, R_3 and R_5 – tied to the bits b_0, b_1, b_2, b_3, b_4 and b_5 respectively of Port A.

14.8.2.7   Have a 3-digit 7-segment display. Display the value of number stored in the selected register. If 'Select Parameter' button is pressed, the bit in high state of Port A register shifts; along with that, the selected register – from amongst R_0, R_1, R_2, R_3, R_4 and R_5 – also shifts. Content of the new register is displayed.

14.8.2.8   A push button is tied to bit b_2 of Port B; when it is pressed, LED connected to bit b_6 of Port B turns ON. It shows an arrow ' '. Every time b_2 is pressed the register tied to the selected bit of Port A, is incremented by one.

14.8.2.9   A push button is tied to bit b_3 of Port B. When it is pressed, LED connected to bit b_7 of Port B turns ON. It shows an arrow ' '. Every time b_3 of Port B is pressed, the register tied to the selected bit of Port A is decremented by one.

All the above activities – changing the parameter selected, increasing the parameter value or decreasing the parameter value and displaying the parameter value are carried out in the ISR. The µC will respond to any of the keys here only when it is in ISR. It enters the ISR only if Port B <bo> is pressed. If none of the keys is pressed for a continuous period of 10s, µC will quit the ISR.

Implement the Soft Selector Switch scheme.

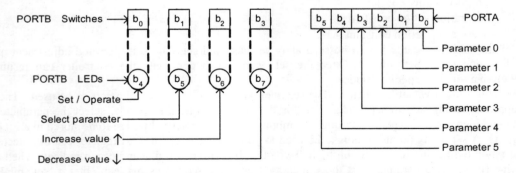

Figure 14.6 Port bit assignments to implement a soft selector type of selector switch

14.9 FUNCTIONS OF SAMPLES

$S_{-0}, S_{-1}, S_{-2} .. S_{-14}$ and S_{-15} represent a set of 16 samples of a signal, S_0 being the latest value. All S's are one byte wide. Write assembly language programs to generate the following outputs [15]. All these are obtained from the backward difference formulae.

Moving average: Y represents the average of last 8 sample values – it also represents the 'sub sampled value' obtained from the Ss. The simplest approximation to Y_M is

$Y_M = [S_0 + S_{-1} + .. + S_{-15}]/16$

Different better approximations to Y_M are available. Express Y_M using each of the following:

Trapezoidal rule:

$$Y_M = (S_0 + 2S_{-1} + 2S_{-2} + 2S_{-3} .... + 2S_{-14} + S_{-15})\frac{1}{32}$$

Simpson's (1/3) rule (for odd number of samples only):

$$Y_M = (S_0 + 4S_{-1} + 2S_{-2} + 4S_{-3} .... + 4S_{-13} + S_{-14})\frac{1}{45}$$

Simpson's (3/8) rule:

$$Y_M = (S_0 + 3S_{-1} + 3S_{-2} + 2S_{-3} + 3S_{-4} .... + 3S_{-14} + S_{-15})\frac{3}{128}$$

Weddle's rule (for 13 samples):

$$Y_M = (S_0 + 5S_{-1} + S_{-2} + 6S_{-3} + S_{-4} + 5S_{-5} + S_{-6} + S_{-6} + 5S_{-7} + S_{-8} + 6S_{-9} + S_{-10} + 2S_{-11} + S_{-12})\frac{3}{130}$$

Derivative: Y_D represents the derivative of the signal obtained from the set of previous sample values. Different approximations are available.

First approximation

$Y_D = S_0 - S_{-1}$

Second approximation

$$Y_D = (3S_0 - 4S_{-1} + 2S_{-2})\frac{1}{2}$$

Third approximation

$$Y_D = (11S_0 - 18S_{-1} + 9S_{-2} + 2S_{-3})\frac{1}{3}$$

Predictor: Y_P represents a predicted value for S_1 (the next sample) based on the sample values so far. Different approximate values for Y_P are

First approximation

$Y_P = 2S_0 - S_{-1}$

Second approximation

$Y_P = 3S_0 - 3S_{-1} + S_{-2}$

Third approximation

$Y_D = 4S_0 - 6S_{-1} + 4S_{-2} - S_{-3}$

Integral: Y_1 – the cumulative sum obtained when computing the moving average m, represents the integral Y_1 of the signal

Composite signal : Using the derivative and integral values generate the composite signal Y_C as

$Y_C = K_P S_0 + K_1 Y_1 + K_d Y_d$

Here Y_C represents the current sample value of the composite signal. The constants K_P, K_1 and K_d are called proportionality constant, integral constant, and derivative constant respectively. Take these as byte values available in EEPROM area. Write a program to compute Y_C.

14.10 PID CONTROLLER

A typical temperature control system is shown in block diagram form in Figure 14.7. T represents the temperature of a furnace and it is to be controlled. The temperature is sensed by the transducer and converted into equivalent voltage V_M (Voltage representing the measured value). The desired temperature is represented by V_s (Voltage representing the set value). The difference $E = V_s - V_M$ represents the error in temperature. Based upon the error value – its samples – the controller generates a command signal used to adjust the furnace temperature. The functional relationship of controller output Y_c to the error E is decided such that the furnace temperature adjusts itself to the set value in 'the best' possible manner. Controllers use different methods and strategies to get best results. The PID (Proportional, Integral and Derivative) controller is a well-established and widely used type of controller. The error E and controller output Y_c in a PID controller are related through the equation.

$$Y_c = K_p E + K_I \int E \, dt + K_d \, (dE/dt)$$

14.10.1   The proportional, integral, and derivative signals are generated as explained in project 14.9 above. The controller interface is organized as shown in Figure 14.8. It has a 4 digit display and a set of keys. Each key has an associated LED; when a key is pressed, the LED associated with it turns ON. The eight keys on the left are to select any one of eight parameters of the controller for adjustment. The seven keys on the right are ON/OFF type command keys.

14.10.2   The key P/R is to select program or run mode for the controller. When in run mode, the controller will respond to closure of keys DS and DM but ignore others. When in program mode, it will respond to all keys being pressed, except DS and DM.

14.10.3   When in run mode pressing key DS is a command to display the set value (of temperature). Pressing Key DM is a command to display the measured value (of temperature).

14.10.4   Consider the adjustments in program mode:

14.10.4.1   A/M key selects between auto and manual modes of control (by toggling). If it is in manual mode, the controller is bypassed; control action is carried out manually, external to the controller. The LED marked 'M' at the bottom lights up signifying that the operation is in manual mode. If control is in auto mode, the LED marked 'A' at the bottom lights up, signifying that control action is automatic and through the PID controller.

14.10.4.2   When a key on the left is pressed, the corresponding parameter is selected for adjustment. The current value of the parameter is displayed. If 'I' key is pressed, the set value increases by one count. If 'D' key is pressed, the set value of the parameter decreases by one count. Repeatedly operating 'I' and 'D' keys, one can set the new value of the selected parameter.

14.10.4.3   The ON/OFF key turns the controller output ON or OFF.

14.10.4.4   Roles of keys for KP, KI and Kd have been explained already.

14.10.4.5   Key S_P is for changing the 'Set Point' value (of temperature).

14.10.4.6   The key 'R_R' is to adjust the 'ramp rate' of control action. If the furnace temperature changes too fast, the furnace wall, heater, or the object heated can get damaged; hence the need to clamp the ramp rate.

14.10.4.7   The key 'L_u' sets an upper limit to the controller output. Safety of equipment is one of the possible considerations here. Similarly the key 'L_l' sets a lower limit to the controller output.

14.10.4.8   The key 'O_s' is to add an offset to the controller output Yc. Normally Yc will have a range 0 to N where N is a scaled positive integer. With the addition of offset, controller output can be shifted to any other convenient range.

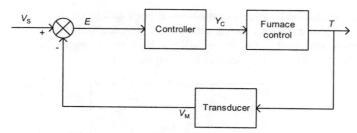

Figure 14.7 A temperature control system in block diagram form

14.10.5 The measured value is an analog input to the controller. It has to be converted to equivalent digital form through an ADC. Similarly the controller output (after adding the off set) is in digital form. It has to be converted into equivalent analog form through a DA Converter.

14.10.6 Realize the PID controller using a µC with the following guidelines.

14.10.6.1 The Display, ADC, DAC and Soft Selector can be configured on the lines discussed earlier.

14.10.6.2 The 'Program Mode' can be realized on the lines described with the Soft Selector Project.

14.10.6.3 The 'Program Mode' can be interrupt driven. Control action can continue in the back ground (if necessary). The changed parameter values can be stored in scratchpad registers during programming. When one exits program mode, parameter values in use can be updated.

14.10.6.4 The updated values can be stored in EEPROM

14.10.7 Sampling and ADC can be in interrupt mode. Calculation of integral value and derivative value can be in separate subroutines. Calculations of Yc can be done in a subroutine; it can call the integral and derivative subroutines. After every ADREG read operation, Yc can be computed and updated.

14.10.8 In some cases, the time available between two successive samples may not be enough to complete the calculation of Yc and updating it. In such a case Yc calculation has to be in 2 or 3 segments extending over 2 or 3 sample periods. Correspondingly Yc need be updated only once in 2 or 3 sample periods.

14.10.9 The sample period need not be rigidly fixed as done here. One can have another programmable parameter as the sampling period – Ts. Its value is normally selected to be 1/5th to 1/20th of the process time constant.

14.10.10 More often sampling is done at a much faster rate than what is required to compute the integral and derivative values of the process variable. In such cases the average of 8 or 16 successive samples is taken and used as one sample for the subsequent computations. The operation is equivalent to 'sub-sampling'. Thus if the basic sampling is done at 1s intervals, a sub-sampling factor of 8 leads to an effective process sampling period of 8s. Instead of the sampling period of Ts, one can have the sub-sampling factor as a programmable parameter. Values as multiples of 2 are preferable here – it reduces software overhead.

14.10.11 Easy visualization of process variable makes the process operators feel comfortable. One can provide two bar graphs in the controller to display the measured and actual values of the process variable.

14.10.12 PID controllers have a serial asynchronous port. Values of all the displayed quantities can be communicated in a structured manner (to a master) through the link. The facility for remote programming from the master too can be added.

The PID controller, its facilities and operation have been described here with temperature as the process variable. In practice the process variable can be different but the PID controller, its functions and operation remain invariant.   Other controllers like fuzzy controller, neural network based controllers etc. can be designed and realized in a similar manner.

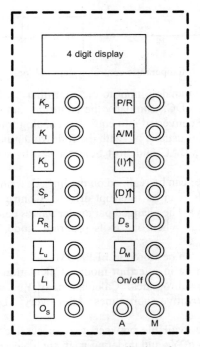

Figure 14.8 PID controller interface

14.11 AUTORANGING DIGITAL VOLTMETER

14.11.1 Consider the 4 digit display example in Section 10.5.

14.11.1.1   Estimate the time for initialization and the first sweep of the processor through its program.

14.11.1.2   Estimate the time for interrupt service in one full cycle of display.

14.11.1.3   What is the % of time the μC spends on interrupt service?

14.11.1.4   What is the % of time μC is idling?

14.11.2 In the example there, the time period for one display cycle has been mentioned as 20mS (50 frames/s). In program-1, it is about 33ms. The program uses Timer0 and allows it to count every time from 00 to 0FFh before interrupt generation. Alter the program by loading a number into the timer register and allowing it to count from there. Along with this, change prescale factor and make display refresh cycle as 50 frames per second.

14.11.2.1   What is the % of time μC spends on interrupt with the changes in 14.11.2?

14.11.2.2   What is the % of idling time of μC with the changes above?

14.11.3 Consider the AD converter example – Program-2, Chapter 12. Obtain the average of the 20 samples acquired. Convert the average into BCD format and display in 4 digits. This requires integration of programs and some modifications. Take voltage reference as 1.024V (external reference). The converted output is to be displayed as 4 digits with the decimal point being to the right of the left-most digit.

14.11.4 Modify task 14.11.3 as follows:

14.11.4.1   Get the average of 20 samples of ADC output – Designate it as S_1. Repeat procedure for successive sets of 20 samples each and get S_2, S_3, . . . S_{15}, and S_{16}. Get C_1 – the average of S_1, S_2, S_3, . . and S_{16} . Continue the procedure and obtain C_1, C_2, . . . C_{29}, and C_{30}. Obtain average of C_1, C_2, C_3, . . C_{29}, C_{30} – designated as C_a. Display C_a for 5s.

14.11.4.2   Obtain the average of $|C_i - C_a|$. Display this quantity for 5s. The scheme we have implemented is a '4 digit Digital Voltmeter'. It displays the average voltage – average taken over 10s (approx). It also displays the average deviation from the average taken for 10s.

14.11.5 The above voltmeter can be converted into an 'auto-ranging voltmeter' with the modification in Figure 14.9. It has a potential divider and a selector switch S. Connection of S to 1 makes Vo = Vi /300. Connections to positions 2, 3, 4, 5, and 6 give potential divider ratios of 1:100, 1:30, 1:10, 1:3, and 1:1 respectively. An auto-ranging voltmeter requires Vo to be given as analog input to the ADC. The potential divider is to be operated as follows:

14.11.5.1   Connect S to position 1 of potential divider. Observe the ADC output. If its value is within the maximum permissible limit (0x3FFh for a 10-bit ADC) and above the 1/3rd of maximum permissible value (0x155h for a 10-bit ADC), the switch position is OK.

14.11.5.2   If the converted value is 0x3FFh, display 'Err' indicating that the meter is out of range and reading is in error.

14.11.5.3   If the converted value is below 0155h, connect S to position 2 and do conversion. Repeat procedure until position 6.

14.11.5.4   When the switch position is shifted from position 2 to 3, the decimal point is to be shifted right by one position. As the switch is shifted from position 4 to 5, decimal point is to be shifted right by one position (second time). When the switch position is shifted from position 6 to 7, the decimal point is to be shifted right by one position once again (third time).

14.11.5.5   An 8-to-1 analog mux IC can be used for the selector switch function. It has 3 selector bit inputs – S_2, S_1, and S_0. For S_2 S_1 S_0 = 000, channel I_0 is selected and connected to output line O (See Figure 14.10). For S_2 S_1 S_0 = 001, Channel I_1 is connected to output line O and so on. S_2, S_1, and S_0 can be connected to 3 port lines of the µC.

Implement the auto-ranging voltmeter with the details as given above. Leave channel I_0 grounded as shown. Connect other inputs as required.

14.11.6 The auto-ranging meter can be refined and accuracy improved to some extent. For this modify the scheme as follows:-

14.11.6.1   Convert the analog input with the I_0 as input. Let his be V_z (zero correction).

14.11.6.2   After every AD conversion, subtract the number representing V_z from the converted output. The net value is the zero-corrected voltage.

14.11.6.3   Do auto-ranging as explained in 14.11.5 above.

14.11.6.4   Complete meter implementation with the 'zero-correction' refinement.

14.11.7 The buffer in Figure 14.9 can be replaced by the buffer / amplifier in Figure 14.11. It has a selector switch S2 with two positions marked a and b. Position a makes the unit, a unit gain buffer. In position b it is an amplifier of gain 10. Use of the amplifier with the above scheme adds two additional ranges with full scale values of 300mV and 100 mV respectively. Implement the modified and enhanced auto-ranging voltmeter.

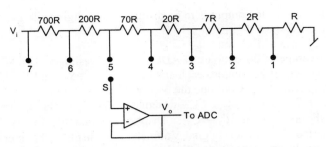

Figure 14.9 Potentiometer scheme for auto-ranging voltmeter

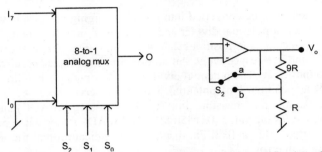

Figure 14.10 Selector switch to be used with the auto-ranging meter

Figure 14.11 Modification of the buffer in Figure 14.9 to extend the range of the auto-ranging meter

14.12 PHASE AND FREQUENCY METER

Frequency Meter:

Consider the task of measuring the frequency of an incoming signal. The signal is taken as a periodic pulse sequence. Connect it to T1CK1 input lead.

14.12.1 Get a pulse of desired width from Timer2 as follows:

14.12.1.1   Select a port bit and configure it in output mode – Let this be signal G.

14.12.1.2   Configure Timer2 with pre-scaler of 1:1 and post-scaler of 1:1. Load a number N to period register PR2. Enable TMR2IF and global interrupt Make G = 1 and start Timer 2 by making TMR2ON = 1.

14.12.1.3   As soon as TMR2IF goes high, make G = 0. Make TM2ON=0 and disable TMR2IF.

14.12.1.4   What is the time period for which G = 1?

14.12.1.5   Considering firmware overheads, adjust N such that G = 1 for 1 ms. Estimate the error in the pulse width.

14.12.2 Set timer 1 in asynchronous mode (TMRCS = 1). Set TMR prescaler value as 1:1. Disable Timer 1 oscillator (T1OSEN = 0). Select external input to timer 1 (TMRICS = 1). Disable TMRIF. With these settings, TMRI is ready for operation. Make TMR1ON=1 and start TIMER1. Timer1 will start incrementing with every rising edge on TICKI. Start Timer2 with the above settings. As soon as TIM2IF goes high, Make TMR1ON = 0 and stop counting incoming pulses.

14.12.3 Read TMR1H and TMRIL. Let it be W. W is a measure of the frequency of incoming signal.

14.12.4 Convert W into a decimal number and display it: – Use the 4 digit display scheme considered in Section 10.4.2.1. Note that the frequency is displayed in kHz [since the pulse period for counting is taken as 1ms].

14.12.5 What is the maximum frequency that can be measured without error (appx. 1 MHz with 50% duty cycle)?

14.12.6 Introduce auto-scaling to the frequency meter as follows:

14.12.6.1 If W in (14.12.3) above is less than 064h, increase the pulse width of Timer2 by a factor or 10. For this alter postscaler to 1:10. Repeat measurement. The display is in terms of × 100hz.

14.12.6.2 If *W* in (14.12.6.1) is less than 064h, increase pulse width by another factor of 10 (appx.). For this use prescaler of 1:16 and postscaler of 1:6 for Timer2; it gives a total scale factor of 96. Adjust period register number PR2 suitably. Fine tune pulse width of Timer2 using NOP instructions. The display is in ×10 Hz.

14.12.6.3 In 3.2, if *W* value is less than 064h, readjust the prescaler and postscaler values in Timer2 to increase pulse width to 1s. The reading here is in Hz. Check whether this is possible. If necessary use a prescale factor 1:8 with Timer1 to get the required pulse width.

14.12.6.4 Have three output bits connected to respective LEDs. Use them to show the scale as KHz, 100Hz and 10Hz, depending on the range used for the meter.

14.12.6.5 Configure three port pins as input pins and provide keys for input for these. Interface the keys to the μC [see Section 10.4.2]

14.12.6.6 Key K1 is to select manual mode or auto mode of scale selection. In auto-mode, implement frequency meter as detailed so far.

14.12.6.7 Key K2 is for going up in scale in manual mode and key 3 for going down in scale in manual mode.

14.12.6.8 If in manual mode, Key K2 is pressed once, it is interpreted as increasing frequency. Change from Hz to ×10Hz or ×10Hz to ×100Hz or 100 Hz to KHz as the case may be. If key K3 is pressed once, it is interpreted as decreasing frequency in a similar manner.

14.12.6.9 In manual mode decide range selection depending on closure of keys K2 and K3.

14.12.6.10 In manual mode decide the LEDs in (14.12.6.4) to be lit, depending on frequency setting as above.

14.12.6.11 Set up a software timer to use input from Timer2. Use it to refresh measurement and display every 60s.

14.12.6.12 Set up a software timer to use input from Timer 2. With this add one more scale to the meter − × 0.1 Hz. Modify the LED set mentioned in (14.12.6.4) to accommodate this enhancement.

Phase Meter:

14.12.7 Configure Timer 1 to function with internal Oscillator and prescaler factor 1:1. Configure a CCP module to have prescale factor of 1 and function on positive edge. CCP1 pin is to be set to input mode.

14.12.8 Phase shift is to be measured between two square waves A and B. Give B as input to CCP1 pin. Give A as input to the other port pin configured as input pin.

14.12.9 Poll inputs A and B successively and wait until both are zero.

14.12.10 Wait until input A goes to high state. Start a timer. Continue timer until input A goes to low state. Read timer value (represents the half period of inputs [T/2]. Timer2 can be used as the timer here. Starting and stopping can be done by setting and resetting theTMR2ON bit.

14.12.11 At the next positive edge on input A, start Timer1. With CCP1 operation when input B goes high next time <TMR1H:TMR1L> will be captured and stored in <CCPR1A:CCPR1L>: Let this time period by TP.

14.12.12 If Tp < T/2, input A leads input B. The required phase shift in degrees is 360 TP/T. If TP > T/2, B leads A. The required phase shift in degrees is (TP-T/2) 360/T. Configure two port bits as outputs – each driving an LED. Use these to display 'A leads B' and 'B leads A'. Convert phase shift into equivalent decimal number and display the same.

14.12.13 If T/2 is large than 07FFFh, use pre-scale factor for the timer. Pre-scale factor up to 8 can be used. If it is insufficient, part of the timer is to be realized in software.

14.12.14 At 50 Hz, for phase shift of 1°, what is the value of TP?

14.13 DOT MATRIX DISPLAY

14.13.1   A matrix of 7×5 dots can be used as a palette to form all alphanumeric characters. Standard tables for the same are available and can be used to form the table of dots to be selected in each column for each of the alphanumeric characters. Figure 14.12 shows the matrix along with two displayed characters as examples. In the dot matrix representation of characters, each dot can be represented by an LED. The matrix of LEDs can be driven by row lines and column lines as was done with the set of four 7 segment characters in Section 10.5.2.

14.13.1.1   Row R_1 can have all LED anodes connected together; the common point can be driven through a port line. A current limiting resistor is to be inserted in series. Row R_2 can be driven by the second port line and so on, up to R_7.

14.13.1.2   Connect all the cathodes of LEDs in column C_1 together. Drive the common cathode through a MOSFET transistor from a port pin. Repeat for columns C_2, C_3, C_4, and C_5.

14.13.1.3   To display one character, activate columns C_1 to C_5 successively. When column C_1 is active, output the row bits as required for column C_1. Repeat with all columns. Repeat the sequence cyclically and display the character.

14.13.1.4   For each character maintain the corresponding row activation table of 5 rows. The ASCII code of the character is to be available in a scratch pad register. Write the program to display the character.

14.13.2   A 4 × 4 keyboard matrix was considered in Section 10.5.4. Expand it to accommodate a normal keyboard in an 8 × 8 matrix. Use two 8-bit ports and do the keyboard interface to identify a key closure.

14.13.2.1   Take a µC with 4 ports. Integrate tasks 14.13.1 and 14.13.2 above. If a key is pressed, its alphanumeric equivalent should be displayed on the 5 × 7 matrix.

14.13.2.2   Consider two µCs. Establish a one-way asynchronous 8-bit link between the two. The first µC has a 7 × 5 matrix display interfaced to it. The second has a keyboard attached to it. If a key in the keyboard is pressed, the corresponding character is to be displayed in the matrix display attached to the first µC.

14.13.2.3   Consider a set of LEDs arranged as a 7 × 23 matrix. It represents a scheme of displaying 4 characters of 7 × 5 type in one line with 3 vertical lines left blank in between each pair of characters. Interface it to µC with 4 ports as shown in Figure 14.13.

14.13.3   Display a set of 4 characters in multiplexed mode.

14.13.3.1   Displace the character set one position to left with a delay of 1s.

14.13.3.2   Continue shifting as in 5.2 above and shift out the whole display in 23s.

14.13.3.3   Instead of shifting out the character-set, rotate the same. Show a cyclically moving display.

14.13.3.4   Consider the display in 5.1. Move it down by one position with a delay of 1s.

14.13.3.5   Repeat shifting as in 5.5 above and shift out the complete character set.

14.13.3.6   Instead of shifting out the character set, cyclically shift it down.

14.13.3.7   Rotate the character set display left 10 times as in 5.4. Subsequently rotate the character set down 10 times.

14.13.4   Figure 14.14 shows two display units of the type in considered above. Configure them to display a set of 8 characters. Modify the same to move the display cyclically in left direction. This can be achieved in two ways.

14.13.4.1   The display is to be shifted left by one position every one second. For this transfer one column data across C – C from μC_b to μC_a after every one second. Simultaneously transfer one column data across d-d from μC_a to μC_b.

14.13.4.2   Have the full set of 8 characters stored in μC_a and μC_b in the required format. Make both units work independently. Synchronize them by μC_a giving output flag which forms an external interrupt input to μC_b; that is, μC_a decides when to shift out; it does shifting within itself and (through the interrupt) commands μC_b to do the same simultaneously. Compare the two alternatives.

14.13.5   Have μC_c interfaced to an ASCII Keyboard as in task 14.13.2 above. Provide one-way serial link from μC_c to μC_a and μC_b in Problem (6) above. Integrate the display scheme. Turn off display completely when it is being refreshed.

14.13.6   Have a 2-line display with each line of 8 characters as in task 14.13.4 above. Let these be interfaced to 4 μCs – μC_a and μC_b in top line and μC_d and μC_e in bottom line. μC_c is interfaced to the keyboard. All other μCs are in serial communication with μC_c in asynchronous mode. μC_c is the master. μC_a provides an output flag which forms interrupt input to μC_b, μC_d, and μC_e. It does the synchronizing for the whole display.

14.13.6.1   Consider the display scheme in task 14.13.6. Have the following programs.
   PM1:   Shift display left
   PM2:   Shift display right
   PM3:   Shift display up
   PM4:   Shift display down
   PM5:   Shift upper line left, lower line right
   PM6:   Blink display at 4s intervals – ON for 2s and OFF for 2s.

14.13.6.2   Do a program which will do all the above cyclically and sequentially; each of PM1, PM2, and PM6 is to be done for 30s

14.13.7   PC interface to display:

14.13.7.1   Set a PC in serial communication with the μC of task 14.13.1. Choose 8-bit asynchronous mode and 9.6 kbaud as data rate. The μC is to be programmed to function in two modes; two bytes – B_1 and B_2 – are to be used for this. If $(B_1 <bo>) = 0$, the display is to be OFF. If $(B_1 <bo>) = 1$, the display has to show the character whose ASCII equivalent is B_2. When reset, the μC is in receive mode and $(B_1 <bo>) = 0$. Two bytes received on serial line are to be read and loaded as B_1 and B_2 in that order. Write a C program which will download two characters to the μC. In the μC, the first character is to be loaded as B_2 and the second one as B_1. The second one is 001h. It starts the display.

14.13.7.2   Modify the C program to display different characters at regular intervals.

14.13.7.3   Repeat tasks so far, replacing μC_a by the PC.

14.13.7.4   Consider the modified version of task 14.13.7 with PC programming. Provide other alternatives of display. Make a program in PC where the user need only fill up data in a user friendly window to set the moving display fully.

14.13.8   Figure 14.15 shows a 2-colour LED in one pack. lC is the common cathode lead, lR and lG are the anode leads of Red and Green LEDs respectively. Each LED in Problem 1 is replaced by a 2-colour LED. Red LED set is driven by μCR and Green LED set is driven by μCG. Repeat Problem 1 with the option to show the character in Green colour or Red colour. The option is to be exercised by making a digital input, dedicated for the purpose high (green) or low (red). This control input is given to both μCs – μCG and μCR.

14.13.8.1   Redo all earlier tasks with the 2 colour display scheme. In the modified versions of tasks 14.13.3 to 14.13.6, include combinations of multi-colour displays.

14.13.8.2   The display scheme in the final form has matrix of 48 × 16 positions with one green LED and a red LED at each position. It can be used to display a variety of figures, shapes and so on.

14.13.8.3   As an example consider the set of LEDs in Figure 14.16. Display the red LED in position 0. Five seconds later display red LEDs in position 1. Ten seconds later display red LEDs in position 2 and so on. Repeat the display with green LEDs. Repeat with different combinations and sequences.

14.13.8.4   Display all LEDs (red) along a radius from the center. Sweep the radius counter-clockwise and show the full matrix lit.

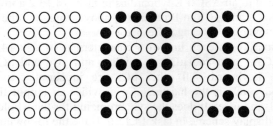

Figure 14.12 The 7 × 5 dot matrix for display: Display of character 'A' and the numeral '1' are shown as examples.

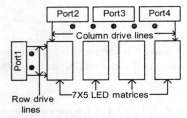

Figure 14.13 LED matrix to display running characters

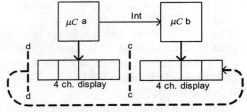

Figure 14.14 Scheme for two display units of the type in Figure 14.13 to function together

Figure 14.15 2-colour LED set with common cathode

2	2	2	2	2
2	1	1	1	2
2	1	0	1	2
2	1	1	1	2
2	2	2	2	2

Figure 14.16 Typical LED display scheme considered in Problem 13

14.14 SIGNALLING FOR TRAFFIC CONTROL

14.14.1 Consider the automatic traffic control at a junction formed by the crossing of two roads A and B as shown in Figure 14.17. The simplest sequence of traffic control at the junction is as follows:-

14.14.1.1   Allow the Up-traffic on road A for 60s – it includes traffic going ahead on road A itself, traffic turning right into Road B as well as the traffic turning left into Road B. It requires the green light for A up-traffic to be ON for 57s, followed by the amber light for A up-traffic to be ON for 3s. All other directions are to keep the red lights ON.

14.14.1.2   Allow the down traffic – all 3 types as the up-traffic on road A explained above – on road A for 60s; for this keep the green light for A down-traffic on for 57s; follow it by the amber light in the same direction to be ON for 3s. During this whole interval all other directions are to have the red lights ON.

14.14.1.3   Allow up-traffic on road B for 60s on the lines explained above. For this keep the green light and the amber light for the direction on for 57s and 3s respectively.

14.14.1.4   Allow down-traffic on road B for 60s as explained above – with the green light concerned being ON for 57s followed by the amber light being on for 3s.

14.14.1.5   Repeat the above sequence cyclically. Lights of 3 colours – green, amber, and red – are to be lit in each direction. Implement the scheme using a µC. Each such light can be driven by a port bit. Altogether 12 bits are to be output. The hardware interface from the port bit to the lamp can be a properly selected solid state relay; the relay output drives the lamp. When a bit is at 1 state, the corresponding light is on; when the bit is in 0 state, the light is off.

14.14.1.6   Four bits of Port A and the 8 bits of Port B can be configured as output pins. Each can be used to keep a light ON or OFF.

14.14.1.7   Configure timer1 to function with external clock and have an oscillator with a standard 32.768 KHz crystal connected between T1OS1 and T1OS0. With a prescaler scale factor of 1:1, the timer overflows every 2s and the TMR1IF is set. ISR can reload the timer with 08000h so that timer 1 interrupt is at every 1s interval.

14.14.1.8   Have two software counters in scratchpad registers – designated C57 and C03.

14.14.1.9   After initialization, the µC shall idle. All activities may be carried out in ISR following an IRQ due to TMR1IF being set. As part of initialization C57 and C03 are cleared. The green lamp in A up-direction is lit. Have a bit in a scratchpad register designated G-A. Initially reset every interrupt. Increment C57 if G-A is at 0 state. Continue until the count in C57 is 57. When it is 57, reset C57, turn off the green light in A up-direction and turn on amber light in A up-direction. Set G-A.

14.14.1.10   Set G-A at 1 state, increment C03 at every interrupt from TMRIF; continue until the count is 3. When it is 3, reset C03, turn off amber light in A-up direction.

14.14.1.11   Continue the above procedure until the full cycle of lamp lighting and turning off is complete. Repeat the cycle.

14.14.1.12   A separate scratchpad register can be used to keep count of the lamp being turned on. Whenever the count reaches 12, the counter can be reset. The count value can be used to select the next lamp in the sequence that is to be turned on.

14.14.1.13   Prepare a flowchart for the program. Implement the traffic control scheme.

14.14.2 Add a set of 8 seven-segment displays – one for each direction. As soon as the green light in A up-direction is turned on, the display near it has to show 60. The count should decrease with every second. It becomes zero when the amber light in A-up direction turns off. The display should get reset to 180 immediately. It should decrement every second and become zero when the green light turns on again. When the green light is on, the digital display shows the time remaining for the on-going traffic. When the red light is on, the digital display shows the time one has to wait for the green light to turn on. Functions of other digital displays are identical. The digital displays can be updated using the software counters C-57 & C03.

14.14.3 Modify the traffic control scheme for a junction of 5 roads.

14.14.4 In the scheme of Figure 14.17, traffic on road A is heavier than on road B. The green lights in both directions on road A are to be kept on for 60s; those on road B are to be kept on for 30s only. Modify the scheme accordingly. Implement the scheme as modified here.

14.14.5 So far we have not made any provision for pedestrian crossings. Make provision for pedestrian crossings for 30s at the end of every cycle. During pedestrian crossings, all vehicular traffic is halted. Pedestrian crossing requires a red light and a green light. Amber light is not provided; instead the green light is made to blink for 6s in the last phase of pedestrian crossing.

14.14.6 Depending on the variations in traffic conditions, the on-timer may have to be adjustable. Further to meet exigencies like traffic jam, accidents, and unexpected road blocks the automatic signalling may have to be bypassed and manual control used. A soft selector can be added to accommodate these requirements. Modify the scheme accordingly.

14.14.7 Consider the up-traffic on road A. The segment which turns right may be small; Similar is the case with the down-traffic on road A. Traffic management can be modified and made more efficient by taking this factor into account.

14.14.7.1 Allow up traffic from 0 to 60s. Out of this allow the traffic that wind to the right only for 30s. This requires separate green lights (of the arrow type) for traffic that proceeds straight ahead and that turns right.

14.14.7.2 From 30th second to 90th second, allow down traffic on road A. Allow the segment that turns right from 60th to 90th second. With the modifications here and the previous steps, the traffic on road A has been handled fully in 90s (instead of the 120s earlier).

14.14.7.3 From 90th to 120th second, allow up-traffic on road B.

14.14.7.4 From 120th to 150th second, allow down-traffic on road B.

14.14.7.5 In every case insert amber lights for 3s, as done earlier.

14.14.7.6 Stop vehicular traffic for 30s. Allow pedestrian crossings during this period. Ensure that from 24th to 30^{th}s the green light blinks.

14.14.8 In the above scheme the traffic on road B can be given as much importance as that on road A and pedestrian crossings accommodated in a better manner. Up traffic on road A was allowed right turn only from 0s to 30s because of its limited size. A similar restriction can be put on left turn – that is right turn and left turn for up traffic on road A are restricted to be during 0s to 30s. From 30th to 60th second, the up- and down-traffic on road A are allowed. At 60th second, up-traffic on road A is stopped. From 60th to 90th second, left and right turns are allowed for the down-traffic on road A. With the modification here, the pedestrian crossings on road B – on either side – can be allowed from 30th to 60ths. Accommodate traffic on road B from 90th to 180th second through the above type of sequence. The pedestrian crossing on road A is accommodated from 120th to 150th second to 150th second.

14.14.8.1 In the scheme as developed so far, each group of lights has 5 lights – three greens, one amber and one red. Their control requires at least 6 wires; the preferred procedure (to improve reliability) is to use 5 pairs of wires. The cumbersome wiring requirements can be dispensed with, by using a separate µC for each group of lamps. A master µC at a traffic control box can do the overall control. It can be in asynchronous serial communication with all the other µCs.

14.14.8.2   Decide roles for master μC and slave μCs by dividing and assigning activities to each. Decide on communication protocol between master μC and the slave μCs. With these preliminaries implement the overall scheme.

14.14.9 Coordination of traffic signal operations in consecutive junctions along busy main roads can ease traffic congestion. Consider two successive road crossings on road A, the two being 0.8 KM apart. From 8 am to 11 am the up-traffic on road A is relatively high. A vehicle with an average speed of 40 km/hr between the two crossings takes 72s to travel the distance. The two signal systems can be coordinated in their operation if the green light for up-traffic in A direction in the second junction turns on 72s after that in the first junction. With this all the up-traffic on road A can smoothly flow through the two junctions if the average speed of 40 Km/hr is sustained. The coordination for such synchronization can be done through a serial link between the two traffic control systems. Operation can be similarly synchronized in the evening hours to ensure smooth flow of down traffic. Implement the synchronized scheme.

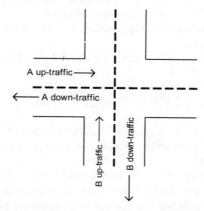

Figure 14.17 Traffic junction of 2 roads considered for signaling

14.15 SLIP DETECTION FOR ELECTRIC LOCOMOTIVE

The tractive power of electric locos can extend up to 5000 KW. With main line traction the maximum permissible tractive effort is used during acceleration and the train brought to full speed. With an acceleration of 0.8 kmph/s, the train can be brought to full speed of 120 kmph in 150s (2.5 minutes). This possibly represents the best possible acceleration to full speed. In practice the maximum value of tractive effort that can be exerted (and hence the maximum permissible acceleration) is limited by the adhesion coefficient. The adhesion coefficient depends on track conditions. If the tractive effort exceeds the limit imposed by adhesion conditions, the wheels will slip; such wheel slippage affects the wear and tear of loco wheels and the track adversely. Automatic wheel slippage detection can enhance quality of electric traction.

An electric loco has 4 traction motors driving the wheels on the 4 sides. Each such motor has its own inverter drive. The 4 inverter drives work in coordination through the electronic controller. Each of the 4 wheel sets has a tacho-generator. Slip detection is done with the help of the 4 tacho-voltages. Let the tacho-voltages be represented by V_1, V_2, V_3 and V_4.

14.15.1 When running at steady speed condition, slippage can be detected as follows:-

14.15.1.1   Let the average speed – V_a – be obtained as

$$V_a = (V_1, + V_2 + V_3 + V_4)/4$$

Sample the speed at 1s intervals. Compute V_a after every sample set. Obtain the moving average – V_{am} – of V $_a$ over 4 successive samples. The wheels – singly or in combinations – slip if

$|V_{am} - V_a|$ 5% of V_{am}

or if

$|V_a - V_1|$, $|V_a - V_2|$, $|V_a - V_3|$, or $|V_a - V_4|$ exceeds 5% of V_a

Instead of the 5% value, one can use 6.25% (1 in 16th) value; note that $(1/16)^{th}$ of a binary number is obtained by right shifting the number 4 times in succession.

14.15.1.2   The signals – V1, V2, V3 and V4 – are analog inputs. Implement wheel slip detection using a μC. Output a bit to indicate wheel slip. The bit is normally at 0 state. In case of a wheel slip, the bit is set. It is reset after 10s or when the wheel slip disappears - whichever is later.

14.15.1.3   Commit 4 bits – b_1, b_2, b_3, and b_4 – of an output port to the four tacho outputs – V_1, V_2, V_3 and V_4 – respectively. Normally all these bits are at 0 state. If four consecutive slips are with the same wheel, set the corresponding bit.

14.15.1.4   Consider the loco during acceleration. Let Δ be the increase in V_a in every second. In the absence of slip we have

$V_a - V_{am} = V_a$ - $[(V_a- Δ)+(V_a - 2 Δ)+ (V_a-3 Δ) + (V_a-4 Δ)]/4$
$= 2.5 Δ$
$V_a = V_{am} + 2.5 Δ$

Slip occurs if $|V_a-V_{am} - 2.5 Δ| ≥$ 5% $(V_{am} + 2.5 Δ)$

Or

If $|V_a - V_1|$, $|V_a - V_2|$, $|V_a-V_3|$, or $|V_a-V_4|$ exceeds 5% of V_a.

14.15.1.5   Modify wheel slip indication to accommodate motion under acceleration. in (3) to accommodate braking.

14.15.2   In rainy weather, in hilly forest terrain, and in places where dry leaves can cover the tract over long stretches, slippage can be high and maximum attainable acceleration may be limited. Still it is desirable to accelerate the loco at the maximum possible rate and bring it to full speed as soon as possible. For this the tractive effort should be lower than the value that causes slippage but as high as possible. Start with a low tractive effort (low acceleration – 30% of maximum nominal value on level terrain) and accelerate for 10s. If there is no slippage, increase it to 40% and accelerate for 10s. Continue until full tractive effort is used (corresponding to the maximum power output) or maximum permissible acceleration or slippage. In case of slippage reduce acceleration to the previous value and maintain it. A similar approach is used during deceleration. Use an output bit of the μC as a command to increase tractive effort and another bit as a command to reduce tractive effort. Implement the scheme using these.

14.16 THERMISTOR BASED THERMOMETER [SEE AN 649]

Thermistor is one of the simplest and most economic of temperature sensors. It can be used to construct a room temperature indicator. With the simple circuit in Figure 14.18 (a), as the temperature increases, thermistor resistance reduces and V_o increases; V_o is a monotonic function of temperature. The unit is to be calibrated and a table of temperature versus V_o prepared. From the temperature versus voltage table, the inverse table of AD converter output versus temperature is to be generated. We may use the LS byte of ADC output with 256 states – 000h to 0FFh. The table will have 256 entries – the output temperature range extending from 0°C to 100°C.

The sensor circuit can be interfaced to a µC as shown in Figure 14.18 (b). V_o is an analog input to the µC; P is a port pin. The display can be a 3 digit 7-segment display. Operation sequence can be as follows:

14.16.1 Configure Timer1 to function as a timer with internal clock source and maximum prescaler setting – 1:8. With 10MHz clock frequency, counting from 00000h to 0FFFFh takes about 200ms. Allow the timer to continuously count from 00000h and overflow. It generates an interrupt at 200ms.Use the interrupt and an internal software timer to have 10s interval generated. At the end of every 10s, do AD conversion and display of temperature.

14.16.2 Configure the AD Converter with one analog input and internal clock for conversion. Result is to be adjusted as left justified.

14.16.3 Configure pin P in Figure 14.18 (b) to be an output pin. Output '0' on it.

14.16.4 Start AD conversion. Wait for conversion to be complete.

14.16.5 Configure pin P to be in input mode; being a high impedance input, the current drawn is zero. Thermistor current is zero.

14.16.6 Read ADRESH byte (Ignore the last 2 bits). If the thermistor is properly selected, the number can have the range 000h to 0FFh.

14.16.7 Using the look up table, convert the number representing temperature into equivalent temperature and display the value.

Observations:

* Normally thermistor tolerance is 20%. For each thermistor, the unit has to be recalibrated.
* The unit can be adapted to function with any other resistance sensor. Depending on the nature of variation of the resistance with the variable, R value has to be selected. If the sensor resistance increases as the value of the variable increases, R and RT may have to be interchanged.
* When used as a meter with other variables, external reference can be used to improve repeatability. The reference has to be stabilized with low temperature coefficient.
* With addition of instrumentation amplifiers and necessary power supplies, resistance bridge type transducer outputs can be used as analog input.
* Additional functions can be incorporated into the instrument.

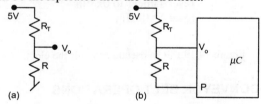

Figure 14.18 Use of thermistor as a temperature sensor (a) basic circuit (b) interface to µC

14.17 Clock [AN649]

A Clock can be designed to function around a timer. An oscillator functioning at 32.768 KHz [standard crystal specifically intended for clocks] can be used for the purpose. Timer1 can be configured to function with input from the oscillator as the clock signal. If initialized to a count of 0800h, the timer will overflow with a delay of 1s causing an interrupt. The interrupt can be used to update time. The update routine (in pseudo-code) can be as in Figure 14.19.

14.17.1 The processor can be configured to be in sleep mode. Timer1 alone will function with the external clock. The timer interrupt will wake up the µC. It will execute the update routine and revert to sleep mode.

14.17.2 Facility to set time manually can be provided through two keys interfaced through PORT B (RB4 & RB5). One is the 'Set' key and the other the 'UP' key. When either is pressed, an interrupt can be generated (interrupt on 'PORT B status change'). The interrupt service routine can identify key pressed and carry out a 'time set' routine.

14.17.3 Program Outline

14.17.4 Set Timer1 by setting T1OSGEN (TICON<3>). Internal oscillator is selected. Use 32.768 KHz crystal and 33pF capacitor to ground [See µC data book on Timer1]. Set prescaler to 1:1.

14.17.5 Configure port B pins PB4 and PB5 as inputs. Use 'interrupt on change' feature to identify key closure.

14.17.6 To update time, have ISR as on the lines of the pseudo code routine in Figure 14.19.

14.17.7 To set time, have ISR. The ISR can sequentially select registers displaying time and increment the content. AM/PM bit too can be changed in a similar manner. When the ISR is active the digit being changed can be displayed in a blinking mode.

14.17.8 Use the PWM facility and add audio beeps at hourly intervals.

```
If(TMR1IF)
{seconds ++;
 if(seconds >59)
   {seconds = 0;
    minutes ++;
   }
   if(minutes > 59)
      {minutes = 0;
       hours ++;
      }
      if(hours > 12)
         swap AM ? PM;
         reset Timer1 to 0x80h;
         reset TMR1IF;
}
```

Figure 14.19 ISR in pseudo code for the clock

14.18 SEQUENCING OF CONVEYOR BELT OPERATIONS

Electrical motors used in industry are characterized by three operations. All the three decide the current through the coil of a contactor which provides supply to the motor or shuts down the supply. The coil is shown in Figure 14.20 (a) along with the concerned switches of operation. SR is the start switch which is normally open. SP is the stop switch which is normally closed. OL is the overload switch which is normally closed.

For starting the motor, switch SR is to be closed. The coil C will get energized and start the motor. Once energized, the coil will turn on another contact designated IL (Interlock). IL will bypass SR; hence even if SR is released, the coil (and hence the motor) will remain energized. For stopping the motor, switch SP is opened; the coil gets de-energized and shuts off the supply to the motor. If the motor gets overloaded, the contact OL will open. The coil will get de-energized and shut off supply to the motor.

14.18.1 Treat SR, SP and OL as logic inputs to a µC. Write a program to output a logic signal to turn ON/OFF the supply to the coil C. The interlock IL is to be realized in software (firmware).

14.18.2 A contact NO1 (Normally open) provided in parallel with SR can be used as an alternate start contact. Remote starting of the motor can be achieved using this. As many remote start facilities as desired can be provided in this manner. A contact NC1 (Normally Closed) in series can be used to achieve 'remote stop'; as many such normally closed contacts as desired, can be provided in series. Additional overload type trip facilities too can be provided in the same manner. Modify the above program to accommodate the additional start and stop facilities.

14.18.3 In process industries, power stations etc., material in the form of lumps, pellets and powder is transported through conveyor belts. Such a conveyor belt scheme is shown in Figure 14.20 (b) in a simplified form. S1 is a silo holding one type of material. It is fed at one end of conveyor belt B1, with motor M1 being the drive for the conveyor. At the other end of B1 the material is fed to belt B2 which feeds it to belt B4. B4 delivers the material to storage bin SB. A second category of material is available in Silo S2. It is fed at one end of belt B3. B3 feeds it to B4 at the other end. B4 feeds it to storage bin SB at the other end. With properly adjusted feed rate of the two materials, the mix in SB can be ensured to be of a desired ratio.

14.18.4 Starting Sequence: If B4 stops but B1, B2, and B3 run, material can get clogged on Belt B4. In general, the belt at the downstream end has to be started first and that at the upstream end started last. Further they have to be started in the proper sequence: – B4   B3/B2   B1.

14.18.5 Stopping Sequence: To avoid material clogging on any belt B1 / B3 have to be stopped first and B4 stopped last. Stopping has to be in the proper sequence.

14.18.6 Protection: If the motor of any belt stops (due to overload), all the related upstream belt motors have to be stopped immediately.

14.18.7 Maintaining Material Ratio: If the material feed from one of the Silos stops, all feed from the other has to be stopped. Otherwise the material ratio in SB cannot be maintained.

14.18.8 Screw feeders from Silos: Each Silo (S1 and S2) will have a screw feeder at the bottom to facilitate material feed to the belt below. If a belt stops, the screw feeder behind has to be stopped too.

14.18.9 Design a µC based scheme to control the operation of the conveyor belt system. The controller will have all start, stop and overload switches as logic inputs. The outputs will be commands to turn ON / OFF the motors as well as the screw conveyors. All protective interlocks discussed above are to be provided.

14.18.10 Provide logic outputs to display the status of each belt / motor. The display can be RED (OFF), GREEN (ON) and AMBER (stopped). The belt / motor that stopped first may be identified by the corresponding amber light blinking.

14.18.11 The system has to stop if S1 or S2 goes empty or SB gets full. The full/empty status can be sensed using level gauges or load cells. A simpler scheme is to estimate the possible run time in advance and feed it into the µC scheme. Add the provision to the µC based scheme to select S1, S2, or SB and enter the allowable run time for each. It should be done before the start of the conveyor scheme. The conveyor scheme run time is to be kept track of. It should stop if it matches any of the 3 time values entered. Incorporate this facility into the system.

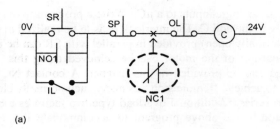

(a)

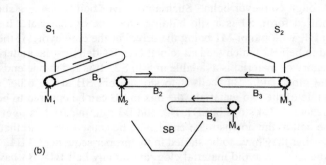

(b)

Figure 14.20 (a) Motor control circuit (a) An elementary conveyor belt scheme

14.19 TARIFF METER FOR HIGH TENSION CONSUMERS

14.19.1 Electricity Supply Organizations supply power to bulk consumers with specific tariffs. A tariff has a number of components to it. The amount paid by a customer on a monthly basis is the sum of all the components. A typical tariff is as follows:

14.19.1.1   Rs.3.50 per kwh for normal load. There is a 5% discount for consumption between 11 pm and 5 am. There is a 20% penalty for consumption between 6 pm and 10 pm.

14.19.1.2   Rs.200/- per KVA of maximum demand in a month

14.19.1.3   Rs.400/- per KVA per month for every KVA of maximum demand over and above the contracted KVA load.

14.19.1.4   Rs.2500/- per month given as incentive for every 0.01 increase in power factor over and above the value of 0.9.

14.19.1.5   An additional charge of Rs.2500/- per month (as penalty) for every 0.01 decrease in power factor from 0.9.

14.19.1.6   Twice the value of the energy consumption over a 30 minute interval (in Kwh) is taken as the KW load. The peak value of such a moving average value of power over a whole month is taken as the maximum demand in KW.

14.19.1.7   Total Kwh energy consumption over a month is computed. Total KVARh is also computed. The power factor is taken as

$$powerfactor = \frac{KWH}{\sqrt{KWH^2 + KVARH^2}}$$

14.19.2 Modify the power meter discussed Project 14.6 and design a tariff meter. The tariff meter
should have in-built clock and calendar. It should have a supervisory mode and normal mode.

14.19.2.1   The facilities desired in the normal mode are the following:-

• Soft key pad and display

• Facility to select KWh, maximum demand, KVARh, and power factor as on date (within the month) and display the same

• Facility to display the charges due under each category and total charge to be paid as on date (within the month)

14.19.2.2   The facilities desired in the supervisory mode are the following:-

• A serial asynchronous link

• Facility to read all data displayed under the normal mode

• Facility to reset maximum demand value, KVARh value, and Kwh value to zero. (Normally Kwh and KVARh are not reset).

• Facility to reset charge payable under different heads to any desired value (to account for part payment).

• Facility to adjust the rates given above to any other desired set of values

14.20 PWM INVERTER

14.20.1 UPS and other power supplies require a sine wave at a specified frequency to be generated. They use a MOSFET bridge with a DC power supply at the power side as shown in Figure 14.21 (a). The MOSFET Bridge is driven by a PWM waveform conforming to specific criteria. One such PWM waveform is shown in Figure 14.21 (b). The angles a1, a2, and a3 can be selected such that the 3rd and 5th harmonics are zero.

14.20.2 Show that the fundamental component of output is proportional to $\cos a1 - \cos a2 + \cos a3$.

14.20.3 Show that the 3rd and 5th harmonics are zero if

14.20.4 $\cos 3a1 - \cos 3a2 + \cos 3a3 = 0$
and
$\cos 5a_1 - \cos 5a_2 + \cos 5a_3 = 0$.

14.20.5 Obtain the different sets of a1, a2 and a3 values satisfying the above two equations. For each set obtain the fundamental component.

14.20.6 For a 50 hz inverter, the quarter cycle duration is 5ms. Convert the a1, a2 and a3 sets of values into equivalent time values for 50 Hz inverter.

14.20.7 Select one a1, a2 and a3 set of values. Generate the waveform in Figure 14.21 (b) using a μC and output the same on a port pin. Use two timers – one to generate 10 ms period and the other to generate the pulses within the 10ms period. After every 10ms, the output is to be inverted.

14.20.8 In a MOSFET inverter shown in Figure 14.21 (a), G1, G2, G3 and G4 are the MOSFET gates. G1 and G3 are to be driven by the signal in Figure 14.21 (b). G2 and G4 are to be driven by a signal which is its complement. However, one constraint is to be observed. Whenever G1 turns off, G4 can be turned on, only after a 1μs delay. This is done to avoid short circuiting the DC source of power.

14.20.9 Similar constraint holds good for all turn-off-turn-on pairs. Commit four output pins of a port to G1, G2, G3 and G4. Output necessary waveforms from the μC.

14.20.10 Any pulse of width less than 4 μs does not contribute significantly to the output. It causes power loss in the MOSEFET power transistor. Modify the scheme to accommodate this.

14.20.11 For every set of angles satisfying the constraint equations above, one can get a fundamental amplitude component. Using such data one can generate the inverse tables; for every fundamental component desired, the set of angle values of a1, a2 and a3 can be identified and made available as look-up tables. Modify the above scheme such that a 10-bit number representing the desired fundamental output is used as reference and the corresponding waveforms for G1, G2, G3 and G4 are generated.

14.20.12 - If the supply voltage changes or the load on the inverter changes, the inverter output voltage can change. However, it is highly desirable to keep the inverter output rms voltage constant. It is achieved by measuring the inverter output voltage, comparing with a set reference and correcting it in a feed back scheme. A possible procedure is as follows:-

14.20.12.1   Select one output voltage value; use corresponding angle values from the look-up tables and output G_1, G_2, G_3 and G_4 signals from the μC.

14.20.12.2   If the actual (measured) output is less than the desired value, shift to the set of entries in the look-up tables, corresponding to the next higher desired output voltage value.

14.20.12.3   If the actual (measured) output is less than the desired value, shift to the set of entries in the look-up tables, corresponding to the next lower desired output voltage value.

14.20.12.4   For satisfactory working, two additional constraints are to be observed:

- The measured output voltage has to be filtered and smoothened over a 50 cycle period prior to every comparison with the reference value. The smoothening can be achieved by taking a 'moving average' over 50 cycles.
- After every change in G_1, G_2, G_3, and G_4, one has to wait for 2s before the next change is brought forth.

14.20.12.5   Modify the inverter scheme to incorporate the above with necessary additions and modifications to software.

Note: The inverter realized above is close enough to a practical one used in UPS schemes; but it requires some additional 'fine tuning' which is not discussed here [1].

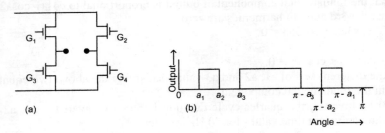

Figure 14.21 (a) Power circuit of the single phase inverter (b) PWM waveform

APPENDIX A

INSTRUCTION SET OF PIC®16F87X SERIES

PIC16F872

12.0 INSTRUCTION SET SUMMARY

The PIC16 instruction set is highly orthogonal and is comprised of three basic categories:

- **Byte-oriented** operations
- **Bit-oriented** operations
- **Literal and control** operations

Each PIC16 instruction is a 14-bit word divided into an **opcode** which specifies the instruction type, and one or more **operands** which further specify the operation of the instruction. The formats for each of the categories is presented in Figure 12-1, while the various opcode fields are summarized in Table 12-1.

Table 13-2 lists the instructions recognized by the MPASM™ Assembler. A complete description of each instruction is also available in the PICmicro™ Mid-Range Reference Manual (DS33023).

For **byte-oriented** instructions, 'f' represents a file register designator and 'd' represents a destination designator. The file register designator specifies which file register is to be used by the instruction.

The destination designator specifies where the result of the operation is to be placed. If 'd' is zero, the result is placed in the W register. If 'd' is one, the result is placed in the file register specified in the instruction.

For **bit-oriented** instructions, 'b' represents a bit field designator, which selects the bit affected by the operation, while 'f' represents the address of the file in which the bit is located.

For **literal and control** operations, 'k' represents an eight- or eleven-bit constant or literal value

One instruction cycle consists of four oscillator periods; for an oscillator frequency of 4 MHz, this gives a normal instruction execution time of 1 µs. All instructions are executed within a single instruction cycle, unless a conditional test is true or the program counter is changed as a result of an instruction. When this occurs, the execution takes two instruction cycles with the second cycle executed as a NOP.

Note:	To maintain upward compatibility with future PIC16F872 products, <u>do not use</u> the OPTION and TRIS instructions.

All instruction examples use the format '0xhh' to represent a hexadecimal number, where 'h' signifies a hexadecimal digit.

12.1 READ-MODIFY-WRITE OPERATIONS

Any instruction that specifies a file register as part of the instruction performs a Read-Modify-Write (R-M-W) operation. The register is read, the data is modified, and the result is stored according to either the instruction or the destination designator 'd'. A read operation is performed on a register even if the instruction writes to that register.

For example, a "CLRF PORTB" instruction will read PORTB, clear all the data bits, then write the result back to PORTB. This example would have the unintended result that the condition that sets the RBIF flag would be cleared.

TABLE 12-1:   OPCODE FIELD DESCRIPTIONS

Field	Description
f	Register file address (0x00 to 0x7F)
W	Working register (accumulator)
b	Bit address within an 8-bit file register
k	Literal field, constant data or label
x	Don't care location (= 0 or 1). The assembler will generate code with x = 0. It is the recommended form of use for compatibility with all Microchip software tools.
d	Destination select; d = 0: store result in W, d = 1: store result in file register f. Default is d = 1.
PC	Program Counter
TO	Time-out bit
PD	Power-down bit

FIGURE 12-1:   GENERAL FORMAT FOR INSTRUCTIONS

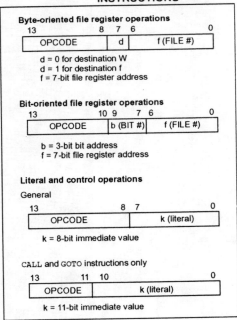

PIC16F872

TABLE 12-2:   PIC16F872 INSTRUCTION SET

Mnemonic, Operands		Description	Cycles	14-Bit Opcode MSb	LSb	Status Affected	Notes
BYTE-ORIENTED FILE REGISTER OPERATIONS							
ADDWF	f, d	Add W and f	1	00 0111 dfff	ffff	C,DC,Z	1,2
ANDWF	f, d	AND W with f	1	00 0101 dfff	ffff	Z	1,2
CLRF	f	Clear f	1	00 0001 1fff	ffff	Z	2
CLRW	-	Clear W	1	00 0001 0xxx	xxxx	Z	
COMF	f, d	Complement f	1	00 1001 dfff	ffff	Z	1,2
DECF	f, d	Decrement f	1	00 0011 dfff	ffff	Z	1,2
DECFSZ	f, d	Decrement f, Skip if 0	1(2)	00 1011 dfff	ffff		1,2,3
INCF	f, d	Increment f	1	00 1010 dfff	ffff	Z	1,2
INCFSZ	f, d	Increment f, Skip if 0	1(2)	00 1111 dfff	ffff		1,2,3
IORWF	f, d	Inclusive OR W with f	1	00 0100 dfff	ffff	Z	1,2
MOVF	f, d	Move f	1	00 1000 dfff	ffff	Z	1,2
MOVWF	f	Move W to f	1	00 0000 1fff	ffff		
NOP	-	No Operation	1	00 0000 0xx0	0000		
RLF	f, d	Rotate Left f through Carry	1	00 1101 dfff	ffff	C	1,2
RRF	f, d	Rotate Right f through Carry	1	00 1100 dfff	ffff	C	1,2
SUBWF	f, d	Subtract W from f	1	00 0010 dfff	ffff	C,DC,Z	1,2
SWAPF	f, d	Swap nibbles in f	1	00 1110 dfff	ffff		1,2
XORWF	f, d	Exclusive OR W with f	1	00 0110 dfff	ffff	Z	1,2
BIT-ORIENTED FILE REGISTER OPERATIONS							
BCF	f, b	Bit Clear f	1	01 00bb bfff	ffff		1,2
BSF	f, b	Bit Set f	1	01 01bb bfff	ffff		1,2
BTFSC	f, b	Bit Test f, Skip if Clear	1 (2)	01 10bb bfff	ffff		3
BTFSS	f, b	Bit Test f, Skip if Set	1 (2)	01 11bb bfff	ffff		3
LITERAL AND CONTROL OPERATIONS							
ADDLW	k	Add literal and W	1	11 111x kkkk	kkkk	C,DC,Z	
ANDLW	k	AND literal with W	1	11 1001 kkkk	kkkk	Z	
CALL	k	Call subroutine	2	10 0kkk kkkk	kkkk		
CLRWDT	-	Clear Watchdog Timer	1	00 0000 0110 0100		$\overline{TO},\overline{PD}$	
GOTO	k	Go to address	2	10 1kkk kkkk	kkkk		
IORLW	k	Inclusive OR literal with W	1	11 1000 kkkk	kkkk	Z	
MOVLW	k	Move literal to W	1	11 00xx kkkk	kkkk		
RETFIE	-	Return from interrupt	2	00 0000 0000 1001			
RETLW	k	Return with literal in W	2	11 01xx kkkk	kkkk		
RETURN	-	Return from Subroutine	2	00 0000 0000 1000			
SLEEP	-	Go into Standby mode	1	00 0000 0110 0011		$\overline{TO},\overline{PD}$	
SUBLW	k	Subtract W from literal	1	11 110x kkkk	kkkk	C,DC,Z	
XORLW	k	Exclusive OR literal with W	1	11 1010 kkkk	kkkk	Z	

Note   1:   When an I/O register is modified as a function of itself (e.g., MOVF PORTB, 1), the value used will be that value present on the pins themselves. For example, if the data latch is '1' for a pin configured as input and is driven low by an external device, the data will be written back with a '0'.

2:   If this instruction is executed on the TMR0 register (and, where applicable, d = 1), the prescaler will be cleared if assigned to the Timer0 module.

3:   If Program Counter (PC) is modified, or a conditional test is true, the instruction requires two cycles. The second cycle is executed as a NOP.

Note: Additional information on the mid-range instruction set is available in the PICmicro™ Mid-Range MCU Family Reference Manual (DS33023).

PIC16F872

12.2 Instruction Descriptions

ADDLW	Add Literal and W
Syntax:	[*label*] ADDLW k
Operands:	$0 \le k \le 255$
Operation:	$(W) + k \rightarrow (W)$
Status Affected:	C, DC, Z
Description:	The contents of the W register are added to the eight-bit literal 'k' and the result is placed in the W register.

ADDWF	Add W and f
Syntax:	[*label*] ADDWF f,d
Operands:	$0 \le f \le 127$ $d \in [0,1]$
Operation:	$(W) + (f) \rightarrow$ (destination)
Status Affected:	C, DC, Z
Description:	Add the contents of the W register with register 'f'. If 'd' is 0, the result is stored in the W register. If 'd' is 1, the result is stored back in register 'f'.

ANDLW	AND Literal with W
Syntax:	[*label*] ANDLW k
Operands:	$0 \le k \le 255$
Operation:	(W) .AND. (k) $\rightarrow$ (W)
Status Affected:	Z
Description:	The contents of W register are AND'ed with the eight-bit literal 'k'. The result is placed in the W register.

ANDWF	AND W with f
Syntax:	[*label*] ANDWF f,d
Operands:	$0 \le f \le 127$ $d \in [0,1]$
Operation:	(W) .AND. (f) $\rightarrow$ (destination)
Status Affected:	Z
Description:	AND the W register with register 'f'. If 'd' is 0, the result is stored in the W register. If 'd' is 1, the result is stored back in register 'f'.

BCF	Bit Clear f
Syntax:	[*label*] BCF f,b
Operands:	$0 \le f \le 127$ $0 \le b \le 7$
Operation:	$0 \rightarrow$ (f<b>)
Status Affected:	None
Description:	Bit 'b' in register 'f' is cleared.

BSF	Bit Set f
Syntax:	[*label*] BSF f,b
Operands:	$0 \le f \le 127$ $0 \le b \le 7$
Operation:	$1 \rightarrow$ (f<b>)
Status Affected:	None
Description:	Bit 'b' in register 'f' is set.

BTFSS	Bit Test f, Skip if Set
Syntax:	[*label*] BTFSS f,b
Operands:	$0 \le f \le 127$ $0 \le b < 7$
Operation:	skip if (f<b>) = 1
Status Affected:	None
Description:	If bit 'b' in register 'f' is '0', the next instruction is executed. If bit 'b' is '1', then the next instruction is discarded and a NOP is executed instead, making this a 2TCY instruction.

BTFSC	Bit Test, Skip if Clear
Syntax:	[*label*] BTFSC f,b
Operands:	$0 \le f \le 127$ $0 \le b \le 7$
Operation:	skip if (f<b>) = 0
Status Affected:	None
Description:	If bit 'b' in register 'f' is '1', the next instruction is executed. If bit 'b', in register 'f', is '0', the next instruction is discarded, and a NOP is executed instead, making this a 2TCY instruction.

PIC16F872

CALL — Call Subroutine

Syntax:	[*label*] CALL k
Operands:	$0 \leq k \leq 2047$
Operation:	(PC)+ 1→ TOS, k → PC<10:0>, (PCLATH<4:3>) → PC<12:11>
Status Affected:	None
Description:	Call Subroutine. First, return address (PC+1) is pushed onto the stack. The eleven-bit immediate address is loaded into PC bits <10:0>. The upper bits of the PC are loaded from PCLATH. CALL is a two-cycle instruction.

CLRF — Clear f

Syntax:	[*label*] CLRF f
Operands:	$0 \leq f \leq 127$
Operation:	00h → (f) 1 → Z
Status Affected:	Z
Description:	The contents of register 'f' are cleared and the Z bit is set.

CLRW — Clear W

Syntax:	[*label*] CLRW
Operands:	None
Operation:	00h → (W) 1 → Z
Status Affected:	Z
Description:	W register is cleared. Zero bit (Z) is set.

CLRWDT — Clear Watchdog Timer

Syntax:	[*label*] CLRWDT
Operands:	None
Operation:	00h → WDT 0 → WDT prescaler, 1 → $\overline{TO}$ 1 → $\overline{PD}$
Status Affected:	$\overline{TO}$, $\overline{PD}$
Description:	CLRWDT instruction resets the Watchdog Timer. It also resets the prescaler of the WDT. Status bits $\overline{TO}$ and $\overline{PD}$ are set.

COMF — Complement f

Syntax:	[*label*] COMF f,d
Operands:	$0 \leq f \leq 127$ $d \in [0,1]$
Operation:	$(\bar{f})$ → (destination)
Status Affected:	Z
Description:	The contents of register 'f' are complemented. If 'd' is 0, the result is stored in W. If 'd' is 1, the result is stored back in register 'f'.

DECF — Decrement f

Syntax:	[*label*] DECF f,d
Operands:	$0 \leq f \leq 127$ $d \in [0.1]$
Operation:	(f) - 1 → (destination)
Status Affected:	Z
Description:	Decrement register 'f'. If 'd' is 0, the result is stored in the W register. If 'd' is 1, the result is stored back in register 'f'.

PIC16F872

DECFSZ	Decrement f, Skip if 0
Syntax:	[label] DECFSZ f,d
Operands:	0 ≤ f ≤ 127 d ∈ [0,1]
Operation:	(f) - 1 → (destination); skip if result = 0
Status Affected:	None
Description:	The contents of register 'f' are decremented. If 'd' is 0, the result is placed in the W register. If 'd' is 1, the result is placed back in register 'f'. If the result is 1, the next instruction is executed. If the result is 0, then a NOP is executed instead, making it a 2TCY instruction.

GOTO	Unconditional Branch
Syntax:	[label] GOTO k
Operands:	0 ≤ k ≤ 2047
Operation:	k → PC<10:0> PCLATH<4:3> → PC<12:11>
Status Affected:	None
Description:	GOTO is an unconditional branch. The eleven-bit immediate value is loaded into PC bits <10:0>. The upper bits of PC are loaded from PCLATH<4:3>. GOTO is a two-cycle instruction.

INCF	Increment f
Syntax:	[label] INCF f,d
Operands:	0 ≤ f ≤ 127 d ∈ [0,1]
Operation:	(f) + 1 → (destination)
Status Affected:	Z
Description:	The contents of register 'f' are incremented. If 'd' is 0, the result is placed in the W register. If 'd' is 1, the result is placed back in register 'f'.

INCFSZ	Increment f, Skip if 0
Syntax:	[label] INCFSZ f,d
Operands:	0 ≤ f ≤ 127 d ∈ [0,1]
Operation:	(f) + 1 → (destination), skip if result = 0
Status Affected:	None
Description:	The contents of register 'f' are incremented. If 'd' is 0, the result is placed in the W register. If 'd' is 1, the result is placed back in register 'f'. If the result is 1, the next instruction is executed. If the result is 0, a NOP is executed instead, making it a 2TCY instruction.

IORLW	Inclusive OR Literal with W
Syntax:	[label] IORLW k
Operands:	0 ≤ k ≤ 255
Operation:	(W) .OR. k → (W)
Status Affected:	Z
Description:	The contents of the W register are OR'ed with the eight-bit literal 'k'. The result is placed in the W register.

IORWF	Inclusive OR W with f
Syntax:	[label] IORWF f,d
Operands:	0 ≤ f ≤ 127 d ∈ [0,1]
Operation:	(W) .OR. (f) → (destination)
Status Affected:	Z
Description:	Inclusive OR the W register with register 'f'. If 'd' is 0, the result is placed in the W register. If 'd' is 1, the result is placed back in register 'f'.

PIC16F872

MOVF	**Move f**
Syntax:	[*label*] MOVF f,d
Operands:	$0 \leq f \leq 127$ $d \in [0,1]$
Operation:	(f) → (destination)
Status Affected:	Z
Description:	The contents of register f are moved to a destination dependant upon the status of d. If d = 0, destination is W register. If d = 1, the destination is file register f itself. d = 1 is useful to test a file register, since status flag Z is affected.

NOP	**No Operation**
Syntax:	[*label*] NOP
Operands:	None
Operation:	No operation
Status Affected:	None
Description:	No operation

MOVLW	**Move Literal to W**
Syntax:	[*label*] MOVLW k
Operands:	$0 \leq k \leq 255$
Operation:	k → (W)
Status Affected:	None
Description:	The eight-bit literal 'k' is loaded into W register. The don't cares will assemble as 0's.

RETFIE	**Return from Interrupt**
Syntax:	[*label*] RETFIE
Operands:	None
Operation:	TOS → PC, 1 → GIE
Status Affected:	None

MOVWF	**Move W to f**
Syntax:	[*label*] MOVWF f
Operands:	$0 \leq f \leq 127$
Operation:	(W) → (f)
Status Affected:	None
Description:	Move data from W register to register 'f'.

RETLW	**Return with Literal in W**
Syntax:	[*label*] RETLW k
Operands:	$0 \leq k \leq 255$
Operation:	k → (W); TOS → PC
Status Affected:	None
Description:	The W register is loaded with the eight-bit literal 'k'. The program counter is loaded from the top of the stack (the return address). This is a two-cycle instruction.

PIC16F872

RLF — Rotate Left f through Carry

Syntax:	[*label*] RLF f,d
Operands:	$0 \le f \le 127$ $d \in [0,1]$
Operation:	See description below
Status Affected:	C
Description:	The contents of register 'f' are rotated one bit to the left through the Carry Flag. If 'd' is 0, the result is placed in the W register. If 'd' is 1, the result is stored back in register 'f'.

$$\boxed{C} \leftarrow \boxed{\text{Register f}}$$

RETURN — Return from Subroutine

Syntax:	[*label*] RETURN
Operands:	None
Operation:	TOS → PC
Status Affected:	None
Description:	Return from subroutine. The stack is POPed and the top of the stack (TOS) is loaded into the program counter. This is a two-cycle instruction.

RRF — Rotate Right f through Carry

Syntax:	[*label*] RRF f,d
Operands:	$0 \le f \le 127$ $d \in [0,1]$
Operation:	See description below
Status Affected:	C
Description:	The contents of register 'f' are rotated one bit to the right through the Carry Flag. If 'd' is 0, the result is placed in the W register. If 'd' is 1, the result is placed back in register 'f'.

$$\boxed{C} \rightarrow \boxed{\text{Register f}}$$

SLEEP

Syntax:	[*label*] SLEEP
Operands:	None
Operation:	00h → WDT, 0 → WDT prescaler, 1 → $\overline{\text{TO}}$, 0 → $\overline{\text{PD}}$
Status Affected:	$\overline{\text{TO}}$, $\overline{\text{PD}}$
Description:	The power-down status bit, $\overline{\text{PD}}$ is cleared. Time-out status bit, $\overline{\text{TO}}$ is set. Watchdog Timer and its prescaler are cleared. The processor is put into SLEEP mode with the oscillator stopped.

SUBLW — Subtract W from Literal

Syntax:	[*label*] SUBLW k
Operands:	$0 \le k \le 255$
Operation:	k - (W) → (W)
Status Affected:	C, DC, Z
Description:	The W register is subtracted (2's complement method) from the eight-bit literal 'k'. The result is placed in the W register.

SUBWF — Subtract W from f

Syntax:	[*label*] SUBWF f,d
Operands:	$0 \le f \le 127$ $d \in [0,1]$
Operation:	(f) - (W) → (destination)
Status Affected:	C, DC, Z
Description:	Subtract (2's complement method) W register from register 'f'. If 'd' is 0, the result is stored in the W register. If 'd' is 1, the result is stored back in register 'f'.

PIC16F872

SWAPF	Swap Nibbles in f
Syntax:	[label] SWAPF f,d
Operands:	0 ≤ f ≤ 127 d ∈ [0,1]
Operation:	(f<3:0>) → (destination<7:4>), (f<7:4>) → (destination<3:0>)
Status Affected:	None
Description:	The upper and lower nibbles of register 'f' are exchanged. If 'd' is 0, the result is placed in the W register. If 'd' is 1, the result is placed in register 'f'.

XORLW	Exclusive OR Literal with W
Syntax:	[label] XORLW k
Operands:	0 ≤ k ≤ 255
Operation:	(W) .XOR. k → (W)
Status Affected:	Z
Description:	The contents of the W register are XOR'ed with the eight-bit literal 'k'. The result is placed in the W register.

XORWF	Exclusive OR W with f
Syntax:	[label] XORWF f,d
Operands:	0 ≤ f ≤ 127 d ∈ [0,1]
Operation:	(W) .XOR. (f) → (destination)
Status Affected:	Z
Description:	Exclusive OR the contents of the W register with register 'f'. If 'd' is 0, the result is stored in the W register. If 'd' is 1, the result is stored back in register 'f'.

APPENDIX B

INSTRUCTION SET OF 8X51 FAMILY

intel.   MCS®-51 PROGRAMMER'S GUIDE AND INSTRUCTION SET

MCS®-51 INSTRUCTION SET

Table 10. 8051 Instruction Set Summary

Interrupt Response Time: Refer to Hardware Description Chapter.

Instructions that Affect Flag Settings(1)

Instruction	Flag C	OV	AC	Instruction	Flag C	OV	AC
ADD	X	X	X	CLR C	O		
ADDC	X	X	X	CPL C	X		
SUBB	X	X	X	ANL C,bit	X		
MUL	O	X		ANL C,/bit	X		
DIV	O	X		ORL C,bit	X		
DA	X			ORL C,bit	X		
RRC	X			MOV C,bit	X		
RLC	X			CJNE	X		
SETB C	1						

(1)Note that operations on SFR byte address 208 or bit addresses 209-215 (i.e., the PSW or bits in the PSW) will also affect flag settings.

Note on instruction set and addressing modes:

Rn   — Register R7–R0 of the currently selected Register Bank.

direct   — 8-bit internal data location's address. This could be an Internal Data RAM location (0–127) or a SFR [i.e., I/O port, control register, status register, etc. (128–255)].

@Ri   — 8-bit internal data RAM location (0–255) addressed indirectly through register R1 or R0.

#data   — 8-bit constant included in instruction.

#data 16 — 16-bit constant included in instruction.

addr 16 — 16-bit destination address. Used by LCALL & LJMP. A branch can be anywhere within the 64K-byte Program Memory address space.

addr 11 — 11-bit destination address. Used by ACALL & AJMP. The branch will be within the same 2K-byte page of program memory as the first byte of the following instruction.

rel   — Signed (two's complement) 8-bit offset byte. Used by SJMP and all conditional jumps. Range is −128 to +127 bytes relative to first byte of the following instruction.

bit   — Direct Addressed bit in Internal Data RAM or Special Function Register.

Mnemonic		Description	Byte	Oscillator Period
ARITHMETIC OPERATIONS				
ADD	A,Rn	Add register to Accumulator	1	12
ADD	A,direct	Add direct byte to Accumulator	2	12
ADD	A,@Ri	Add indirect RAM to Accumulator	1	12
ADD	A,#data	Add immediate data to Accumulator	2	12
ADDC	A,Rn	Add register to Accumulator with Carry	1	12
ADDC	A,direct	Add direct byte to Accumulator with Carry	2	12
ADDC	A,@Ri	Add indirect RAM to Accumulator with Carry	1	12
ADDC	A,#data	Add immediate data to Acc with Carry	2	12
SUBB	A,Rn	Subtract Register from Acc with borrow	1	12
SUBB	A,direct	Subtract direct byte from Acc with borrow	2	12
SUBB	A,@Ri	Subtract indirect RAM from ACC with borrow	1	12
SUBB	A,#data	Subtract immediate data from Acc with borrow	2	12
INC	A	Increment Accumulator	1	12
INC	Rn	Increment register	1	12
INC	direct	Increment direct	2	12
INC	@Ri	Increment direct RAM	1	12
DEC	A	Decrement Accumulator	1	12
DEC	Rn	Decrement Register	1	12
DEC	direct	Decrement direct byte	2	12
DEC	@Ri	Decrement indirect RAM	1	12

All mnemonics copyrighted © Intel Corporation 1980

intel. MCS®-51 PROGRAMMER'S GUIDE AND INSTRUCTION SET

Table 10. 8051 Instruction Set Summary (Continued)

Mnemonic		Description	Byte	Oscillator Period
ARITHMETIC OPERATIONS (Continued)				
INC	DPTR	Increment Data Pointer	1	24
MUL	AB	Multiply A & B	1	48
DIV	AB	Divide A by B	1	48
DA	A	Decimal Adjust Accumulator	1	12
LOGICAL OPERATIONS				
ANL	A,Rn	AND Register to Accumulator	1	12
ANL	A,direct	AND direct byte to Accumulator	2	12
ANL	A,@Ri	AND indirect RAM to Accumulator	1	12
ANL	A, #data	AND immediate data to Accumulator	2	12
ANL	direct,A	AND Accumulator to direct byte	2	12
ANL	direct, #data	AND immediate data to direct byte	3	24
ORL	A,Rn	OR register to Accumulator	1	12
ORL	A,direct	OR direct byte to Accumulator	2	12
ORL	A,@Ri	OR indirect RAM to Accumulator	1	12
ORL	A, #data	OR immediate data to Accumulator	2	12
ORL	direct,A	OR Accumulator to direct byte	2	12
ORL	direct, #data	OR immediate data to direct byte	3	24
XRL	A,Rn	Exclusive-OR register to Accumulator	1	12
XRL	A,direct	Exclusive-OR direct byte to Accumulator	2	12
XRL	A,@Ri	Exclusive-OR indirect RAM to Accumulator	1	12
XRL	A, #data	Exclusive-OR immediate data to Accumulator	2	12
XRL	direct,A	Exclusive-OR Accumulator to direct byte	2	12
XRL	direct, #data	Exclusive-OR immediate data to direct byte	3	24
CLR	A	Clear Accumulator	1	12
CPL	A	Complement Accumulator	1	12

Mnemonic		Description	Byte	Oscillator Period
LOGICAL OPERATIONS (Continued)				
RL	A	Rotate Accumulator Left	1	12
RLC	A	Rotate Accumulator Left through the Carry	1	12
RR	A	Rotate Accumulator Right	1	12
RRC	A	Rotate Accumulator Right through the Carry	1	12
SWAP	A	Swap nibbles within the Accumulator	1	12
DATA TRANSFER				
MOV	A,Rn	Move register to Accumulator	1	12
MOV	A,direct	Move direct byte to Accumulator	2	12
MOV	A,@Ri	Move indirect RAM to Accumulator	1	12
MOV	A, #data	Move immediate data to Accumulator	2	12
MOV	Rn,A	Move Accumulator to register	1	12
MOV	Rn,direct	Move direct byte to register	2	24
MOV	Rn, #data	Move immediate data to register	2	12
MOV	direct,A	Move Accumulator to direct byte	2	12
MOV	direct,Rn	Move register to direct byte	2	24
MOV	direct,direct	Move direct byte to direct	3	24
MOV	direct,@Ri	Move indirect RAM to direct byte	2	24
MOV	direct, #data	Move immediate data to direct byte	3	24
MOV	@Ri,A	Move Accumulator to indirect RAM	1	12

All mnemonics copyrighted © Intel Corporation 1980

intel.   MCS®-51 PROGRAMMER'S GUIDE AND INSTRUCTION SET

Table 10. 8051 Instruction Set Summary (Continued)

Mnemonic		Description	Byte	Oscillator Period
DATA TRANSFER (Continued)				
MOV	@Ri,direct	Move direct byte to indirect RAM	2	24
MOV	@Ri,≠data	Move immediate data to indirect RAM	2	12
MOV	DPTR,≠data16	Load Data Pointer with a 16-bit constant	3	24
MOVC	A,@A+DPTR	Move Code byte relative to DPTR to Acc	1	24
MOVC	A,@A+PC	Move Code byte relative to PC to Acc	1	24
MOVX	A,@Ri	Move External RAM (8-bit addr) to Acc	1	24
MOVX	A,@DPTR	Move External RAM (16-bit addr) to Acc	1	24
MOVX	@Ri,A	Move Acc to External RAM (8-bit addr)	1	24
MOVX	@DPTR,A	Move Acc to External RAM (16-bit addr)	1	24
PUSH	direct	Push direct byte onto stack	2	24
POP	direct	Pop direct byte from stack	2	24
XCH	A,Rn	Exchange register with Accumulator	1	12
XCH	A,direct	Exchange direct byte with Accumulator	2	12
XCH	A,@Ri	Exchange indirect RAM with Accumulator	1	12
XCHD	A,@Ri	Exchange low-order Digit indirect RAM with Acc	1	12

Mnemonic		Description	Byte	Oscillator Period
BOOLEAN VARIABLE MANIPULATION				
CLR	C	Clear Carry	1	12
CLR	bit	Clear direct bit	2	12
SETB	C	Set Carry	1	12
SETB	bit	Set direct bit	2	12
CPL	C	Complement Carry	1	12
CPL	bit	Complement direct bit	2	12
ANL	C,bit	AND direct bit to CARRY	2	24
ANL	C,/bit	AND complement of direct bit to Carry	2	24
ORL	C,bit	OR direct bit to Carry	2	24
ORL	C,/bit	OR complement of direct bit to Carry	2	24
MOV	C,bit	Move direct bit to Carry	2	12
MOV	bit,C	Move Carry to direct bit	2	24
JC	rel	Jump if Carry is set	2	24
JNC	rel	Jump if Carry not set	2	24
JB	bit,rel	Jump if direct Bit is set	3	24
JNB	bit,rel	Jump if direct Bit is Not set	3	24
JBC	bit,rel	Jump if direct Bit is set & clear bit	3	24
PROGRAM BRANCHING				
ACALL	addr11	Absolute Subroutine Call	2	24
LCALL	addr16	Long Subroutine Call	3	24
RET		Return from Subroutine	1	24
RETI		Return from interrupt	1	24
AJMP	addr11	Absolute Jump	2	24
LJMP	addr16	Long Jump	3	24
SJMP	rel	Short Jump (relative addr)	2	24

All mnemonics copyrighted © Intel Corporation 1980

intel.   **MCS®-51 PROGRAMMER'S GUIDE AND INSTRUCTION SET**

Table 10. 8051 Instruction Set Summary (Continued)

Mnemonic		Description	Byte	Oscillator Period
PROGRAM BRANCHING (Continued)				
JMP	@A+DPTR	Jump indirect relative to the DPTR	1	24
JZ	rel	Jump if Accumulator is Zero	2	24
JNZ	rel	Jump if Accumulator is Not Zero	2	24
CJNE	A,direct,rel	Compare direct byte to Acc and Jump If Not Equal	3	24
CJNE	A,≠data,rel	Compare immediate to Acc and Jump if Not Equal	3	24

Mnemonic		Description	Byte	Oscillator Period
PROGRAM BRANCHING (Continued)				
CJNE	Rn,≠data,rel	Compare immediate to register and Jump if Not Equal	3	24
CJNE	@Ri,≠data,rel	Compare immediate to indirect and Jump if Not Equal	3	24
DJNZ	Rn,rel	Decrement register and Jump if Not Zero	2	24
DJNZ	direct,rel	Decrement direct byte and Jump if Not Zero	3	24
NOP		No Operation	1	12

All mnemonics copyrighted © Intel Corporation 1980

APPENDIX C

ASCII Table (7- bit)

(ASCII = American Standard Code for Information Interchange)

Decimal	Octal	Hex	Binary	Value		
000	000	000	00000000	NUL		(Null char.)
001	001	001	00000001	SOH		(Start of Header)
002	002	002	00000010	STX		(Start of Text)
003	003	003	00000011	ETX		(End of Text)
004	004	004	00000100	EOT		(End of Transmission)
005	005	005	00000101	ENQ		(Enquiry)
006	006	006	00000110	ACK		(Acknowledgment)
007	007	007	00000111	BEL		(Bell)
008	010	008	00001000	BS		(Backspace)
009	011	009	00001001	HT		(Horizontal Tab)
010	012	00A	00001010	LF		(Line Feed)
011	013	00B	00001011	VT		(Vertical Tab)
012	014	00C	00001100	FF		(Form Feed)
013	015	00D	00001101	CR		(Carriage Return)
014	016	00E	00001110	SO		(Shift Out)
015	017	00F	00001111	SI		(Shift In)
016	020	010	00010000	DLE		(Data Link Escape)
017	021	011	00010001	DC1	(XON)	(Device Control 1)
018	022	012	00010010	DC2		(Device Control 2)
019	023	013	00010011	DC3	(XOFF)	(Device Control 3)
020	024	014	00010100	DC4		(Device Control 4)
021	025	015	00010101	NAK		(Negative Acknowledgement)
022	026	016	00010110	SYN		(Synchronous Idle)
023	027	017	00010111	ETB		(End of Trans. Block)
024	030	018	00011000	CAN		(Cancel)
025	031	019	00011001	EM		(End of Medium)
026	032	01A	00011010	SUB		(Substitute)
027	033	01B	00011011	ESC		(Escape)
028	034	01C	00011100	FS		(File Separator)
029	035	01D	00011101	GS		(Group Separator)
030	036	01E	00011110	RS		(Request to Send) (Record Separator)
031	037	01F	00011111	US		(Unit Separator)
032	040	020	00100000	SP		(Space)
033	041	021	00100001	!		(exclamation mark)

034	042	022	00100010	"	(double quote)
035	043	023	00100011	#	(number sign)
036	044	024	00100100	$	(dollar sign)
037	045	025	00100101	%	(percent)
038	046	026	00100110	&	(ampersand)
039	047	027	00100111	'	(single quote)
040	050	028	00101000	(	(left/opening parenthesis)
041	051	029	00101001	)	(right/closing parenthesis)
042	052	02A	00101010	*	(asterisk)
043	053	02B	00101011	+	(plus)
044	054	02C	00101100	,	(comma)
045	055	02D	00101101	-	(minus or dash)
046	056	02E	00101110	.	(dot)
047	057	02F	00101111	/	(forward slash)
048	060	030	00110000	0	
049	061	031	00110001	1	
050	062	032	00110010	2	
051	063	033	00110011	3	
052	064	034	00110100	4	
053	065	035	00110101	5	
054	066	036	00110110	6	
055	067	037	00110111	7	
056	070	038	00111000	8	
057	071	039	00111001	9	
058	072	03A	00111010	:	(colon)
059	073	03B	00111011	;	(semi-colon)
060	074	03C	00111100	<	(less than)
061	075	03D	00111101	=	(equal sign)
062	076	03E	00111110	>	(greater than)
063	077	03F	00111111	?	(question mark)
064	100	040	01000000	@	(AT symbol)
065	101	041	01000001	A	
066	102	042	01000010	B	
067	103	043	01000011	C	
068	104	044	01000100	D	
069	105	045	01000101	E	
070	106	046	01000110	F	
071	107	047	01000111	G	
072	110	048	01001000	H	
073	111	049	01001001	I	
074	112	04A	01001010	J	
075	113	04B	01001011	K	
076	114	04C	01001100	L	
077	115	04D	01001101	M	
078	116	04E	01001110	N	
079	117	04F	01001111	O	
080	120	050	01010000	P	
081	121	051	01010001	Q	
082	122	052	01010010	R	
083	123	053	01010011	S	
084	124	054	01010100	T	
085	125	055	01010101	U	

086	126	056	01010110	V	
087	127	057	01010111	W	
088	130	058	01011000	X	
089	131	059	01011001	Y	
090	132	05A	01011010	Z	
091	133	05B	01011011	[	(left/opening bracket)
092	134	05C	01011100	\	(back slash)
093	135	05D	01011101	]	(right/closing bracket)
094	136	05E	01011110	^	(caret/cirumflex)
095	137	05F	01011111	_	(underscore)
096	140	060	01100000	`	
097	141	061	01100001	a	
098	142	062	01100010	b	
099	143	063	01100011	c	
100	144	064	01100100	d	
101	145	065	01100101	e	
102	146	066	01100110	f	
103	147	067	01100111	g	
104	150	068	01101000	h	
105	151	069	01101001	i	
106	152	06A	01101010	j	
107	153	06B	01101011	k	
108	154	06C	01101100	l	
109	155	06D	01101101	m	
110	156	06E	01101110	n	
111	157	06F	01101111	o	
112	160	070	01110000	p	
113	161	071	01110001	q	
114	162	072	01110010	r	
115	163	073	01110011	s	
116	164	074	01110100	t	
117	165	075	01110101	u	
118	166	076	01110110	v	
119	167	077	01110111	w	
120	170	078	01111000	x	
121	171	079	01111001	y	
122	172	07A	01111010	z	
123	173	07B	01111011	{	(left/opening brace)
124	174	07C	01111100			(vertical bar)
125	175	07D	01111101	}	(right/closing brace)
126	176	07E	01111110	~	(tilde)
127	177	07F	01111111	DEL	(delete)

APPENDIX D

ABBREVIATIONS

ADC	Analog to Digital Converter
ALP	Assembly Language Program
ALU	Arithmetic Logic Unit
BCD	Binary Coded Decimal
C	Carry (bit)
CD	Compact Disc
CISC	Complex Instruction Set Computer
CLK	CLocK (signal)
CS	Chip Select (signal)
CT	Current Transformer
DC	Digit Carry (bit)
DEMUX	Demultiplexer
DM	Data Memory
DRAM	Dynamic RAM
DTPR	Data Pointer
EEPROM	Electrically Erasable Programmable Memory
EPROM	Erasable Programmable Memory
FIFO	First Input First Output (register)
FOC	Fibre Optic Cable
FSR	File Select Register
GIE	Global Interrupt Enable
HDD	Hard Disc Drive
HT	High Tension
IDE	Integrated Development Environment
IR	Instruction Register
IRQ	Interrupt ReQuest
ISR	Interrupt Service Routine
LCD	Liquid Crystal Display
LED	Light Emitting Diode
LIFO	First Input First Output (register)
LT	Low Tension
LUT	Look Up Table
MS	Master Slave (Flip flop)
MAR	Memory Address Register
MR	WRite (signal)
MUX	Multiplexer

PC	program Counter
PID	Proportional Integral and Derivative
PM	Program Memory
PROM	Programmable Read Only Memory
PSW	Program Status Word
PSU	Power Supply Unit
RAM	Random Access Memory
RD	ReaD (signal)
RISC	Reduced Instruction Set Computer
ROM	Read Only Memory
SP	Stack Pointer
SRAM	Static RAM
SR	Shift Register
SFR	Special function Register
USART	Universal Serial Asynchronous Receiver and Transmitter
WDT	Watch Dog Timer
Z	Zero (bit)

REFERENCES

1. Bose, B. K., "Modern Power Electronics and AC Drives", Pearson Education, Singapore, 2002.

2. Brown, S., and Z. Vranesic., "Fundamentals of Digital Logic with Verilog Design", McGraw Hill, NY (USA), 2002.

3. Hayes, J. P., "Computer Architecture and Organization", 3rd Ed.., McGraw Hill, Singapore 1998.

4. Hayes, J. P., "Introduction to Digital Logic Design", Addison Wesley, Reading MA (USA), 1993.

5. Hill F. J., and G. R. Peterson., "Introduction to Switching Theory and Logic Design", 3rd Ed., McGraw-Hill, NY (USA), 1981.

6. Intel, "MCS51 Microcontroller Family User's Manual", 1994.

7. Jaeger, R. C., and T. N. Blalock., "Microelectronic Circuit Design", 2nd Ed., McGraw Hill, NY (USA), 2003.

8. Kamal, R., "Embedded Systems – Architecture, Programming, and Design", Tata McGraw_Hill, New Delhi, 2003.

9. Leach, D. P., and A. P. Malvino., "Digital Principles and Applications", 5th Ed., McGraw Hill, NY (USA), 1994.

10. Mano, M. M., "Digital Design", Pearson Education Inc., Singapore, 2002.

11. Microchip, "PIC16F87X Data Sheet", 2001

12. Microchip, "PICmicro Mid-Range Family Reference Manual", 1997.

13. Noergaard, T., "Embedded System Architecture", Elsevier Inc., Oxford (UK), 2005.

14. Null, L., and J Lobur., "Computer Organization and Architecture", Narosa, New Delhi, 2004.

15. Padmanabhan, T. R., "Industrial Instrumentation – Principles and Design", Springer Verlag, London, 2000.

16. Ray, A. K., and K. M. Bhurchandi.;"Advanced Microprocessors and Peripherals – Architecture, Programming and Interfacing", 2nd Ed., Tata McGraw-Hill, New Delhi, 2006.

17. Tinder R. F., "Engineering Digital Design", 2nd Ed., Academic Press, CA (USA), 2000.

18. Tocci, R. J., and N. S. Widmer., "Digital Systems, Principles and Applications", 8th Ed., Prentice Hall, NJ (USA), 2001.

19. Vahid, F., and T Givargis, "Embedded System Design – A unified Hardware / Software Introduction", John Wiley, NY (USA), 2002.

20. Valvano, J. W., "Embedded Microcomputer Systems", Thomson Learning Inc., Singapore, 2002.

21. Wakerly, J. F., "Digital Systems, Principles and Applications", 3rd Ed., Prentice Hall, NJ (USA). 2002.

INDEX